普通高等教育“十三五”规划教材
新工科建设之路·计算机类规划教材

Web 应用开发
——基于 Spring MVC + MyBatis + Maven

韩　冬　编著

電子工業出版社
Publishing House of Electronics Industry
北京 • BEIJING

内 容 简 介

本书是目前国内较少的系统讲解 Spring MVC+MyBatis+Maven 这一组合的教材。全书分为 Java Web 开发基础、MyBatis 和 Spring MVC 三部分，共 20 章。Java Web 开发基础（1～9 章），主要包括搭建 Java Web 开发环境、Servlet 编程、JSP 语言基础、会话管理、EL 和 JSTL、过滤器与监听器、AJAX 技术、Maven、jQuery 基础和 jQuery EasyUI 等内容。MyBatis（10～14 章），主要包括 MyBatis 入门、配置和映射、动态 SQL、MyBatis 注解方式、MyBatis 缓存配置和 MyBatis 应用等内容。Spring MVC（15～20 章），主要包括 Spring 基础、Spring MVC 入门、基于注解的控制器、拦截器、数据转换和格式化、Spring MVC 的表单标签库、Spring MVC 的文件上传和下载、Spring MVC 的国际化和 Spring MVC+MyBatis 应用等内容。

本书内容丰富、翔实，实用性强，适用面广，既可作为高等学校计算机相关专业学生学习 Java Web 应用开发的教材，又可作为软件培训机构和编程人员的参考用书。

图书在版编目（CIP）数据

Web 应用开发：基于 Spring MVC+MyBatis+Maven / 韩冬编著. 一北京：电子工业出版社，2018.9

ISBN 978-7-121-34891-4

Ⅰ. ①W… Ⅱ. ①韩… Ⅲ. ①JAVA 语言－程序设计－高等学校－教材 Ⅳ. ①TP312.8

中国版本图书馆 CIP 数据核字（2018）第 184629 号

策划编辑：戴晨辰
责任编辑：裴 杰
印 刷：北京捷迅佳彩印刷有限公司
装 订：北京捷迅佳彩印刷有限公司
出版发行：电子工业出版社
北京市海淀区万寿路173信箱 邮编 100036
开 本：787×1 092 1/16 印张：21.75 字数：640千字
版 次：2018年9月第1版
印 次：2023年1月第9次印刷
定 价：59.00元

凡所购买电子工业出版社图书有缺损问题，请向购买书店调换。若书店售缺，请与本社发行部联系，联系及邮购电话：（010）88254888，88258888。

质量投诉请发邮件至 zlts@phei.com.cn，盗版侵权举报请发邮件至 dbqq@phei.com.cn。

本书咨询联系方式：dcc@phei.com.cn。

前　言

"蒹葭苍苍，白露为霜。所谓伊人，在水一方……"，新技术、新知识宛若一位美貌典雅的妙龄女子，让人怦然心动。这位妙龄女子就是 Spring MVC + MyBatis + Maven。

Spring MVC 是一个基于动作的 MVC 框架。该框架突出了 HTTP 中的请求/响应特性，在该框架中，用户的每一个请求都声明了一个需要执行的动作。而这主要是通过将每个请求 URI 映射到一个可执行的方法来实现的。同时，其也将请求参数映射到对应方法的参数上。

Spring MVC 与 Spring 框架集成在一起（如 IoC 容器、AOP 等），具有非常灵活的数据校验、数据转换和格式化，以及数据绑定机制，支持 RESTful，提供强大的约定优于配置的契约式编程支持，能够让开发者进行更简洁的 Web 层的开发。

MyBatis 是支持定制化 SQL、存储过程以及高级映射的优秀的持久层框架。MyBatis 避免了几乎所有的 JDBC 代码和手动设置参数及获取结果集。MyBatis 可以对配置和原生 Map 使用简单的 XML 或注解，将接口和 Java 的 POJO 映射成数据库中的记录。

MyBatis 框架的优点在于：①与 JDBC 相比，减少了 50%以上的代码量；②MyBatis 是最简单的持久化框架，小巧且简单易学；③MyBatis 相当灵活，不会对应用程序或者数据库的现有设计强加任何影响，SQL 写在 XML 里，从程序代码中彻底分离，降低了耦合度，便于统一管理和优化，并可重用；④提供了 XML 标签，支持编写动态 SQL 语句；⑤提供了映射标签，支持对象与数据库的 ORM 映射。

为什么有 Maven？构建是程序员每天要做的工作，而且相当长的时间花在了这方面，而 Maven 可使这一系列的工作完全自动化。人们一直在寻找避免重复的方法，这里的重复有：设计的重复、文档的重复、编码的重复、构建的重复等。而 Maven 是跨平台的，使用它最大限度地消除了构建的重复。

Maven 不仅是构建工具，还是依赖管理工具和项目管理工具，Maven 提供了中央仓库，能够帮用户自动下载构件。使用 Maven 可以进行项目高度自动化构建、依赖管理和仓库管理等。而使用 Maven 最大的好处就是可以实现依赖管理。

在移动互联网兴起的时代，特别是针对后台开发，越来越多的企业喜欢使用 Spring MVC + MyBatis + Maven 的组合，Maven 也替代 Ant 成为构建 Java Web 项目的流行工具。

一般来说，对于性能要求较高的互联网项目，通常会选用 SSM 框架。

本书的编写思路

本书基于以下教学理念编写而成。

1．注重基础

要把 Spring MVC + MyBatis + Maven 学好，必须要有扎实的基本功，这也就是本书第一部分的内容——Java Web 开发基础，主要包括 Servlet 编程、JSP 基础、会话管理、EL 和 JSTL、过滤器与监听器、AJAX 技术等内容。同样，要想把"Java Web 开发基础"学好，就要有扎实的 Java 基础。在学习的旅途中，我们要循序渐进。

2．注重知识（或者技术）格局（框架）

注重技术细节是个好事，但学以致用。这样才会更有学习的动力。无论是 Spring MVC 、MyBatis 、Maven，或者 Java Web 开发基础，其中的哪一部分拿出来都可以是一本书的内容，或者是一学期的课程。但一旦我们的技术格局有了，骨架有了，血肉是可以逐渐丰满起来的。学生一旦产生兴趣，技术（或者知识）细节方面的，他自己会努力学会并加以运用的。这样（同时）还会培养他自己的自学能力。这也是本书集“Java Web 开发基础 + MyBatis + Spring MVC + Maven”于一体，作为《Web 应用开发》课程主要教学内容的原因。

3．注重编程实践

本书的编程实例很多，限于篇幅，书中的一些示例代码往往只是核心代码，并不是全部的源程序。读者需要到源码包里仔细地阅读代码，以掌握相应的知识（或技能）。要成为优秀的程序员，编写相当数量的代码还是必要的。设计模式是在代码重构的过程中凸现出来的。所以，编写大量代码、注重编程实践，也为以后成为优秀的架构师打下了基础。

4．注重学习者自身的努力和悟性

“师傅领进门，修行在个人”，要成为优秀的软件工程师以致架构师，尤其需要个人的勤奋努力及悟性。如果在学习的过程中，学习者态度消极、被动、有惰性，那么谁也帮不上忙。

5．注重能力培养

“授人以鱼不如授人以渔”，但就编者的教学体会而言，实际上大多数学生喜欢的是“鱼”，而不是“渔”。所以，在教学实践中，恰如其分的引导是非常必要的。学生的“娇气”对学生以后的发展并没有多大好处。

“Rome was not build in a day!”，能力的培养需要一个过程，虽然未必是“宝剑锋从磨砺出，梅花香自苦寒来”，但还是需要大量的编程实践、代码重构和反思（“悟”），经过长时间的潜移默化来逐渐习得。

在教学过程中，更多的需要是“随风潜入夜，润物细无声”。

要说明的是，本书的定位是——卓越工程师培养创新教材，不仅面向在校学生，还面向广大 Java 程序员、技术人员和培训机构等。

给授课教师的教学建议

如下表所示，有星号的章节可作为可选章节。如果要给高校学生实施教学，则正常教学课时可安排为 48~72 课时，教师可根据具体情况来灵活安排。

具体章节	建议课时	是否可选章节	可选章节建议课时
第 1 章 搭建 Java Web 开发环境	2		
第 2 章 Servlet 编程	6		
第 3 章 JSP 语言基础	2		
第 4 章 会话管理	4		
第 5 章 EL 和 JSTL	4		
第 6 章 过滤器与监听器	4		
第 7 章 AJAX 技术	2		

续表

具体章节	建议课时	是否可选章节	可选章节建议课时
第 8 章 Maven	2		
*第 9 章 jQuery EasyUI		可选	2
第 10 章 MyBatis 入门	2		
第 11 章 配置和映射	6		
第 12 章 动态 SQL	4		
*第 13 章 MyBatis 其他		可选	4
第 14 章 MyBatis 应用	2		
第 15 章 Spring 基础	4		
第 16 章 Spring MVC 入门	4		
第 17 章 基于注解的控制器	4		
*第 18 章 拦截器、数据转换和格式化		可选	6
*第 19 章 Spring MVC 其他		可选	4
*第 20 章 Spring MVC+MyBatis 应用		可选	4
合计	52		20

致 谢

本书由韩冬（苏州大学）负责编制提纲及主要撰写工作，参与本书编写工作的还有张建、曹国平、肖广娣、李炜、李家伟、胡楠等，在此一并表示感谢。

在编写本书过程中，编者参考、借鉴了很多 IT 技术专家、学者的相关著作，对于引用的段落或文字尽可能一一列出，谨向各位 IT 技术专家、学者一并表示感谢！

本书的配套资源中提供了所有章节程序的源代码、配套 PPT、习题解答和拓展学习资源（包括 HTML、CSS、JavaScript 基础内容等）。读者可在华信教育资源网进行下载，具体下载地址为 http://www.hxedu.com.cn/。

鉴于编者水平有限，书中难免存在不足和错误之处，敬请专家和读者提出宝贵意见和建议，以便再版时改进。

韩 冬

目　　录

第一部分　Java Web 开发基础

第二部分　MyBatis

第三部分 Spring MVC

第一部分

Java Web 开发基础

第1章 搭建Java Web开发环境

本章导读

目前两种流行的软件体系结构是客户端/服务器端（Client/Server，C/S）体系结构和浏览器端/服务器端（Browser/Server，B/S）体系结构。对开发人员来说，在项目开发过程中针对不同项目选择恰当的软件体系结构非常重要。适当的软件体系结构与软件的安全性、可维护性等密切相关。B/S 是 Web 兴起后的一种网络结构模式，Web 浏览器是客户端最主要的应用软件。这种模式统一了客户端，将系统功能实现的核心部分集中到服务器上，简化了系统的开发、维护和使用。本章主要内容有：（1）基于 B/S 架构的 Web 应用；（2）JDK 安装与配置；（3）服务器 Tomcat 下载与安装；（4）Eclipse 安装与使用；（5）MySQL 安装配置与使用；（6）PostgreSQL 安装与使用；（7）Navicat Premium 安装与使用。

1.1 基于B/S结构的Web应用

C/S 架构也可以看作胖客户端架构。因为客户端需要实现绝大多数的业务逻辑和界面展示。在这种架构中，作为客户端的部分需要承受很大的压力，因为显示逻辑和事务处理都包含在其中，通过与数据库的交互（通常是 SQL 或存储过程的实现）来达到持久化数据，以此满足实际项目的需要。

C/S 架构的优点是安全性较好，而突出的缺点是开发和维护成本都比 B/S 高，而且客户端负载重。在 2000 年以前，C/S 架构占据了网络程序开发领域的主流。

B/S 架构是随着 Internet 技术的兴起，对 C/S 架构的一种变化或者改进的结构。在这种结构下，用户工作界面是通过 WWW 浏览器来实现的，极少部分事务逻辑在前端（Browser）实现，主要事务逻辑在服务器端（Server）实现。

基于 B/S 架构的 Web 应用，通常由客户端浏览器、Web 服务器和数据库服务器三部分组成。用户通过客户端浏览器向服务器端发送请求；服务器收到请求后，需要对用户发送过来的数据进行业务逻辑处理，多数还伴随对数据库的存取操作；最后，服务器将处理结果返回给客户端浏览器。

1.2 JDK安装与配置

在 http://www.oracle.com/technetwork/java/javase/downloads/index.html 下载最新版本的 JDK，这里是 JDK8（jdk-8u92-windows-i586.exe），按默认安装路径进行 JDK 安装即可。

设置环境变量如下。

```
JAVA_HOME= C:\Program Files (x86)\Java\jdk1.8.0_92
CLASSPATH=.;%JAVA_HOME%\lib\dt.jar;%JAVA_HOME%\lib\tools.jar
PATH=%PATH%;%JAVA_HOME%\bin
```

注意：%PATH%为原来的环境变量值，添加";"和后面的内容到原来值的后面。

验证是否配置成功，可在命令行窗口中输入 java -version，显示版本为 1.8.0_92，说明 JDK 安装及环境变量配置成功。

1.3　服务器 Tomcat 下载与安装

在 http://tomcat.apache.org/ 下载最新版本的 Tomcat，这里下载的是 tomcat-8.0.45 的解压缩版本（apache-tomcat-8.0.45-windows-x86.zip），解压缩后，tomcat 根目录可改名为 tomcat8，把 tomcat8 文件夹复制至 E 盘下（其他磁盘亦可）。

tomcat8 目录层次如图 1-1 所示，注意，目录名不能有中文和空格。目录介绍如下。

（1）bin 目录：二进制执行文件。其中最常用的文件是 startup.bat，如果是 Linux 或 Mac 系统，则启动文件为 startup.sh。

（2）conf 目录：配置目录。其中最核心的文件是 server.xml，可以在其中修改端口号等。默认端口号是 8080，也就是说，此端口号不能被其他应用程序所占用。

（3）lib 目录：库文件。lib 目录为 Tomcat 运行时需要的 JAR 包所在的目录。

（4）logs 目录：日志。

（5）temp 目录：临时产生的文件，即缓存。

（6）webapps 目录：Web 应用程序。Web 应用放置到此目录下，浏览器可以直接访问。

（7）work 目录：编译以后的 class 文件。

在命令行窗口中进入 E:\tomcat8\bin 目录，运行 startup，启动 Tomcat 服务器。打开浏览器，在浏览器地址栏中输入 http://localhost:8080，则进入如图 1-2 所示页面，说明 Tomcat 安装成功。如果想关闭 Tomcat 服务器，在命令行窗口的 E:\tomcat8\bin 路径下输入 shutdown，即可关闭 Tomcat 服务器。

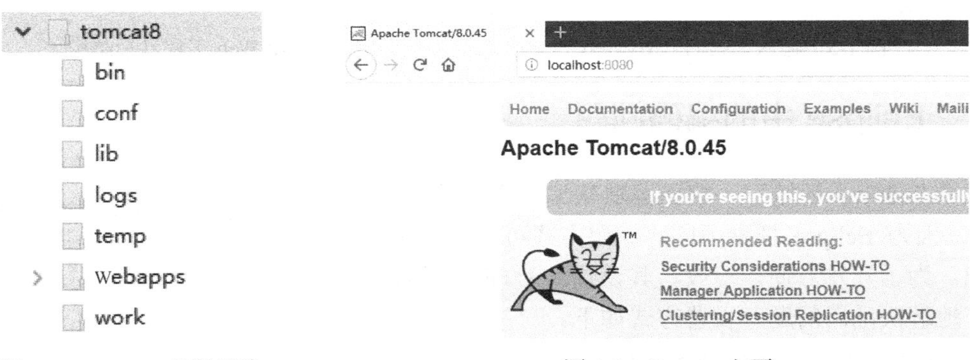

图 1-1　tomcat8 目录层次　　　图 1-2　Tomcat 主页

1.4　Eclipse 安装与使用

1.4.1　Eclipse 下载及创建 Dynamic Web Project

用户可以在 https://www.eclipse.org/downloads/下载最新版本的 Eclipse，注意是 Java EE 版。此处，使用的 Eclipse 版本为 eclipse-jee-oxygen-R-win32.zip，详见本书教学资源包 tools 文件

夹，解压缩后，双击 eclipse.exe，选择工作区路径，如图 1-3 所示，便可打开 Eclipse。

在 Eclipse 中选择“File→New→Project…”选项，在弹出的“New Project”对话框中选择 Web 中的“Dynamic Web Project”节点，如图 1-4 所示。在弹出的“New Dynamic Web Project”对话框中，输入工程名“MyWeb”，如图 1-5 所示，得到 MyWeb 工程目录结构，如图 1-6 所示。

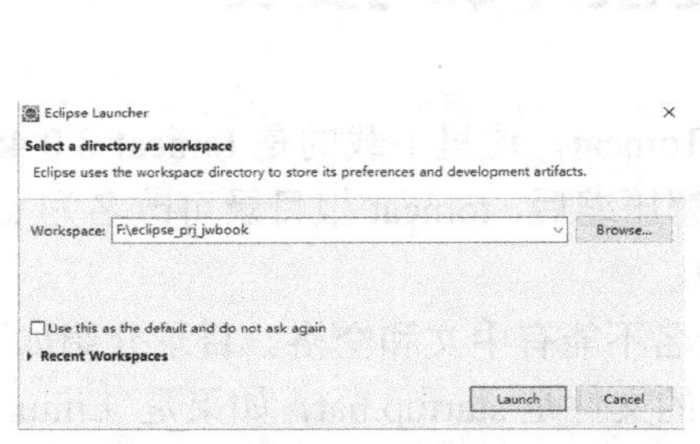

图 1-3 选择工作区路径

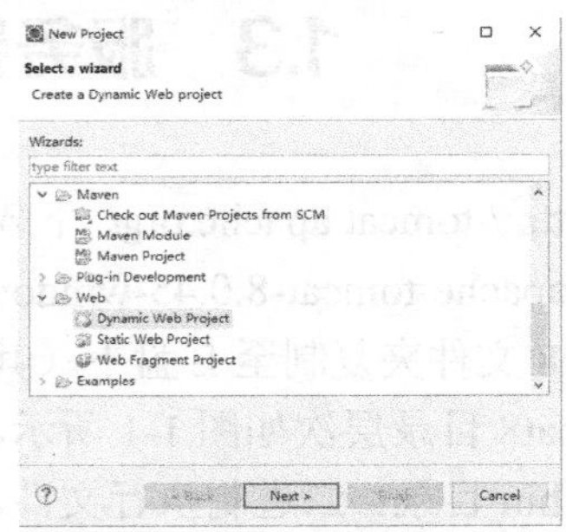

图 1-4 在 Eclipse 中创建 Dynamic Web Project

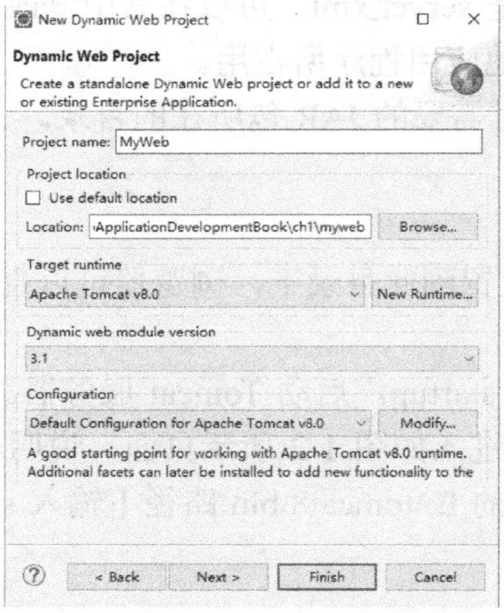

图 1-5 输入工程名“MyWeb”

图 1-6 MyWeb 工程目录结构

1.4.2 Eclipse 中的编码问题

在 Windows 环境下，编译器默认编码方式并不是 UTF-8，因为 Eclipse 插件不同，编码方式可能是 GBK 或其他。为了开发方便，一般将其设置为 UTF-8，这样能更好地解决乱码问题，设置方式如下。

1. 改变整个工作空间的编码格式

分别进行 Eclipse->Window->Preferences->General->Workspace->Other->UTF-8->OK 操作，如图 1-7 所示（汉化版：Eclipse->窗口->首选项->常规->工作空间->其他->UTF-8->确定操作）。

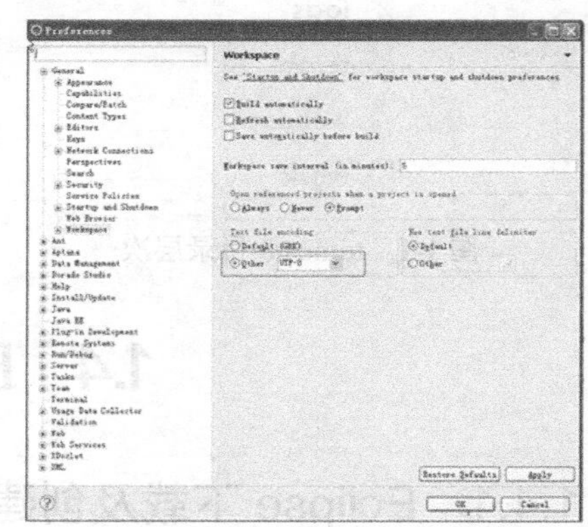

图 1-7 工作空间编码格式设置

2. 项目范围的编码格式设置

分别进行 Project->Properties->General->Resource->Inherited from container(UTF-8)->OK 操作，如图 1-8 所示。

3．某类型文件的编码格式设置

分别进行 Eclipse->Window->Preferences->General->Content Types->在右侧找到要修改的文件的类型（Java 等）->在下面的 Default encoding 输入框中输入 UTF-8->Update->OK 操作。

（汉化版：分别进行 Eclipse->窗口->首选项->常规->内容类型->在右侧选择要修改的文件的类型->下面省略编码输入->更新操作。）

譬如，Java 文件的编码格式设置：选择“Window->Preferences…选项，弹出首选项对话框，在左侧导航树中找到 General->Content Types，在右侧 Context Types 树中展开 Text，选择“Java Source File”节点，在下面的“Default encoding”输入框中输入“UTF-8”，单击“Update”按钮，即可设置 Java 文件编码为 UTF-8，如图 1-9 所示。

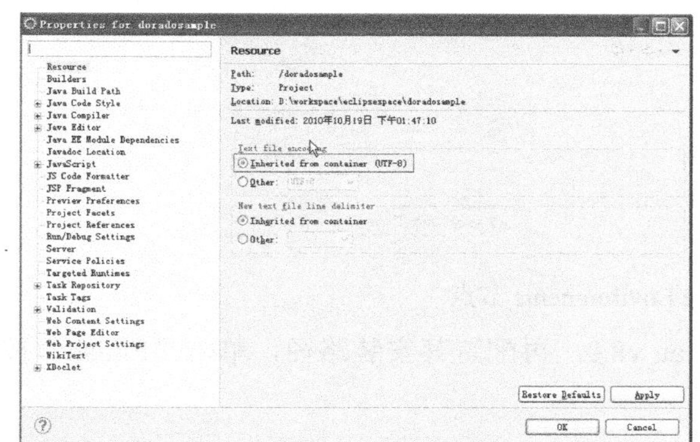

图 1-8　项目范围的编码格式设置

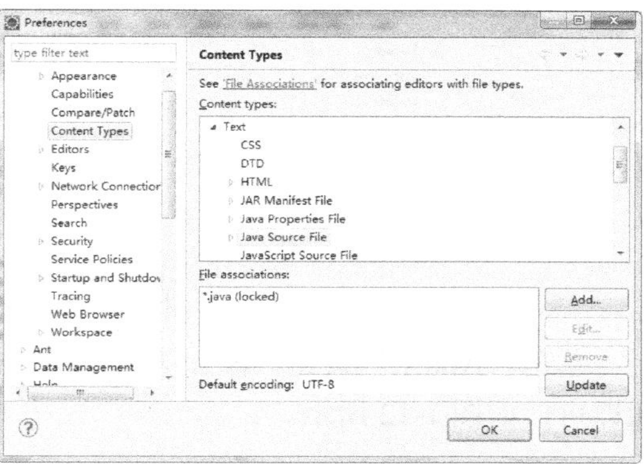

图 1-9　Java 文件的编码格式设置

又如，JSP 文件的编码格式设置：选择“Window->Preferences...选项，弹出首选项对话框，在左侧导航树中找到 Web->JSP Files，在右侧的“Encoding”下拉列表中选择“ISO 10646/Unicode(UTF-8)”，依次单击“Apply”和“OK”按钮，即可设置 JSP 文件编码为 UTF-8，如图 1-10 所示。

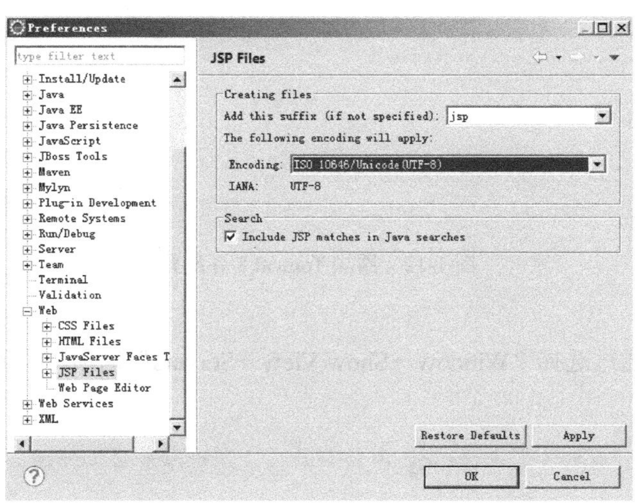

图 1-10　JSP 文件的编码格式设置

4．单个文件编码格式设置

在包资源管理器中右击文件，选择属性选项，改变文本文件编码格式为 UTF-8。

（注意：改变编码格式前应做备份，另外，可以用编码格式批量转换工具处理。）

1.4.3 将 Tomcat 和 Eclipse 相关联

1．配置 Server 的 Runtime Environments，添加 Apache Tomcat v8.0

打开 Eclipse，选择“Window→Preferences”选项，再选择左侧 Server 下的“Runtime Environments”节点，如图 1-11 所示。

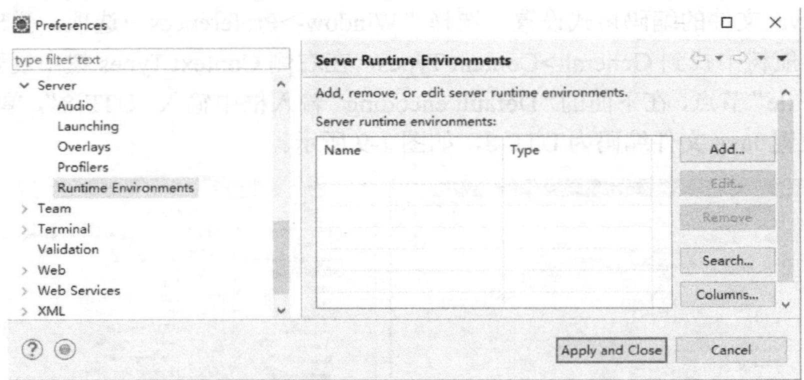

图 1-11 Runtime Environments 节点

单击“Add…”按钮，添加 Apache Tomcat v8.0，再配置其安装路径，单击“Finish”按钮即可，如图 1-12 所示。

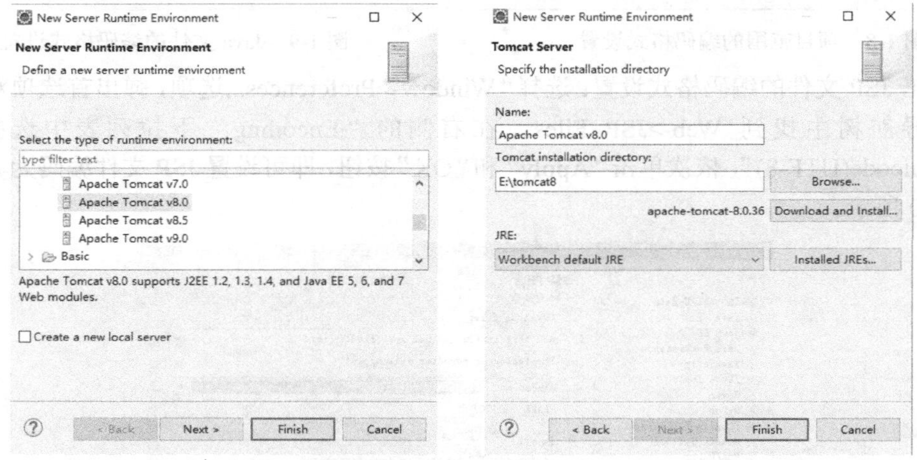

图 1-12 添加 Tomcat 8 并配置

2．创建 Server

在 Eclipse 中，通过选择“Window→Show View→Servers”选项，进入“Servers”选项卡，如图 1-13 所示。

图 1-13 “Servers”选项卡

单击图 1-13 中的超链接，弹出创建 Server 对话框，如图 1-14 所示，选择“Tomcat v8.0 Server”；单击“Next”按钮后，再单击“Finish”按钮即可。此时进入如图 1-15 所示的界面，得到已创建的“Tomcat v8.0 Server at localhost”。

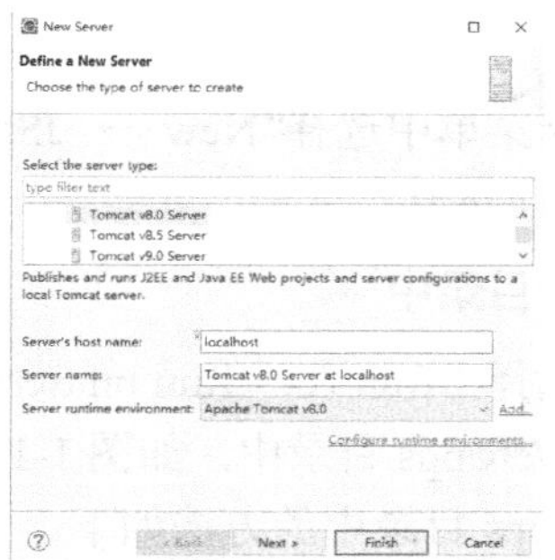

图 1-14　选择“Tomcat v8.0 Server”

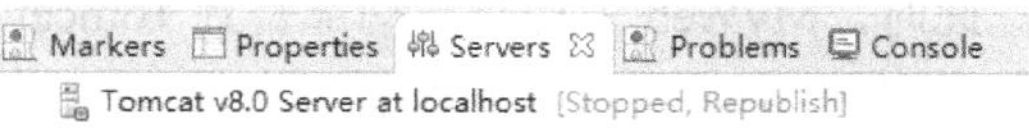

图 1-15　已创建的“Tomcat v8.0 Server at localhost”

1.4.4　Eclipse 自动部署项目到 Tomcat 的 webapps 目录

1．对 Server Locations 和 Server Options 进行配置

双击“Tomcat v8.0 Server at localhost”，或在其上右击，在打开的快捷菜单中选择“Open”选项。在进入的界面中，对 Server Locations 和 Server Options 进行配置，如图 1-16 所示，保存设置即可。

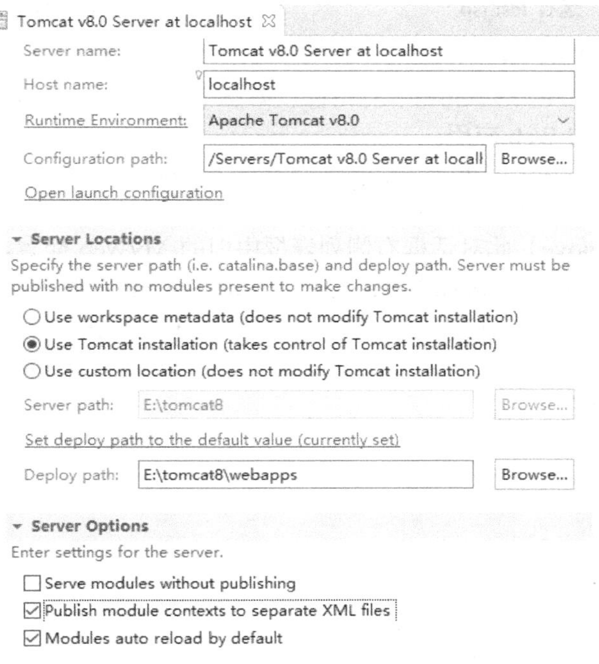

图 1-16　对 Server Locations 和 Server Options 进行配置

Server Locations 的配置中有三处需做修改。

（1）选中“Use Tomcat installation（takes control of Tomcat installation）”单选按钮。

（2）选中“Use Tomcat installation（takes control of Tomcat installation）”单选按钮后，Server Path 自动修改为 Tomcat8 实际安装位置。

（3）在“Deploy path”处修改为“Tomcat 根目录\webapps”。

在 Server Options 的配置中，选中如图 1-16 所示的“Server Options”选项组中的第 2 项

和第 3 项复选框即可。

2. 创建 JSP

右击 MyWeb 工程中的 WebContent 文件夹，在弹出的快捷菜单中选择“New → JSP File”，创建 test.jsp 页面，在 test.jsp 页面的<body>部分输入：This is a test!。

3. 将 Web 项目自动部署到 Tomcat 服务器的 Webapps 目录中

右击 MyWeb 工程的 test.jsp 文件，在弹出的快捷菜单中选择“Run As→Run on Server”选项即可。此时，MyWeb 工程会自动部署到 Tomcat 服务器的 Webapps 目录中，如图 1-17 所示。

另外一种将 Web 项目自动部署到 Tomcat 服务器的 webapps 目录中的方法如下。

先关闭前面运行的服务器。可通过单击“Servers”标签页的红色的“Stop the server”按钮，来停止 Tomcat 服务器，如图 1-18 所示。

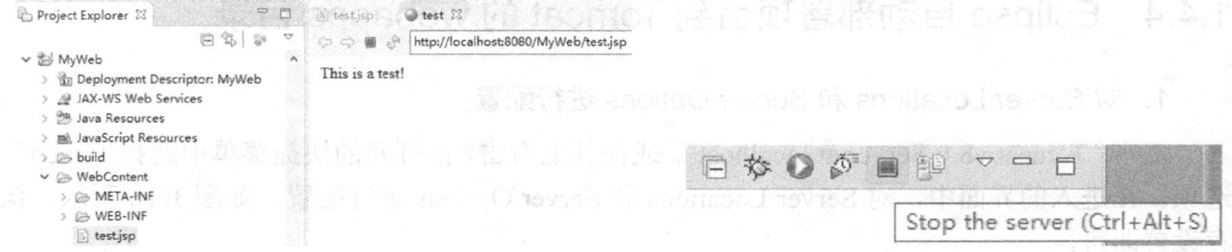

图 1-17 运行 test.jsp　　图 1-18 停止 Tomcat 服务器

右击“Tomcat v8.0 Server at localhost”，在弹出的快捷菜单中选择“Add and Remove…”选项，此时便弹出“Add and Remove”对话框，如图 1-19 所示。在这个对话框中配置部署发布到 Tomcat 服务器上的 Web 应用。

譬如，MyWeb 工程发生了更改，test.jsp 页面的<body>部分多了以下内容：

```
<br>发生更改，再次部署发布到服务器!<hr>
```

我们可删除（Remove）此对话框右侧列表框中旧的 MyWeb 配置，再从对话框左侧列表框中重新添加（Add），单击“Finish”按钮后重新启动 Tomcat 服务器（可通过单击“Servers”标签页的绿色的“Start the server”按钮来重新启动 Tomcat 服务器，在图 1-18 中也可以看到）。

启动外部的浏览器，这里使用的是火狐，访问 http://localhost:8080/MyWeb/test.jsp，便进入如图 1-20 所示的页面。

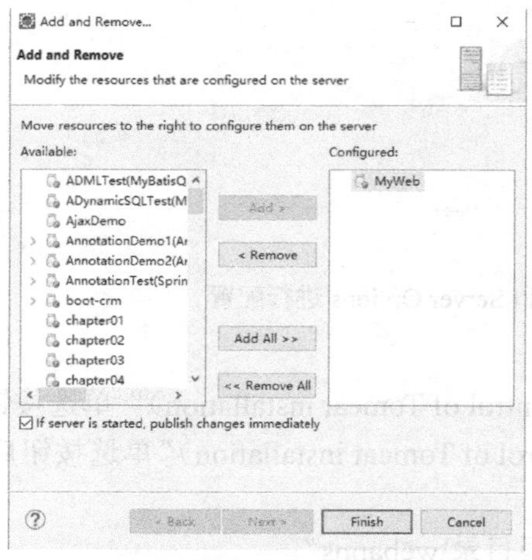

图 1-19 部署服务器上的 Web 应用

图 1-20 MyWeb 工程重新部署发布

1.5　MySQL 安装配置与使用

1. MySQL 下载、安装与配置

在 MySQL 官方网站 https://www.mysql.com/downloads/ 可下载 MySQL 最新版本。这里使用的 MySQL 版本是 mysql-5.5.19-win32.msi。

双击打开 mysql-5.5.19-win32.msi，在图 1-21（a）中单击“Next”按钮，进入图 1-21（b），选中“I accept the terms in the License Agreement”复选框，进入图 1-21（c）所示界面。

在图 1-21(c)中，选择安装类型，有“Typical”(默认)、“Custom”(用户自定义)、“Complete”（完全）三个选项，这里选择的是 Custom（定制）。

在图 1-21（d）中，可以选择安装路径及定制安装的组件，这里使用的是默认安装路径，定制安装的组件也是默认的。单击“Next”按钮后进入图 1-21（e）所示界面，单击“Install”按钮，开始安装，如图 1-21（f）和图 1-21（g）所示。

在图 1-21（h）中，选中“Launch the MySQL Instance Configuration Wizard”复选框，单击“Finish”按钮，完成 MySQL 的安装并启动 MySQL 配置向导。

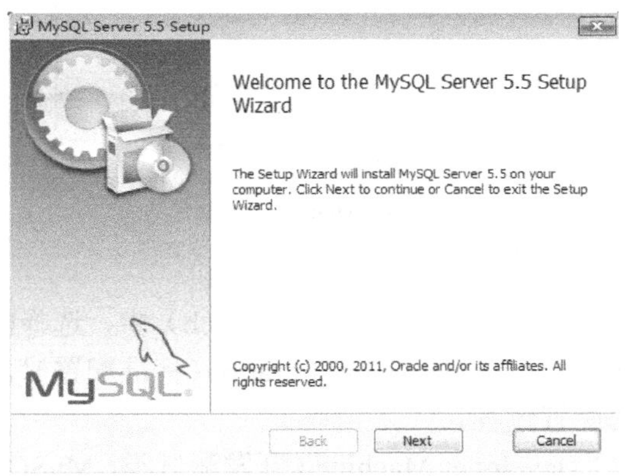

（a）MySQL 安装①

（b）MySQL 安装②

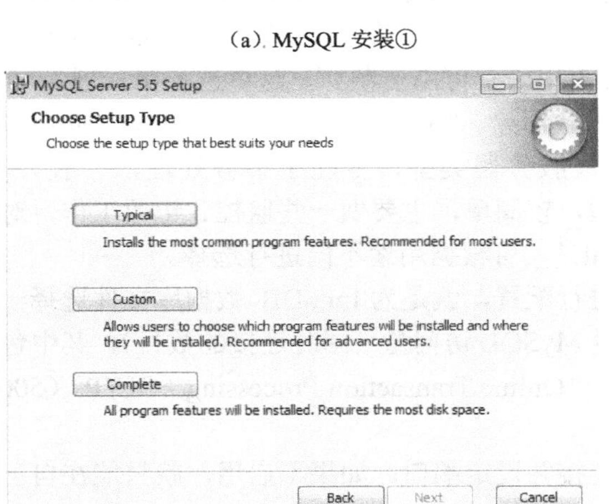

（c）MySQL 安装③

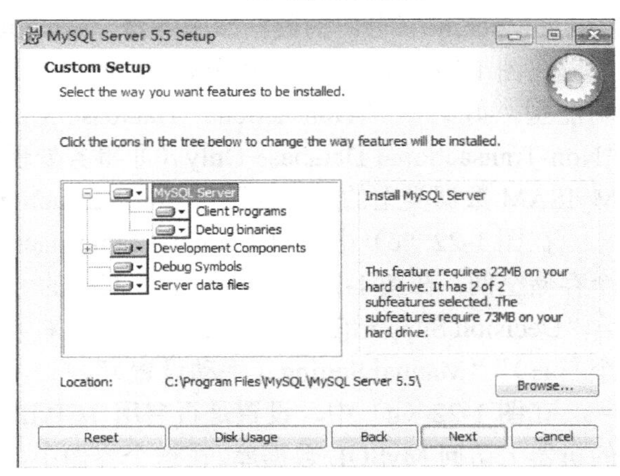

（d）MySQL 安装④

图 1-21　MySQL 安装

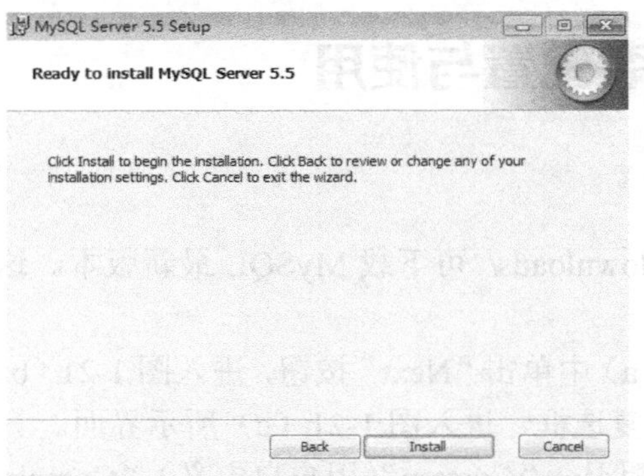

（e）MySQL 安装⑤

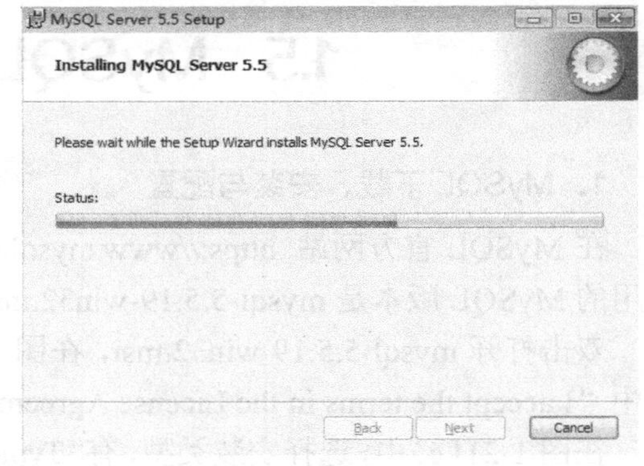

（f）MySQL 安装⑥

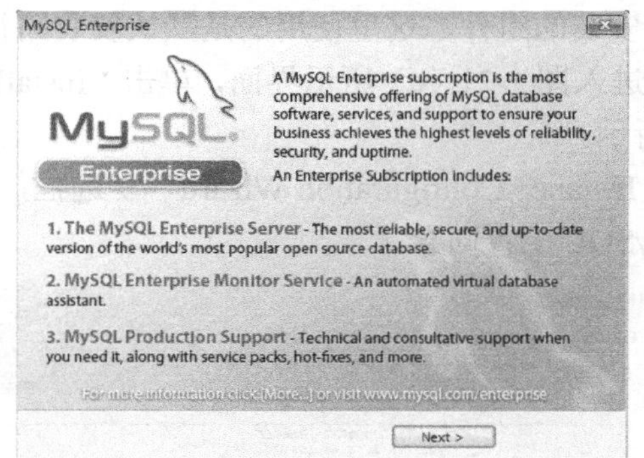

（g）MySQL 安装⑦

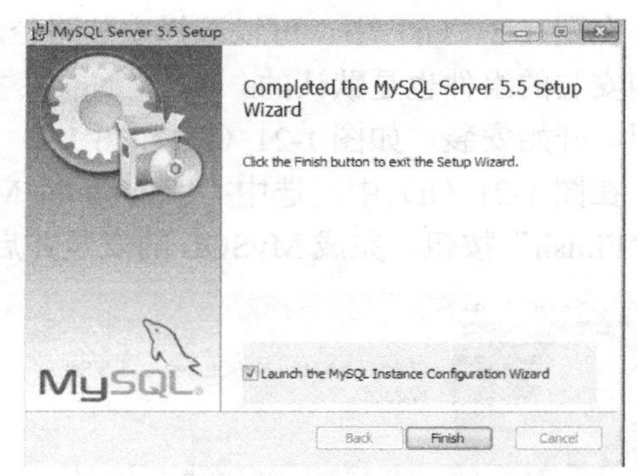

（h）MySQL 安装⑧

图 1-21　MySQL 安装（续）

下面开始配置 MySQL，在图 1-22（a）中单击“Next”按钮，在图 1-22（b）中，选择配置方式，包含“Detailed Configuration（手动精确配置）”“Standard Configuration（标准配置）”，这里选中“Detailed Configuration”，方便读者熟悉配置过程。

在图 1-22（c）中，选择服务器类型，其中包含“Developer Machine（开发测试类，MySQL 占用很少资源）”“Server Machine（服务器类型，MySQL 占用较多资源）”“Dedicated MySQL Server Machine（专门的数据库服务器，MySQL 占用所有可用资源）”。这里选择第一项。

在图 1-22（d）中，选择 MySQL 数据库的大致用途。“Multifunctional Database（通用多功能型，好）”　“Transactional Database Only（服务器类型，专注于事务处理，一般）”“Non-Transactional Database Only（非事务处理型，较简单，主要做一些监控、记数工作，对 MyISAM 数据类型的支持仅限于 non-transactional）”，可根据用途不同进行选择。

在图 1-22（e）中，对 InnoDB Tablespace 进行配置，就是为 InnoDB 数据库文件选择一个存储空间。在图 1-22（f）中，选择网站的一般 MySQL 访问量（同时连接的数目）。其中包含“Decision Support(DSS)/OLAP（20 个左右）”“Online Transaction Processing（OLTP）（500 个左右）”“Manual Setting（手动设置）”。

在图 1-22（g）中，设置是否启用 TCP/IP 连接并设定端口。如果不启用，就只能在自己的机器上访问 MySQL 数据库。在这个界面中，还可以选择启用标准模式（Enable Strict Mode），这样 MySQL 就不会允许细小的语法错误。如果使用者是新手，则建议取消标准模式以减少麻烦。但熟悉 MySQL 以后，应尽量使用标准模式，因为它可以降低有害数据进入数据库的可能性。

在图 1-22（h）中，对 MySQL 默认数据库语言编码进行设置，这比较重要。这里选择的编码是“utf8”。

在图 1-22（i）中，选择是否将 MySQL 安装为 Windows 服务，还可以指定 Service Name（服务标识名称），以及是否将 MySQL 的 bin 目录加入 Windows PATH。这里全部选中，Service Name 不变。

如果选中“Include Bin Directory in Windows PATH”复选框，则可以直接调用 bin 下的可执行文件，而不用指出可执行文件的完整路径。例如，连接 MySQL 服务器，在 Windows 的命令行窗口中输入命令“mysql.exe –u username –p”，再输入 password 即可，不用指出 mysql.exe 的完整路径，这样比较方便。

在图 1-22（j）中，询问是否要修改默认 root 用户（超级管理员）的密码（默认为空），如果要修改，就在“New root password”文本框中填入新密码（如果是重装，并且之前已经设置了密码，则在这里更改密码可能会出错，应留空，并取消选中“Modify Security Settings”复选框，安装配置完成后另行修改密码即可），在“Confirm（确认）”文本框中再输一遍密码，防止输错。“Enable root access from remote machines”表示是否允许 root 用户在其他机器上登录。如果要保证安全，就不要选中；如果要方便，就选中。“Create An Anonymous Account”表示是否新建一个匿名用户，匿名用户可以连接数据库，不能操作数据，包括查询，一般不选中，设置完毕，单击“Next”按钮。进入如图 1-22（k）所示的界面，在这个界面中，确认设置无误，如果有误，单击“Back”按钮可返回检查。最后单击“Execute”按钮使设置生效。

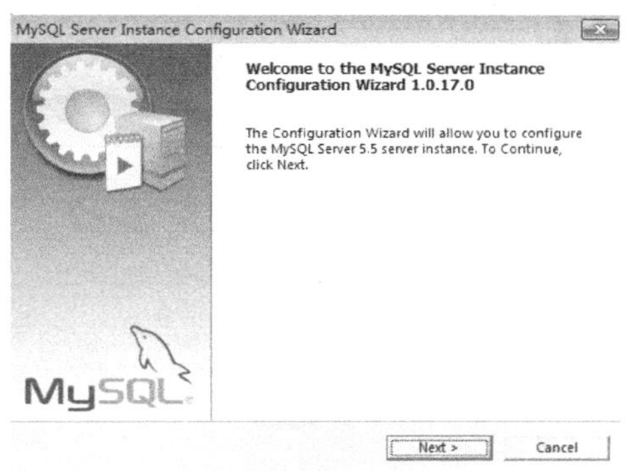

（a）MySQL 配置①

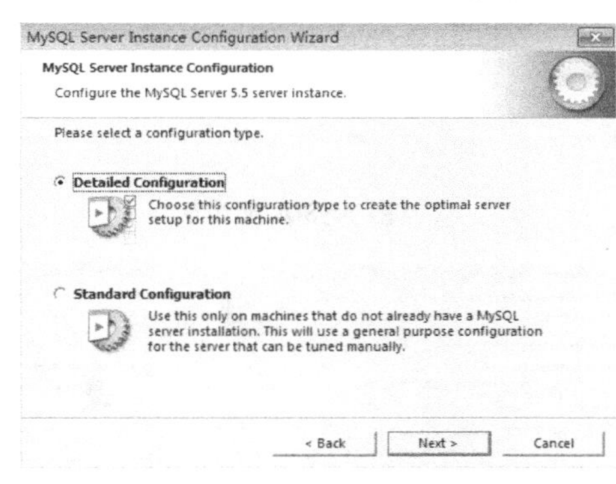

（b）MySQL 配置②

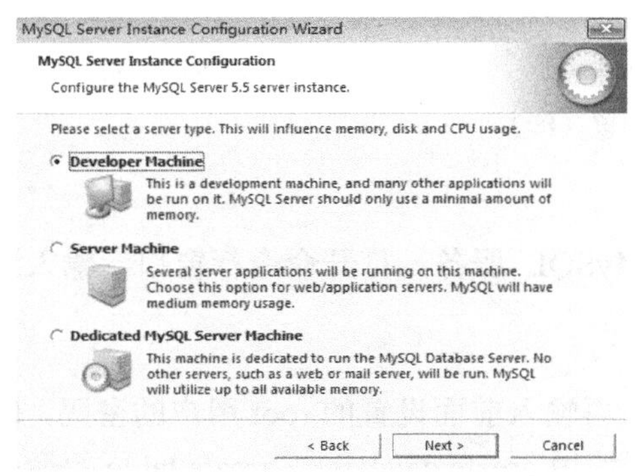

（c）MySQL 配置③

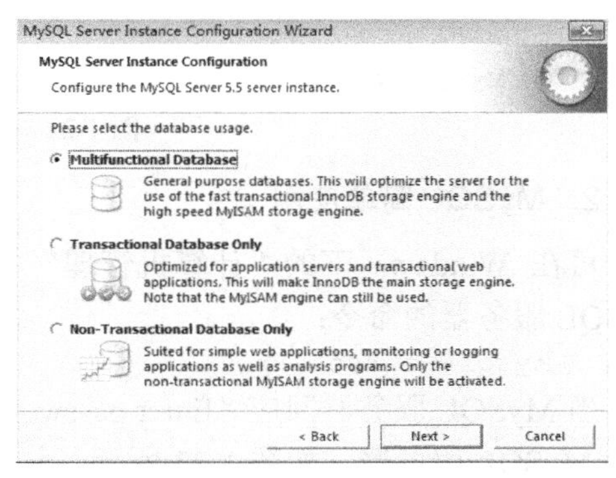

（d）MySQL 配置④

图 1-22　MySQL 配置

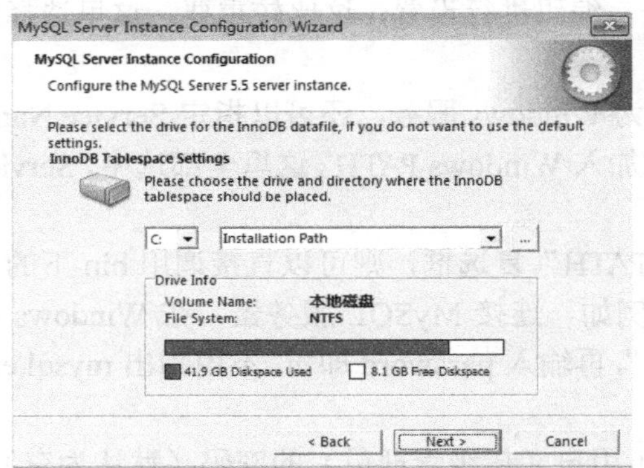

（e）MySQL 配置⑤

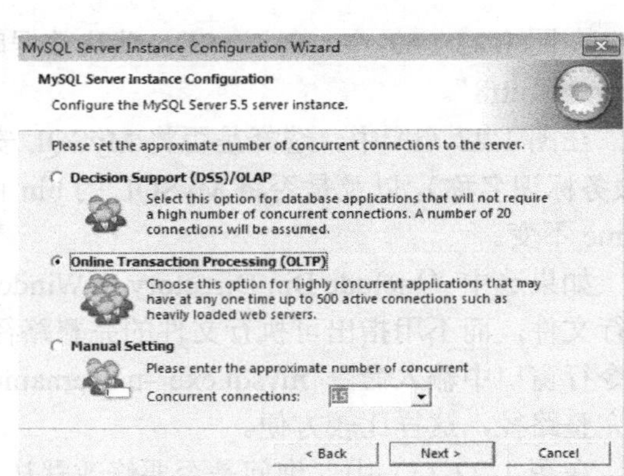

（f）MySQL 配置⑥

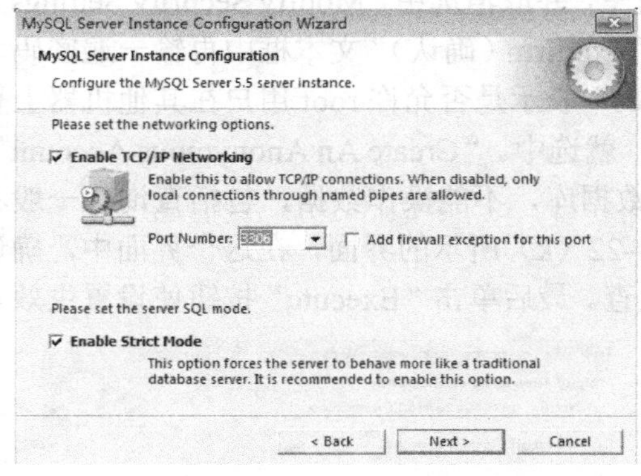

（g）MySQL 配置⑦

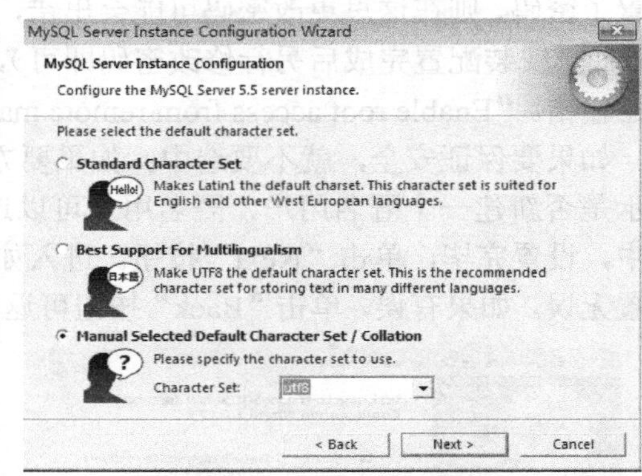

（h）MySQL 配置⑧

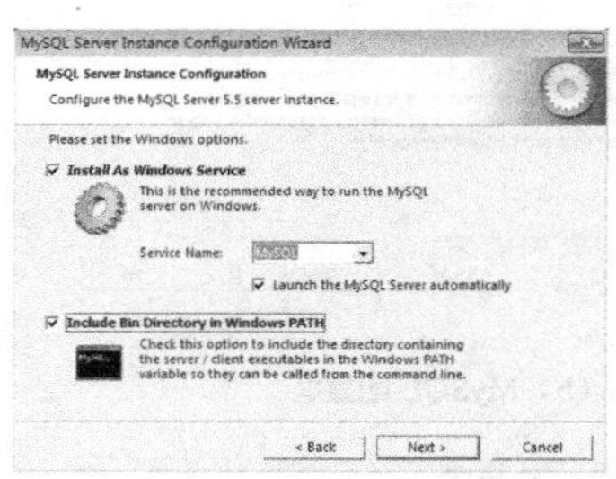

（i）MySQL 配置⑨

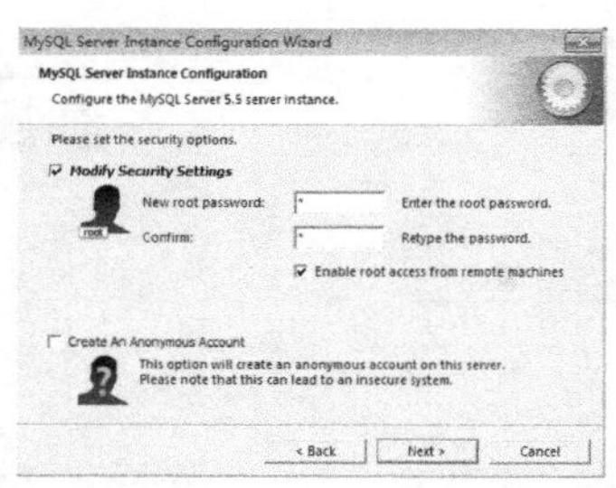

（j）MySQL 配置⑩

（k）MySQL 配置⑪

图 1-22　MySQL 配置（续）

2．MySQL 简单使用

可在 Windows 下的“计算机管理”中启动 MySQL 服务，打开命令行窗口，输入连接 MySQL 服务器的命令：

```
mysql -u root -p
```

当 MySQL 服务器回应“Enter password：”时，再输入前面设置的 root 用户的密码，即可连接 MySQL 服务器，如图 1-23 所示。此时就可以使用 create database、create table 等命令创建数据库和数据表了。

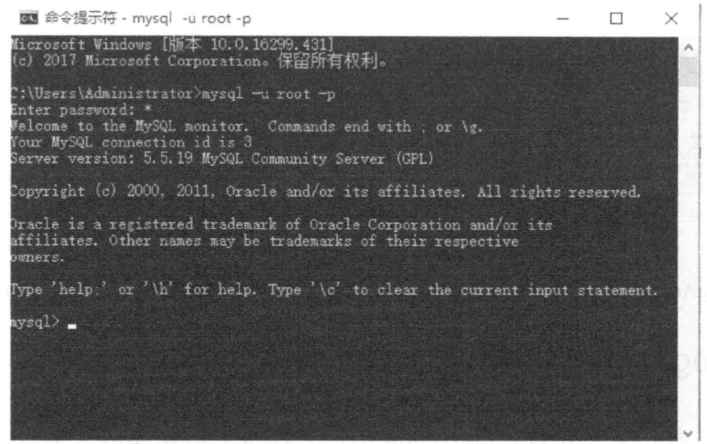

图 1-23　使用“mysql”命令连接 MySQL 服务器

1.6　PostgreSQL 安装与使用

PostgreSQL 由加州大学伯克利分校计算机系开发，是一个自由的对象-关系数据库服务器，它提供了相对其他开放源代码数据库系统（如 MySQL 和 Firebird）和专有系统（如 Oracle、Sybase、IBM 的 DB2 和 Microsoft SQL Server）之外的另一种选择。

1.6.1　PostgreSQL 下载与安装

使用浏览器打开网址 http://www.postgresql.org/download/，这是 PostgreSQL 官方下载地址，选择安装平台，本书以 Windows 版 PostgreSQL 为例，所以单击【Windows】超链接。在进入的新页面中，单击【Download】按钮，找到最新版本下方的图标，可以分别下载 Windows 32 位、64 位版本的 PostgreSQL，请根据自己的系统选择对应版本。

双击安装程序图标，默认安装即可。

1.6.2　使用 pgAdmin III 连接 PostgreSQL

pgAdmin III 是 PostgreSQL 默认的图形化管理软件，通过此管理软件可以连接本地和远程的 PostgreSQL 数据库，支持增、删、查、改、数据库及其表、视图等操作。

下面介绍如何使用 pgAdmin III 连接默认的 PostgreSQL 数据库。

（1）打开 pgAdmin III。从“开始”菜单中找到 PostgreSQL 程序组，找到并打开 pgAdmin III。打开后的软件主界面如图 1-24 所示。软件界面功能区分为以下几个部分：菜单栏、快捷按钮栏、对象浏览器、标签栏、SQL 窗口栏、状态栏。

（2）找到对象浏览器下的【服务器组】→【服务器】→【PostgreSQL 9.4(localhost：5432)】并双击，弹出如图 1-25 所示的对话框。输入安装时设置的密码，单击【确定】按钮进行连接，此处可以勾选【保存密码】复选框，这样下次连接时可以不用再次输入密码。

（3）如果密码输入正确，此时的对象浏览器会更新，展开后进入如图 1-26 所示的界面。至此，已经成功连接到默认的数据库引擎，并且可以看到此时的数据库引擎中有一个数据库，且名为 postgres，这个数据库下无任何表、触发器或者视图等内容。

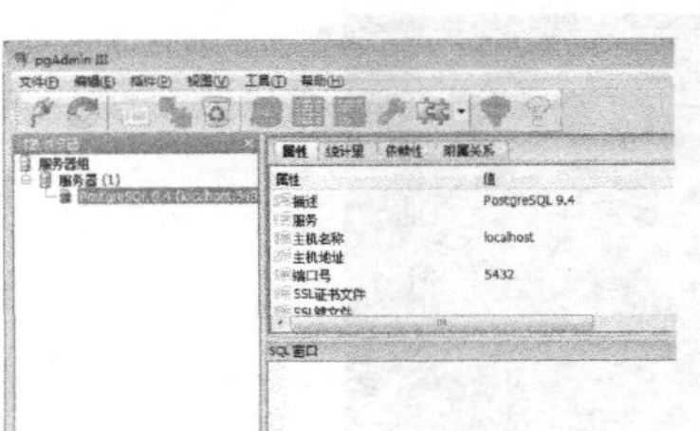

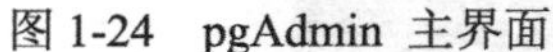

图 1-24 pgAdmin 主界面

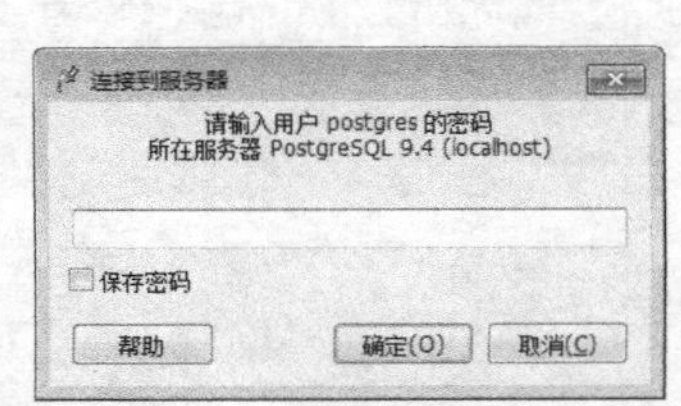

图 1-25 输入密码

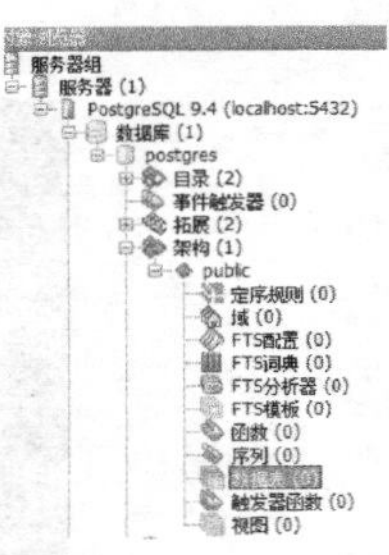

图 1-26 对象浏览器

1.6.3 使用 pgAdmin III 创建数据库

在创建数据库前，首先要打开 pgAdmin III，并连接上数据库服务器。

【示例】使用 pgAdmin III 创建数据库。

（1）右击对象浏览器中的“数据库”，在弹出的快捷菜单中选择“新建数据库...”选项。

（2）设置“新建数据库”对话框中的“属性”标签。“名称”输入“my_db”，“所有者”选择“postgres”，在“注释”中填写任意文本即可。

（3）设置“新建数据库”对话框中的“定义”标签。所有参数可以不用修改。“字符排序”和“字符分类”可以选择“Chinese (Simplified)_People's Republic of China.936”。“连接数限制”的“-1”代表了不限制连接数。

（4）设置“新建数据库”对话框中的“变量”标签。用户如有需要，可以自定义一些变量及其值，变量名必须是菜单中已经存在的变量。

（5）设置“新建数据库”对话框中的“权限”标签。用户如有需要，可以修改不同用户或组对本数据库的权限。

（6）设置“新建数据库”对话框中的“SQL”标签。用户可以在这里查看创建数据的 SQL 语句。单击【确定】按钮创建数据库，创建成功后，在对象浏览器中即可查看到相关信息。

1.7 Navicat Premium 安装与使用

Navicat Premium 是一套数据库管理工具，其结合其他 Navicat 成员，支持单一程序同时连接到 MySQL、MariaDB、SQL Server、SQLite、Oracle 和 PostgreSQL 数据库。它不仅符合专业开发人员的所有需求，对数据库服务器的新手来说学习起来也相当容易。

1. Navicat Premium 下载与安装

在 Navicat 官方授权经销商 http://www.formysql.com/xiazai.html，可下载 Navicat Premium 试用版：navicat120_premium_cs_x86.exe 和 navicat120_premium_cs_x64.exe。

假设操作系统是 32 位的，则双击 navicat120_premium_cs_x86.exe，便可以开始 Navicat Premium 的安装。Navicat Premium 安装图示如图 1-27 所示。

在图 1-27（c）中，可设置 Navicat Premium 的安装路径。在图 1-27（e）中，可设置是否创建 Navicat Premium 的桌面快捷方式图标。

（a）Navicat Premium 安装① 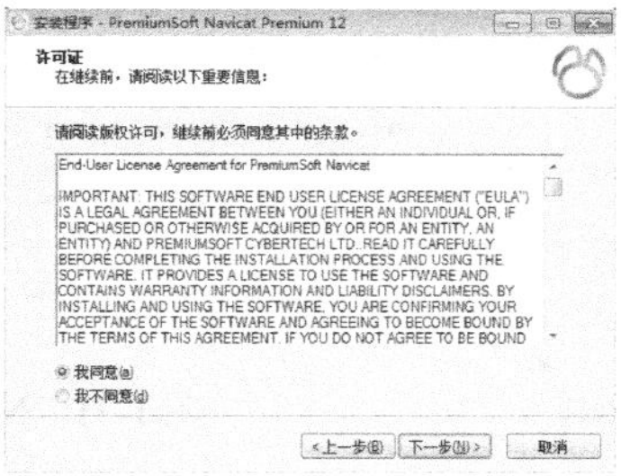　（b）Navicat Premium 安装② 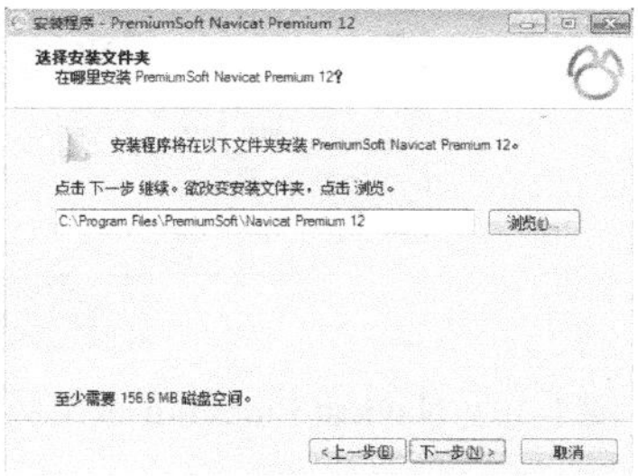　（c）Navicat Premium 安装③

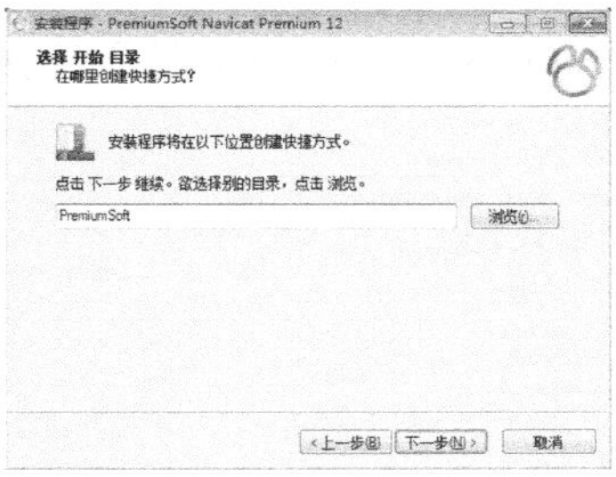

（d）Navicat Premium 安装④ 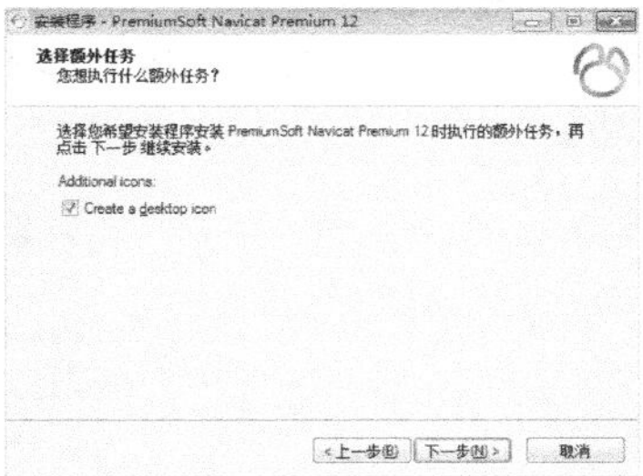　（e）Navicat Premium 安装⑤ 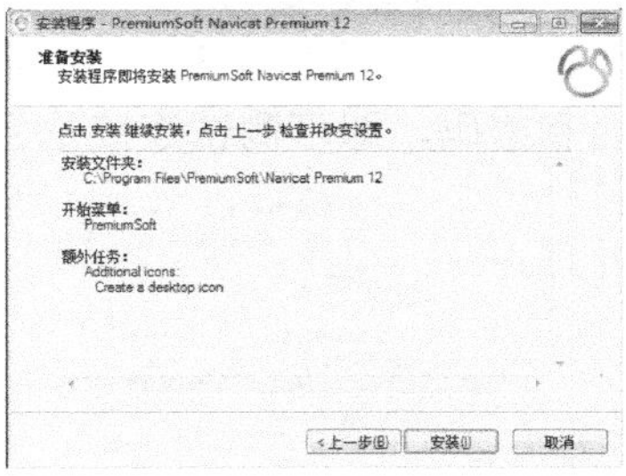　（f）Navicat Premium 安装⑥

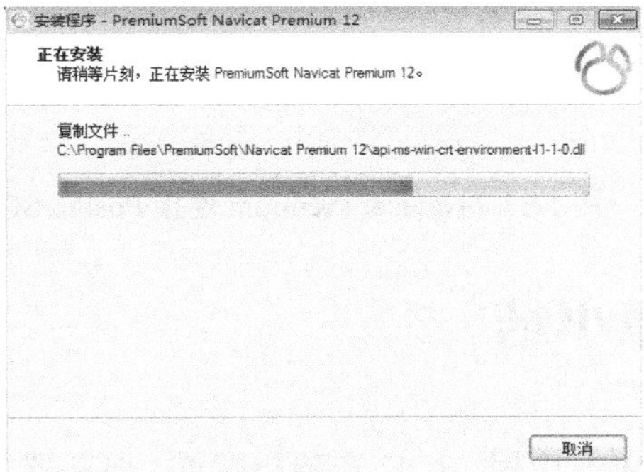

（g）Navicat Premium 安装⑦ 　（h）Navicat Premium 安装⑧

图 1-27　Navicat Premium 安装

也可以下载 Navicat Premium 11.0+破解补丁的安装包，根据自己的计算机是 32 位操作系统还是 64 位操作系统，选择相对应的安装包进行安装，如图 1-28 所示。

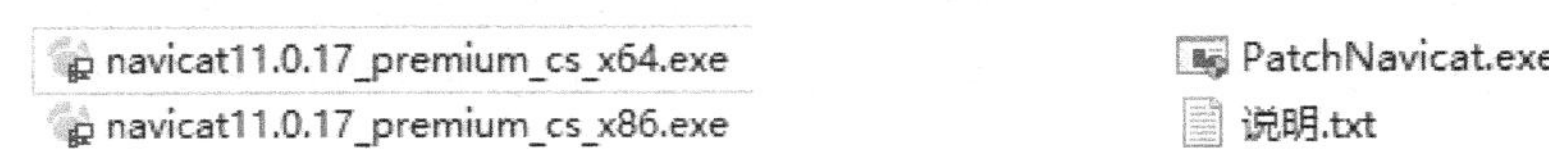

图 1-28　Navicat Premium 11.0 安装包　　图 1-29　Navicat Premium 11.0 破解版

打开破解补丁，把 PatchNavicat.exe 文件放在安装目录下运行，就可以成功激活软件。如图 1-29 所示。

1．Navicat Premium 简单使用

双击 Navicat Premium 桌面快捷方式，如果是试用版，则会有相应的提示，如图 1-30 所示。单击“试用”按钮，进入 Navicat Premium 主界面。

Navicat Premium 主界面如图 1-31 所示。

图 1-30　Navicat Premium 试用版提示

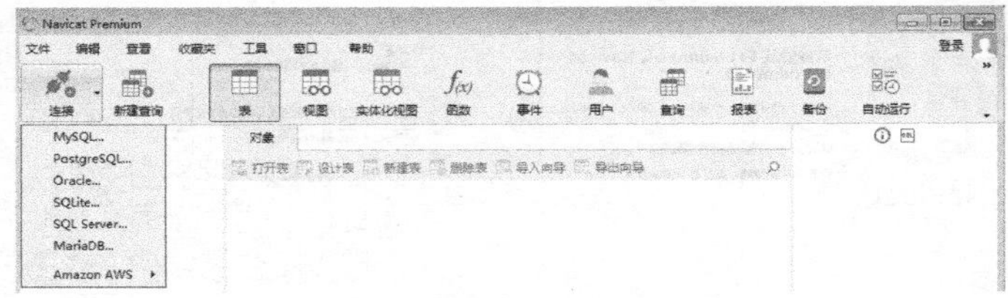

图 1-31　Navicat Premium 主界面

在 Navicat Premium 主界面中，单击工具栏中的“连接”图标，可选择连接到 MySQL 服务器、PostgreSQL 服务器以及其他数据库服务器。

图 1-32 和图 1-33 分别是 Navicat Premium 连接 MySQL 和 PostgreSQL 的图示。输入连接名和密码，即可测试连接是否成功。

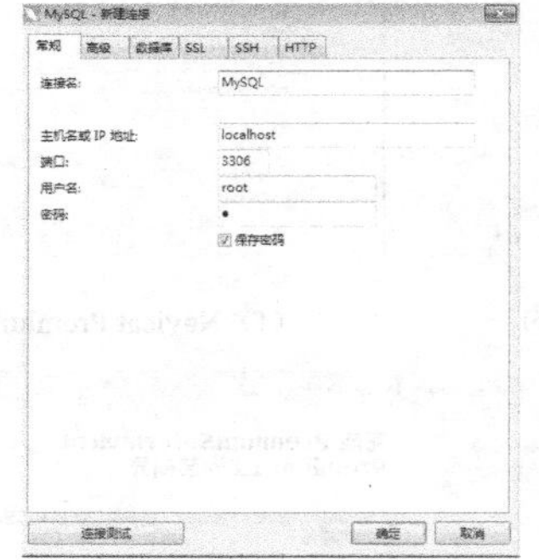

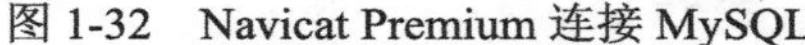

图 1-32　Navicat Premium 连接 MySQL

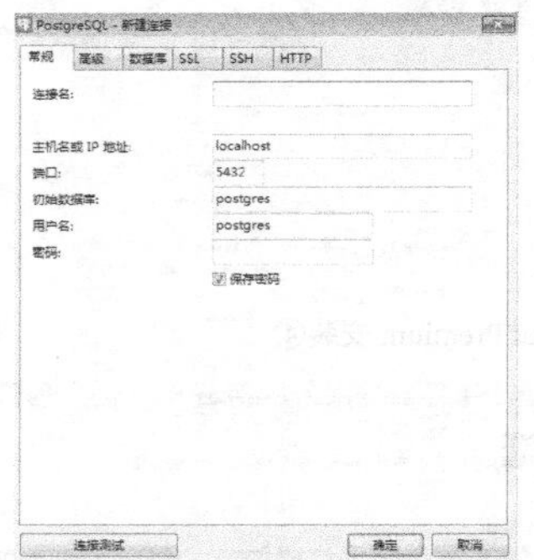

图 1-33　Navicat Premium 连接 PostgreSQL

1.8　本章小结

本章主要介绍了以下内容：基于 B/S 架构的 Web 应用、JDK 安装与配置、服务器 Tomcat 下载与安装、Eclipse 安装与使用、MySQL 安装配置与使用、PostgreSQL 安装与使用、Navicat Premium 安装与使用。

C/S 架构和 B/S 架构是当今世界网络程序开发体系结构的两大主流。

C/S 架构是一种典型的两层架构，即客户端/服务器端架构，其客户端包含一个或多个在用户的电脑上运行的程序；而服务器端有两种，一种是数据库服务器端，客户端通过数据库连接访问服务器端的数据，另一种是 Socket 服务器端，服务器端的程序通过 Socket 与客户端的程序通信。

B/S 架构是随着 Internet 技术的兴起，对 C/S 架构的一种变化或者改进的结构。在这种结构下，用户工作界面是通过 WWW 浏览器来实现的，极少部分事务逻辑在前端（Browser）实现，主要事务逻辑在服务器端（Server）实现。

基于 B/S 架构的 Web 应用，通常由客户端浏览器、Web 服务器和数据库服务器三部分组成。用户通过客户端浏览器向服务器端发送请求；服务器收到请求后，需要对用户发送过来的数据进行业务逻辑处理，多数还伴随对数据库的存取操作；最后，服务器将处理结果返回给客户端浏览器。

习题 1

1. B/S 架构和 C/S 架构有什么不同？什么是基于 B/S 架构的 Web 应用？
2. 在 Eclipse 中如何设置编码？
3. 如何将 Tomcat 和 Eclipse 相关联？
4. 在 Eclipse 中如何自动部署项目到 Tomcat 的 webapps 目录中？
5. 在 MySQL 中如何创建数据库和数据表？
6. 在 PostgreSQL 中如何创建数据库和数据表？
7. 如何安装和破解 Navicat Premium？
8. 如何使用 Navicat Premium 连接 PostgreSQL 数据库服务器？
9. 如何使用 Navicat Premium 连接 MySQL 数据库服务器？

第2章 Servlet编程

本章导读

Java Servlet 是运行在 Web 服务器或应用服务器上的程序，它是来自 Web 浏览器或其他 HTTP 客户端的请求和 HTTP 服务器上的数据库或应用程序之间的中间层。使用 Servlet，可以收集来自网页表单的用户输入，呈现来自数据库或者其他源的记录，还可以动态创建网页。本章主要内容有：（1）Servlet 简介；（2）Servlet 基础；（3）Servlet API 编程常用接口和类；（4）Servlet 处理表单数据；（5）Servlet 重定向和请求转发；（6）Servlet 数据库访问；（7）Servlet 异常处理；（8）异步 Servlet。

2.1 Servlet 简介

Servlet 在 Web 服务器的地址空间内执行。这样，其没有必要再创建一个单独的进程来处理每个客户端请求。在传统的 CGI 中，每个请求都要启动一个新的进程，如果 CGI 程序本身的执行时间较短，则启动进程所需要的开销很可能反而超过了实际执行时间。而在 Servlet 中，每个请求由一个轻量级的 Java 线程（而不是重量级的操作系统进程）处理。

在传统 CGI 中，如果有 *N* 个并发的对同一 CGI 程序的请求，则该 CGI 程序的代码在内存中重复装载了 *N* 次；而对于 Servlet，处理请求的是 *N* 个线程，只需要一份 Servlet 类代码。在性能优化方面，Servlet 也比 CGI 有着更多的选择。

Servlet 是独立于平台的，因为它们是用 Java 编写的。

Java 类库的全部功能对 Servlet 来说都是可用的。

Servlet 对请求的处理和响应过程可分为以下几个步骤。

（1）客户端发送请求至服务器端。

（2）服务器将请求信息发送至 Servlet。

（3）Servlet 生成响应内容并将其传给服务器。响应内容动态生成，通常取决于客户端的请求。

（4）服务器将响应返回给客户端。

1．Servlet 的存在就是要为客户端服务

Servlet 的任务是得到一个客户端的请求，再发回一个响应。请求可能很简单——“请给我一个欢迎页面。”，也可能很复杂——“为我的购物车结账。”这个请求携带着一些重要的数据，Servlet 代码必须知道怎么找到和使用这个请求。响应也携带着一些信息，浏览器需要这些信息来显示一个页面，Servlet 代码必须知道怎么发送这些信息。

2．Servlet 3.0

Servlet 3.0 作为 JavaEE 6 规范体系中的一员，随着 Java EE 6 规范一起发布。该版本在前一版本（Servlet 2.5）的基础上提供了若干新特性，以用于简化 Web 应用的开发和部署。

（1）**异步处理支持**：有了该特性，Servlet 线程不再需要一直阻塞，直到业务处理完毕才

能再输出响应，最后才能结束该 Servlet 线程。在接收到请求之后，Servlet 线程可以将耗时的操作委派给另一个线程来完成，自己在不生成响应的情况下返回至容器。针对业务处理较耗时的情况，这将大大减少服务器资源的占用，并且提高并发处理速度。

（2）**新增的注解支持**：该版本新增了若干注解，用于简化 Servlet、过滤器（Filter）和监听器（Listener）的声明，**这使得 Web.xml 部署描述文件从该版本开始不再是必选的了**。

3. Servlet 与 Tomcat 版本支持

以目前主流的 Web 服务器 Tomcat（包含 Servlet 容器）为例，Tomcat 对 Servlet 版本的支持关系如表 2-1 所示。

表 2-1 Servlet 与 Tomcat 版本支持

Servlet 版本号	Tomcat 版本号	Servlet 版本号	Tomcat 版本号
Servlet 2.4	Tomcat 5.x	Servlet 3.0	Tomcat 7.x
Servlet 2.5	Tomcat 6.x	Servlet 3.1	Tomcat 8.x

2.2 Servlet 基础

2.2.1 用记事本写一个 Servlet

1. 编写 HelloServlet.java

使用记事本编写 HelloServlet.java 文件，代码如下：

```
import javax.servlet.ServletException;
import java.io.*;
import javax.servlet.http.*;
public class HelloServlet extends HttpServlet
{
 //重写了父类 HttpServlet 中的 doGet()方法，用于对 GET 请求方法做出响应
   public void doGet(HttpServletRequest req, HttpServletResponse resp)
          throws ServletException,IOException
   {
       PrintWriter out = resp.getWriter();  //得到 PrintWriter 对象
       out.println("Hello Servlet");  //向客户端发送字符数据
       out.close();
   }
}
```

2. 编译 HelloServlet.java

把 Tomcat 自带的 servlet-api.jar 的路径加到 classpath 中，或者在命令行窗口中运行命令：set classpath=%classpath%;E:\tomcat8\lib\servlet-api.jar。

（这里，本书 Tomcat 的安装路径是 E:\tomcat8。）

进入 HelloServlet.java 所在目录，运行命令：javac HelloServlet.java 。编译后得到 HelloServlet.class 字节码文件。

3. 部署 Servlet

在 E:\tomcat8\webapps 目录下，创建一个文件夹 hello，在 hello 目录下创建 Web-INF 文件

夹和 Web.xml 文件。Web.xml 代码如下（其中<!-- --> 部分表示注释）：

```
<?xml version="1.0" encoding="UTF-8"?>                    <!--XML声明-->
<Web-app xmlns:xsi="http://www.w3.org/2001/XMLSchema-instance"
 xmlns="http://xmlns.jcp.org/xml/ns/javaee"
 xsi:schemaLocation="http://xmlns.jcp.org/xml/ns/javaee
 http://xmlns.jcp.org/xml/ns/javaee/Web-app_3_1.xsd"
 id="WebApp_ID" version="3.1">       <!--声明XML Schema的名称空间-->
 <servlet>
     <servlet-name>helloworld</servlet-name>          <!--Servlet名称-->
     <servlet-class>HelloServlet</servlet-class> <!--Servlet完全限定名-->
   </servlet>
 <servlet-mapping>
     <servlet-name>helloworld</servlet-name>          <!--Servlet名称-->
     <url-pattern>/myhello</url-pattern>          <!--Servlet的URL路径-->
   </servlet-mapping>
</Web-app>
```

在 Web-INF 目录下创建 classes 文件夹，把前面编译得到的 HelloServlet.class 复制至 E:\tomcat8\webapps\hello\Web-INF\classes 目录下。

在这里，hello 文件夹已成为一个 Web 应用，只不过是手工创建的。

4. 访问 HelloServlet

启动 Tomcat 服务器，在浏览器中输入：http://localhost:8080/hello/myhello。

（注意，这里通过调用</servlet-mapping>中的<url-pattern>值来找寻相应的 Servlet。）

网页输出：

```
Hello Servlet
```

2.2.2 Servlet 体系结构

Servlet 3.1 API 有以下 4 个 Java 包。

① javax.servlet：其中包含定义 Servlet 和 Servlet 容器之间契约的类和接口。

② javax.servlet.http：其中包含定义 HTTP Servlet 和 Servlet 容器之间契约的类和接口。

③ javax.servlet.annotation：其中包含标注 Servlet、Filter、Listener 的标注。它还为被标注元件定义元数据。

④ javax.servlet.descriptor：其中包含提供程序化登录 Web 应用程序的配置信息的类型。

Servlet 技术的核心是 Servlet，它是所有 Servlet 类必须直接或间接实现的一个接口。Servlet 接口定义了 Servlet 与 Servlet 容器之间的契约，Servlet 容器将 Servlet 类载入内存，并在 Servlet 实例上调用具体的方法。用户请求致使 Servlet 容器调用 Servlet 的 Service 方法，并传入一个 ServletRequest 实例和一个 ServletResponse 实例。ServletRequest 中封装了当前的 HTTP 请求。ServletResponse 则表示当前用户的 HTTP 响应，使得将响应发回给用户变得十分容易。

2.2.3 Servlet 接口

在 Tomcat 的有关 Servlet API 的文档中有关于 Servlet 接口的介绍。可以通过运行 Tomcat

服务器，在 http://localhost:8080/docs/servletapi/index.html 页面看到。

javax.servlet.Servlet 的定义如下：

```
public interface Servlet
```

该接口中定义了以下 5 种方法。

① init()：在 Servlet 实例化后，Servlet 容器会调用该方法，初始化该对象。init()方法有一个类型为 ServletConfig 的参数，Servlet 容器通过这个参数向 Servlet 传递配置信息。Servlet 使用 ServletConfig 对象从 Web 应用程序的配置信息中获取以名-值对形式提供的初始化参数。

② service()：容器调用 service()方法来处理客户端的请求。

③ destroy()：Servlet 的销毁方法。容器在终止 Servlet 服务前调用此方法。

④ getServletConfig()：该方法返回容器调用 init()方法时传递给 Servlet 对象的 ServletConfig 对象，ServletConfig 对象包含了 Servlet 的初始化参数。

⑤ getServletInfo()：返回一个 String 类型的字符串，其中包括了关于 Servlet 的信息。例如，作者、描述信息等。

2.2.4 Servlet 生命周期

Servlet 生命周期可被定义为从创建直到毁灭的整个过程。以下是 Servlet 遵循的过程。

（1）Servlet 通过调用 init ()方法进行初始化。

（2）Servlet 调用 service()方法来处理客户端的请求。

（3）Servlet 通过调用 destroy()方法终止（结束）。

（4）Servlet 是由 JVM 的垃圾回收器进行垃圾回收的。

1. init() 方法

init()方法被设计成只调用一次。它在第一次创建 Servlet 时被调用，在后续每次用户请求时不再调用。Servlet 创建于用户第一次调用对应于该 Servlet 的 URL 时，但也可以指定 Servlet 在服务器第一次启动时被加载。

当用户调用一个 Servlet 时，就会创建一个 Servlet 实例，每一个用户请求都会产生一个新的线程，适当的时候移交给 doGet()或 doPost()方法。init()方法简单地创建或加载一些数据，这些数据将被用于 Servlet 的整个生命周期。

init 方法的定义如下：

```
public void init(ServletConfig config) throws ServletException {
  // 初始化代码...
}
```

2. service() 方法

service()方法是执行实际任务的主要方法。Servlet 容器调用 service()方法来处理来自客户端（浏览器）的请求，并把格式化的响应写回客户端。

每次服务器接收到一个 Servlet 请求时，服务器会产生一个新的线程并调用服务。service()方法检查 HTTP 请求类型（GET、POST、PUT、DELETE 等），根据请求类型不同分别调用 doGet、doPost、doPut、doDelete 等方法。

```
public void service(ServletRequest request, ServletResponse response)
      throws ServletException, IOException{
}
```

service()方法由容器调用，只需要根据来自客户端的请求类型来重写 doGet()或 doPost()方法即可。doGet()和 doPost()方法是每次服务请求中最常用的方法。

3. doGet() 方法

GET 请求来自一个 URL 的正常请求，或者来自一个未指定 METHOD 的 HTML 表单，它由 doGet()方法处理。

```
public void doGet(HttpServletRequestrequest,HttpServletResponse response)
    throws ServletException, IOException {
    // Servlet 代码
}
```

4. doPost() 方法

POST 请求来自一个特别指定了 METHOD 为 POST 的 HTML 表单，它由 doPost()方法处理。

```
public void doPost(HttpServletRequest request,
                   HttpServletResponse response)
    throws ServletException, IOException {
    // Servlet 代码
}
```

5. destroy() 方法

destroy()方法只会被调用一次，在 Servlet 生命周期结束时被调用。destroy()方法可以让 Servlet 关闭数据库连接、停止后台线程、把 Cookie 列表或单击计数器写入磁盘中，并执行其他类似的清理活动。在调用 destroy()方法之后，servlet 对象被标记为垃圾回收。destroy()方法定义如下所示：

```
public void destroy() {
    //清理代码...
}
```

2.2.5 Servlet 生命周期示例

Servlet 实质上就是 Java 类，我们将使用 Eclipse 来开发 Servlet，以进行 Servlet 的学习。

1. 创建 Dynamic Web Project

在 Eclipse 中选择“File→New→Project...”选项，在弹出的“New Project”对话框中选择 Web 中的“Dynamic Web Project”，如图 2-1 所示。在弹出的“New Dynamic Web Project”对话框中，设定工程名为“ServletDemo”，如图 2-2 所示（此处所使用的 Eclipse 版本详见本书教学资源 tools 文件夹下 eclipse-jee-oxygen-R-win32.zip，将其解压缩后双击 eclipse.exe 即）。

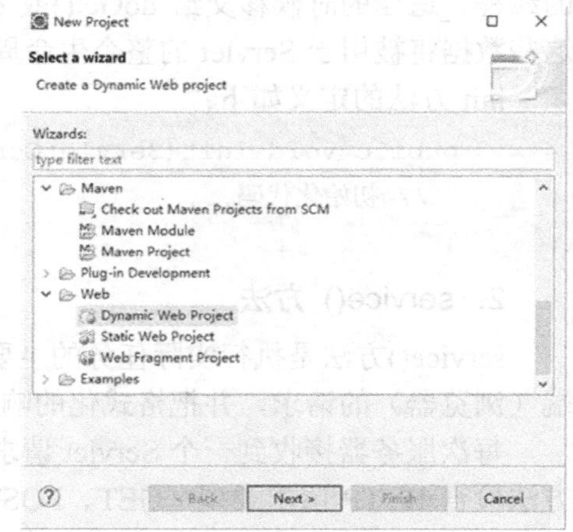

图 2-1 在 Eclipse 中创建 Dynamic Web Project

在如图 2-3 所示的 ServletDemo 工程目录结构中，右击 src 文件夹，创建 Package 名为“com.mialab.servlet_demo”，再在 com.mialab.servlet_demo 包中创建 Servlet。

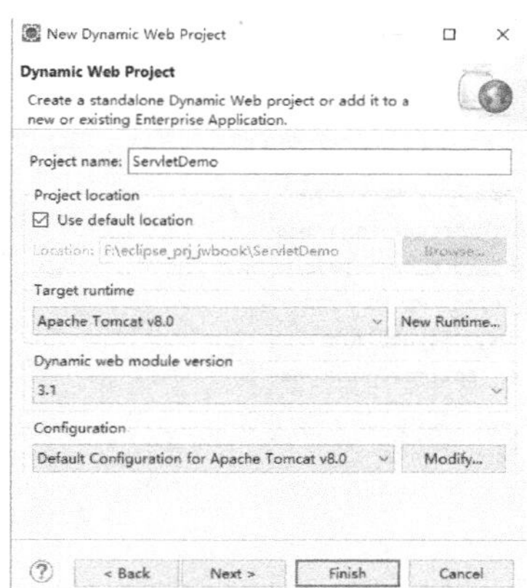

图 2-2　输入工程名“ServletDemo”

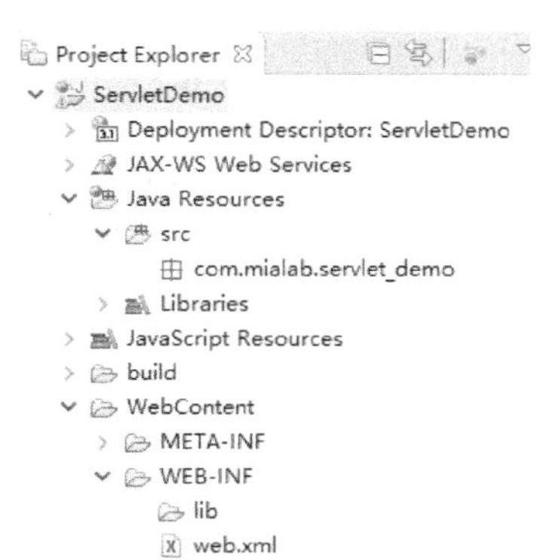

图 2-3　ServletDemo 工程目录结构

2. 开发 Servlet

我们在 com.mialab.servlet_demo 包中创建 LifeServlet。右击 com.mialab.servlet_demo 包，选择“New→Servlet”选项，弹出“Create Servlet”对话框，在此对话框中输入 Servlet 的名称“LifeServlet”，如图 2-4 所示。单击“Next”按钮后，选中“init”、“destroy”等 method，如图 2-5 所示。

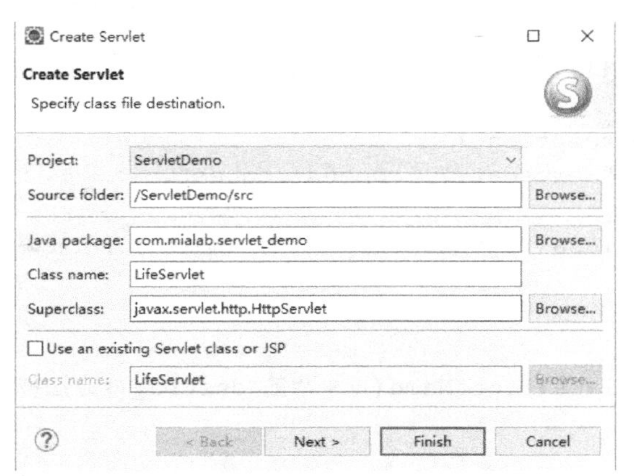

图 2-4　输入 Servlet 的名称“LifeServlet”

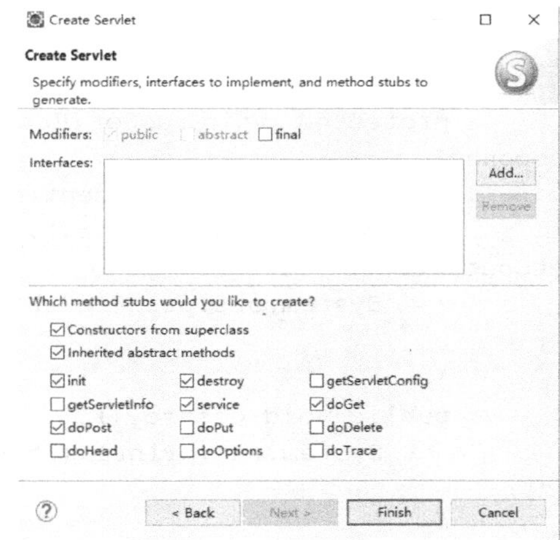

图 2-5　勾选欲生成的 method 代码框架

如果 servlet-api.jar 未加入 ClassPath（或者 Java Build Path），便会出现如图 2-6 所示的错误。我们只需把鼠标指针移入代码错误处，如把鼠标指针移入“HttpServlet”中，稍停片刻，便会弹出一个错误信息提示框，如图 2-7 所示，单击“Fix project setup…”超链接。

```
/**
 * Servlet implementation class LifeServlet
 */
@WebServlet("/LifeServlet")
public class LifeServlet extends HttpServlet
```

图 2-6　servlet-api.jar 未加入 Java Build Path

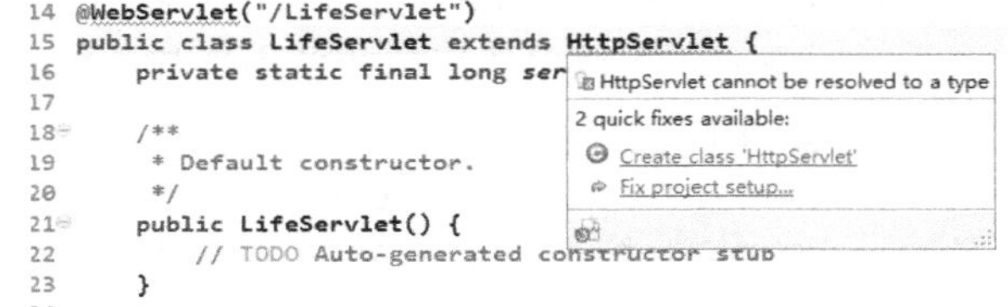

图 2-7　错误信息提示框

在弹出的“Project Setup Fixes”对话框中，把 E:\tomcat8\lib 中的 servlet-api.jar 加入 ServletDemo 工程的 Build Path，如图 2-8 所示。

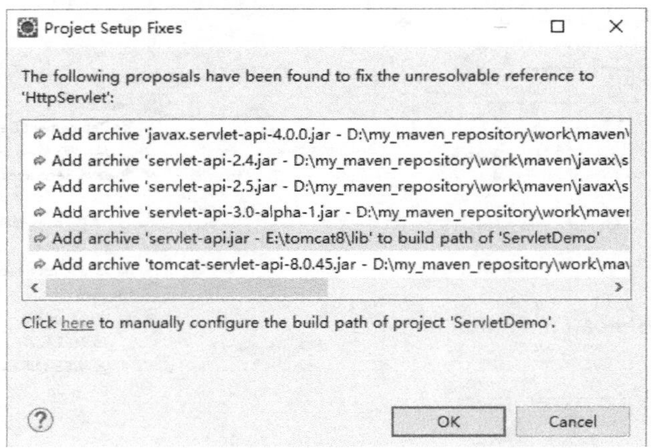

图 2-8 servlet-api.jar 加入 Build Path

LifeServlet.java 主要代码如下：

```
@WebServlet("/LifeServlet")
public class LifeServlet extends HttpServlet {
  public LifeServlet() {
      System.out.println(this.getClass().getName() + "的构造方法被调用");
    }
  public void init(ServletConfig config) throws ServletException {
      System.out.println(this.getClass().getName() + "的 init()方法被调用");
  }
  protected void doGet(HttpServletRequest request, HttpServletResponse
response)
      throws ServletException, IOException {
      response.getWriter().append("Servedat:").append(request.
getContextPath());
      System.out.println(this.getClass().getName() + "的 doGet()方法被调用
");
  }
  public void destroy() {
      System.out.println(this.getClass().getName() + "的 destroy()方法被调
用");
  }
}
```

3. 部署并测试

在 Eclipse 中部署并运行 LifeServlet。启动 Eclipse 中配置好的 Tomcat，访问 LifeServlet 的 URL 为 http://localhost:8080/ServletDemo/LifeServlet。

（可参见第 1 章：Eclipse 自动部署项目到 Tomcat 的 webapps 目录。也可以右击 LifeServlet.java，在弹出的快捷菜单中选择“Run As→Run on Server”选项。）

Eclipse 中的控制台将输出：

```
com.mialab.servlet_demo.LifeServlet 的构造方法被调用
com.mialab.servlet_demo.LifeServlet 的 init()方法被调用
com.mialab.servlet_demo.LifeServlet 的 doGet()方法被调用
```

打开命令行窗口，在 E:\tomcat8\bin 路径下输入 shutdown 命令，关闭 Tomcat 服务器，如图 2-9 所示。

Eclipse 中的控制台将输出：

com.mialab.servlet_demo.LifeServlet 的 destroy()方法被调用

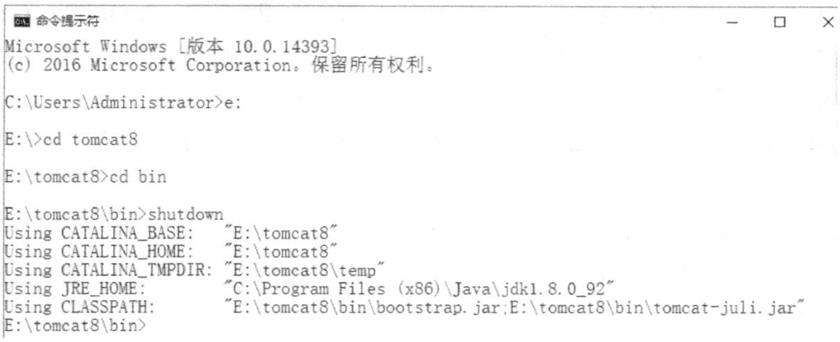

```
命令提示符
Microsoft Windows [版本 10.0.14393]
(c) 2016 Microsoft Corporation。保留所有权利。

C:\Users\Administrator>e:

E:\>cd tomcat8

E:\tomcat8>cd bin

E:\tomcat8\bin>shutdown
Using CATALINA_BASE:   "E:\tomcat8"
Using CATALINA_HOME:   "E:\tomcat8"
Using CATALINA_TMPDIR: "E:\tomcat8\temp"
Using JRE_HOME:        "C:\Program Files (x86)\Java\jdk1.8.0_92"
Using CLASSPATH:       "E:\tomcat8\bin\bootstrap.jar;E:\tomcat8\bin\tomcat-juli.jar"
E:\tomcat8\bin>
```

图 2-9　关闭 Tomcat 服务器

2.3　Servlet API 编程常用接口和类

2.3.1　GenericServlet 类

GenericServlet 类的部分源码如下：

```
package javax.servlet;
…
public abstract class GenericServlet implements Servlet, ServletConfig,
        java.io.Serializable {
    …
    //用 transient 关键字标记的成员变量不参与序列化过程
    private transient ServletConfig config;

    @Override
    public ServletConfig getServletConfig() {
        return config;
    }
    @Override
    public ServletContext getServletContext() {
        return getServletConfig().getServletContext();
    }
    @Override
    public String getInitParameter(String name) {
        return getServletConfig().getInitParameter(name);
    }
    @Override
    public void init(ServletConfig config) throws ServletException {
        this.config = config;
        this.init();
    }
    public void init() throws ServletException {
        // NOOP by default
    }
    …
}
```

GenericServlet 类是一个抽象类，实现了 Servlet 和 ServletConfig 接口，其作用如下。

（1）将 init 方法中的 ServletConfig 赋给一个类中成员（ServletConfig config），以便可以通过调用 get ServletConfig 来获取。（关于 transient 关键字的说明：Java 的 serialization 提供了一种持久化对象实例的机制。当持久化对象时，可能有一个特殊的对象数据成员，我们不想用 serialization 机制来保存它。为了在一个特定对象的域上关闭 serialization，可以在这个域前加上关键字 transient。当一个对象被序列化的时候，transient 型变量的值不包括在序列化的表示中，然而，非 transient 型的变量是被包括进去的。）

（2）为 Servlet 接口中的所有方法提供默认的实现。

（3）可以通过覆盖没有参数的 init 方法来编写初始化代码，ServletConfig 则仍然由 GenericServlet 实例保存。

（4）开发者可以在不用获得 ServletConfig 对象的情况下直接调用 ServletConfig 的方法，如上述代码中的 getServletContext()方法。

【示例】使用 GenericServlet 类。

关于使用 GenericServlet 类的示例，可参见 MyGenericServlet.java 的代码（源码包见本书资源第 2 章 ServletDemo 工程中的 src/com.mialab.servlet_demo 包），访问 MyGenericServlet 的 URL 是 http://localhost:8080/ServletDemo/generic。

2.3.2 HttpServlet 类

HttpServlet 类扩展了 GenericServlet 类，其部分代码如下：

```
package javax.servlet.http;
…
public abstract class HttpServlet extends GenericServlet {
    …
    private static final String METHOD_DELETE = "DELETE";
    private static final String METHOD_HEAD = "HEAD";
    private static final String METHOD_GET = "GET";
    private static final String METHOD_OPTIONS = "OPTIONS";
    private static final String METHOD_POST = "POST";
    private static final String METHOD_PUT = "PUT";
    private static final String METHOD_TRACE = "TRACE";

    protected void doGet(HttpServletRequest req, HttpServletResponse resp)
        throws ServletException, IOException {
        …
    }
    protected void doPost(HttpServletRequest req, HttpServletResponse resp)
        throws ServletException, IOException {
        ...
    }
    //添加新的 Service 方法
    protected void service(HttpServletRequestreq,HttpServletResponseresp)
        throws ServletException, IOException {
        String method = req.getMethod();
        if (method.equals(METHOD_GET)) {
          …
           doGet(req, resp);
```

```
        } else if (method.equals(METHOD_HEAD)) {
            …
            doHead(req, resp);
        } else if (method.equals(METHOD_POST)) {
            doPost(req, resp);
        }…
        else {…
        }
    }

    //覆盖的 Service 方法
    @Override
    public void service(ServletRequest req, ServletResponse res)
        throws ServletException, IOException {
        HttpServletRequest  request;
        HttpServletResponse response;
        try {
            request = (HttpServletRequest) req;
            response = (HttpServletResponse) res;
        } catch (ClassCastException e) {
            throw new ServletException("non-HTTP request or response");
        }
        service(request, response);
    }
}
```

HttpServlet 覆盖了 GenericServlet 中的 Service 方法，并添加了一个新 Service 方法。

新 Service 方法接收的参数是 HttpServletRequest 和 HttpServletResponse，而不是 ServletRequest 和 ServletResponse。

原始的 Service 方法将 Servlet 容器的 request 和 response 对象分别转换成 HttpServletRequest 和 HttpServletResponse，并调用新的 Service 方法。

HttpServlet 中的 Service 方法会检验用来发送请求的 HTTP 方法（通过调用 request.getMethod 来实现）并调用以下方法之一：doGet、doPost、doHead、doPut、doTrace、doOptions 和 doDelete。在这 7 种方法中，doGet 和 doPost 是最常用的。所以，不再需要覆盖 Service 方法了，只须覆盖 doGet 或者 doPost 即可。

2.3.3 ServletConfig 接口

Tomcat 初始化一个 Servlet 时，会将该 Servlet 的配置信息封装到一个 ServletConfig 对象中，通过调用 init(ServletConfig config)方法将 ServletConfig 对象传递给 Servlet。

ServletConfig 接口的常用方法如表 2-2 所示。

表 2-2 ServletConfig 接口的常用方法

方法说明	功能描述
ServletContext getServletContext()	获取 ServletContext 对象
String getServletName()	返回当前 Servlet 的名称
String getInitParameter(String name)	根据初始化参数名称返回对应的初始化参数值
Enumeration<String> getInitParameterNames()	返回一个 Enumeration 对象，其中包含了所有的初始化参数名

【示例】ServletConfig 接口的使用。

ServletConfigDemoServlet.java 主要代码如下：

```
public class ServletConfigDemoServlet extends HttpServlet {
  private ServletConfig servletConfig;
  @Override
    public void init(ServletConfig config) throws ServletException {
      this.servletConfig = config;
      System.out.println("-----------" + servletConfig + "-----------");
  }
    @Override
    public ServletConfig getServletConfig() {
       return servletConfig;
    }
    @Override
  protected void doGet(HttpServletRequest request, HttpServletResponse response)
       throws ServletException, IOException {
      // 使用 ServletConfig 对象获取初始化参数
      ServletConfig servletConfig = getServletConfig();
      System.out.println("-----------" + servletConfig + "-----------");
      String poet = servletConfig.getInitParameter("poet");
      String poem = servletConfig.getInitParameter("poem");
      // 设置响应到客户端的文本类型为 HTML
      response.setContentType("text/html;charset=UTF-8");
      // 获取输出流
      PrintWriter out = response.getWriter();
      out.print("<p>获取 ServletConfigDenoServlet 的初始化参数：");
      out.println("</p><p>poet 参数的值:" + poet);
      out.println("</p><p>poem 参数的值:" + poem + "</p>");
      out.append("Served at: ").append(request.getContextPath());
  }
}
```

Web.xml 中关于 ServletConfigDemoServlet 的配置如下：

```
<servlet>
  <servlet-name>myServletConfig</servlet-name>
  <servlet-class>com.mialab.servlet_demo.ServletConfigDemoServlet</servlet-class>
  <init-param>
     <param-name>poet</param-name>
     <param-value>纳兰容若</param-value>
  </init-param>
  <init-param>
     <param-name>poem</param-name>
     <param-value>我是人间惆怅客</param-value>
  </init-param>
 </servlet>
<servlet-mapping>
  <servlet-name>myServletConfig</servlet-name>
  <url-pattern>/myServletConfig</url-pattern>
 </servlet-mapping>
```

访问 ServletConfigDemoServlet 的 URL：

http://localhost:8080/ServletDemo/myServletConfig

输出结果：

```
获取 ServletConfigDemoServlet 的初始化参数：
poet 参数的值：纳兰容若
poem 参数的值：我是人间惆怅客
Served at: /ServletDemo
```

2.3.4 HttpServletRequest 接口

HttpServletRequest 接口继承自 ServletRequest 接口，专门用来封装 HTTP 请求消息。由于 HTTP 请求消息分为请求行、请求消息头和请求消息体 3 部分，故而在 HttpServletRequest 接口中定义了获取请求行、请求消息头和请求消息体的相关方法，以及存取请求域属性的方法。

1. 获取请求行信息

HTTP 请求报文的请求行由请求方法、请求 URL 和请求协议及版本组成。HttpServletRequest 接口对请求行各部分信息的获取方法如表 2-3 所示。

表 2-3 获取请求行的相关方法

方法说明	功能描述
String getMethod()	用于获取 HTTP 请求消息中的请求方式（如 GET、POST 等）
String getRequestURI()	用于获取请求行中资源名称部分，即位于 URL 的主机和端口之后、参数部分之前的部分
String getQueryString()	用于获取请求行中的参数部分，也就是资源路径后面（？）以后的所有内容。其只对 GET 有效
String getProtocol()	获取请求行中的协议名和版本
String getContextPath()	获取请求 URL 中属于 Web 应用程序的路径。这个路径以“/”开头，表示整个 Web 站点的根目录，路径结尾不含“/”
String getServletPath()	获取 Servlet 所映射的路径
String getRemoteAddr()	用于获取请求客户端的 IP 地址
String getRemoteHost()	用于获取请求客户端的完整主机名
String getScheme()	用于获取请求的协议名，如 HTTP 等
StringBuffer getRequestURL()	获取客户端发出请求时的完整 URL，包括协议、服务器名、端口号、资源路径等信息，但不包括后面的查询参数部分

【示例】使用 HttpServletRequest 接口获取请求行信息。

关于使用 HttpServletRequest 接口获取请求行信息的示例，可参见 RequestLineServlet.java 的代码（源码包见本书资源第 2 章 ServletDemo 工程中的 src/com.mialab.servlet_demo 包），访问 Request- LineServlet 的 URL 是 http://localhost:8080/ServletDemo/RequestLineServlet。

2. 获取请求消息头

当请求 Servlet 时，需要通过请求头向服务器传递附加信息。HTTP 常见的请求头如表 2-4 所示。

表 2-4 HTTP 常见的请求头

请求头名称	说　明
Host	初始 URL 中的主机和端口
Connection	表示是否需要持久连接
Accept	浏览器可接收的 MIME 类型
Accept-Charset	浏览器发给服务器，声明浏览器支持的编码类型

续表

请求头名称	说　明
Content-Length	Content-Length 用于描述 HTTP 消息实体的传输长度。在 HTTP 协议中，消息实体长度和消息实体的传输长度是有区别的，例如，在 GZIP 压缩下，消息实体长度是压缩前的长度，消息实体的传输长度是 GZIP 压缩后的长度
Cache-Control	指定请求和响应遵循的缓存机制
User-Agent	User Agent 的中文名为用户代理，简称 UA，它是一个特殊字符串头，使得服务器能够识别客户使用的操作系统及版本、CPU 类型、浏览器类型及版本、浏览器渲染引擎、浏览器语言、浏览器插件等
Content-Type	表示请求内容的 MIME 类型
Cookie	表示客户端的 Cookie 信息，即告诉服务器，上次访问时，服务器写了哪些 cookie 到客户端

HttpServletRequest 接口中定义的用于获取 HTTP 请求头字段的方法，如表 2-5 所示。

表 2-5　获取请求消息头的方法

请求头名称	说　明
String getHeader(String name)	用于获取一个指定头字段的值
Enumeration getHeaderNames()	用于获取一个包含所有请求头字段的 Enumeration 对象
Enumeration getHeaders(String name)	返回一个 Enumeration 集合对象，该集合对象由请求消息中出现的某个指定名称的所有头字段值组成
int getIntHeader(String name)	用于获取指定名称的头字段，并将其值转换为 int 类型
Cookie[] getCookies()	返回一个数组包含客户端请求的所有 Cookie 对象，如果没有发送任何信息，则这个方法返回 null
String getContentType()	用于获取 Content-Type 头字段的值
int getContentLength()	用于获取 Content-Length 头字段的值
String getCharacterEncoding()	用于返回请求消息的实体部分的字符集编码，通常是从 Content-Type 头字段中提取

【示例】使用 HttpServletRequest 接口获取请求头信息。

使用 HttpServletRequest 接口获取请求头信息的示例，可参见 RequestHeadInfoServlet.java 的代码（源码包见本书资源第 2 章 ServletDemo 工程 src/com.mialab.servlet_demo 包），访问 RequestLineServlet 的 URL 是 http://localhost:8080/ServletDemo/RequestHeadInfoServlet。

3．获取请求参数

在实际开发中，经常需要获取用户提交的表单数据。为了方便获取表单中的请求参数，在 HttpServletRequest 接口的父类 ServletRequest 接口中定义了一系列获取请求参数的方法，如表 2-6 所示。

表 2-6　获取请求参数的方法

方法说明	功能描述
String getParameter(String name)	用于获取某个指定名称的参数值
Enumeration getParameterNames()	返回一个包含请求消息中所有参数名的 Enumeration 对象
String[] getParameterValues(String name)	根返回由 name 指定的用户请求参数对应的一组值
Map getParameterMap()	用于将请求消息中的所有参数名和值装进一个 Map 对象中返回

【示例】使用 HttpServletRequest 接口获取请求参数。

关于使用 HttpServletRequest 接口获取请求参数的示例，可参见 RequestParaServlet.java 和 person_info.html（源码包见本书资源第 2 章的 ServletDemo 工程），先访问 person_info.html，填写个人信息，再提交给 RequestParaServlet 处理。

4．存取请求域属性

存储在 HttpServletRequest 对象中的对象称为请求域属性，属于同一个请求的多个处理组件之间可以通过请求域属性来传递对象数据。

HttpServletRequest 接口与请求域属性相关的方法如表 2-7 所示。

表 2-7　存取请求域属性的方法

方法说明	功能描述
Object getAttribute(String name)	从请求域（Request Scope）中获取 name 属性的值
Enumeration getAttributeNames()	获取所有的请求域（Request Scope）中的属性
void setAttribute(String name, Object value)	设定 name 属性的值为 value，保存在请求域中
void removeAttribute(String name)	从请求域（Request Scope）中移除 name 属性的值

【示例】存取请求域属性。

HttpServletRequest 接口存取请求域属性的示例，可参见 RequestScopeAttrServlet.java 的代码（源码包见本书资源第 2 章 ServletDemo 工程的 src/com.mialab.servlet_demo 包），访问 RequestScope AttrServlet 的 URL 是 http://localhost:8080/ServletDemo/RequestScopeAttrServlet。

2.3.5　HttpServletResponse 接口

HttpServletResponse 接口继承自 ServletResponse 接口，专门用来封装 HTTP 响应消息。由于 HTTP 响应消息分为状态行、响应消息头、消息体三部分，于是在 HttpServletResponse 接口中也相应定义了向客户端发送响应状态码、响应消息头、响应消息体的方法。

1．响应状态码

HTTP 协议响应报文的响应行，由报文协议和版本，以及状态码和状态描述构成。

当 Servlet 向客户端回送响应消息时，需要在响应消息中设置状态码。常见的响应状态码如下：200 表示请求成功；500 表示服务器内部错误；404 表示请求的资源（网页等）不存在；302 表示资源（网页等）暂时转移到其他 URL 等。

HttpServletResponse 接口提供的设置状态码并生成响应状态行的方法有以下几种。

（1）setStatus(int status)方法：setStatus(int status)方法用于设置 HTTP 响应消息的状态码，并生成响应状态行。正常情况下，Web 服务器会默认产生一个状态码为 200 的状态行。

（2）sendError(int sc)方法和 sendError(int sc, String msg)方法：第 1 个方法只是发送错误信息的状态码；而第 2 个方法除了发送状态码外，还可以增加一条用于提示说明的文本信息，该文本信息将出现在发送给客户端的正文内容中。

要说明的是，在实际开发中，一般不需要人为地修改设置状态码，容器会根据程序的运行状况自动响应发送响应的状态码。

2．响应消息头

HttpServletResponse 接口中定义了一些设置 HTTP 响应头字段的方法，如表 2-8 所示。

表 2-8 设置响应消息头字段的方法

方法	功能描述
void setContentType(String type)	设置 Servlet 输出内容的 MIME 类型
void setContentLength(int len)	设置响应消息的实体内容的大小，单位为字节
void setHeader(String name, String value)	设置 HTTP 协议的响应头字段。name 指定响应头字段的名称，value 指定响应头字段的值
void addHeader(String name, String value)	
void setCharacterEncoding(String charset)	用于设置输出内容使用的字符编码
void addCookie(Cookie cookie)	为 Set-Cookie 消息头增加一个值，Set-Cookie 消息头表示应该记录下来的 Cookie，即把 Cookie 发送给客户端
void setLocale(Locale loc)	用于设置响应消息的本地化信息。对 HTTP 来说，就是设置 Content-Language 响应头字段和 Content-Type 头字段中的字符集编码部分

【示例】响应消息头。

关于响应消息头示例，可参见 ResponseHeadServlet.java 的代码（源码包见本书资源第 2 章 ServletDemo 工程的 src/com.mialab.servlet_demo 包），访问 ResponseHeadServlet 的 URL 是 http://localhost:8080/ServletDemo/ResponseHeadServlet。

3. 响应消息体

HttpServletResponse 接口提供了两个获取不同类型输出流对象的方法，如表 2-9 所示。

表 2-9 HttpServletResponse 接口获取不同类型输出流对象的方法

方法	功能描述
ServletOutputStream getOutputStream()	返回字节输出流对象 ServletOutputStream
PrintWriter getWriter()	返回字符输出流对象

【示例】响应消息体。

响应消息体示例，可参见 ResponsePicServlet.java 的代码（源码包第 2 章 ServletDemo 工程的 src/com.mialab.servlet_demo 包），访问 ResponsePicServlet 的 URL 是 http://localhost:8080/ServletDemo/ResponsePicServlet。

2.3.6 ServletContext 接口

Servlet 容器在启动一个 Web 应用时，会为该应用创建一个唯一的 ServletContext 对象供该应用中的所有 Servlet 对象共享。Servlet 对象可以通过 ServletContext 对象来访问容器中的各种资源。获得 ServletContext 对象可以使用以下两种方式。

① 通过 ServletConfig 接口的 getServletContext()方法获得 ServletContext 对象。

② 通过 GenericServlet 抽象类的 getServletContext()方法获得 ServletContext 对象，实质上该方法也调用了 ServletConfig 的 getServletContext()方法。

1. 获取 Web 应用的初始化参数

ServletContext 接口中定义了获取 Web 应用范围的初始化参数的方法。

（1）方法声明：public String getInitParameter(String name)。

作用：返回 Web 应用范围内指定的初始化参数值。在 Web.xml 中使用<context-param>元素表示应用范围内的初始化参数。

（2）方法声明：public Enumeration<String> getInitParameterNames()。

作用：返回一个包含所有初始化参数名称的 Enumeration 对象。

【示例】使用 ServletContext 接口获取 Web 应用的初始化参数。

获取 Web 应用初始化参数的示例，可参见 GetWebInitParamServlet.java 的代码（源码包见本书资源第 2 章 ServletDemo 工程的 src/com.mialab.servlet_demo 包）。

GetWebInitParamServlet.java 中最重要的 doGet 方法代码如下：

```
    protected void doGet(HttpServletRequest request, HttpServletResponse response)
            throws ServletException, IOException {
        // 设置响应到客户端的文本类型为 HTML
        response.setContentType("text/html;charset=UTF-8");
        // 得到 ServletContext 对象
        ServletContext context = this.getServletContext();
        // 得到包含所有初始化参数名的 Enumeration 对象
        Enumeration<String> paramNames = context.getInitParameterNames();
        // 获取输出流
        PrintWriter out = response.getWriter();
        // 遍历所有的初始化参数名，得到相应的参数值，打印到控制台
        out.print("<h2>当前 Web 应用的所有初始化参数：</h2>");
        // 遍历所有的初始化参数名，得到相应的参数值并打印
        while (paramNames.hasMoreElements()) {
            String name = paramNames.nextElement();
            String value = context.getInitParameter(name);
            out.println(name + ": " + value);
            out.println("<br>");
        }
        out.close();
    }
```

Web 应用的初始化参数在 Web.xml 中，主要代码如下：

```
    <!-- Web 应用的初始化参数 -->
    <context-param>
      <param-name>username</param-name>
      <param-value>admin888</param-value>
    </context-param>
    <context-param>
      <param-name>password</param-name>
      <param-value>123</param-value>
    </context-param>
    <context-param>
      <param-name>driverClassName</param-name>
      <param-value>org.postgresql.Driver</param-value>
    </context-param>
    <context-param>
```

```
    <param-name>url</param-name>
    <param-value>jdbc:postgresql://127.0.0.1:5432/postgres</param-value>
  </context-param>
```

访问 GetWebInitParamServlet 的 URL 是 http://localhost:8080/ServletDemo/GetWebInitParamServlet。

这里使用 Firefox 浏览器来访问，输出结果如图 2-10 所示。

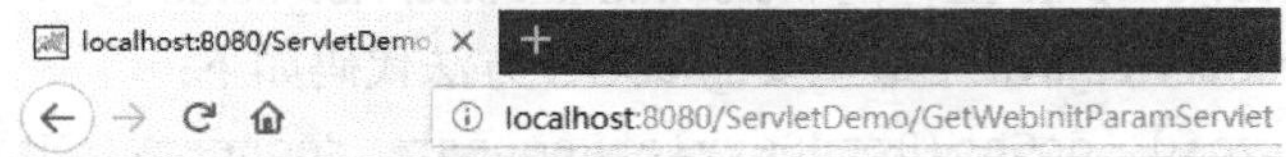

当前Web应用的所有初始化参数：

password : 123
driverClassName : org.postgresql.Driver
url : jdbc:postgresql://127.0.0.1:5432/postgres
username : admin888

图 2-10 获取 Web 应用的初始化参数

2. 实现多个 Servlet 对象共享数据

ServletContext 对象的域属性可以被 Web 应用中的所有 Servlet 访问。在 ServletContext 接口中定义了存取 ServletContext 域属性的 4 个方法，如表 2-10 所示。

表 2-10 存取 ServletContext 域属性的 4 个方法

方法	功能描述
Object getAttribute(String name)	根据参数指定的属性名返回一个与之匹配的域属性值
Enumeration<String> getAttributeNames()	返回一个 Enumeration 对象，该对象包含了所有存放在 ServletContext 中的域属性名
void setAttribute(String name, Object object)	设置 ServletContext 的域属性
void removeAttribute(String name)	根据参数指定的域属性名，从 ServletContext 中删除匹配的域属性

【示例】实现多个 Servlet 对象共享数据。

在 PutContextDataServlet.java 的 doGet 方法中写入以下代码：

```
ServletContext context = this.getServletContext();
// 通过 setAttribute()方法设置属性值
context.setAttribute("contextData", "Here is contexData");
```

先通过访问 http://localhost:8080/ServletDemo/PutContextDataServlet 设置共享数据，再通过调用 GetContextDataServlet 来得到共享数据。

GetContextDataServlet.java 中 doGet 方法的代码如下：

```
PrintWriter out = response.getWriter();
ServletContext context = this.getServletContext();
// 通过 getAttribute()方法获取属性值
String data = (String) context.getAttribute("contextData");
out.println(data);
```

此示例代码可详见 ServletDemo 工程的 PutContextDataServlet.java 和 GetContextDataServlet.java。访问 http://localhost:8080/ServletDemo/GetContextDataServlet 的输出结果是 Here is contexData。

3. 读取 Web 应用下的资源文件

ServletContext 接口中定义了一些读取 Web 资源的方法，如表 2-11 所示。

表 2-11　ServletContext 接口访问 Web 资源的方法

方法	功能描述
String getRealPath(String path)	返回资源文件在服务器文件系统上的真实路径（文件的绝对路径）。参数 path 表示资源文件的虚拟路径，它应该以正斜线“/”开始，“/”表示当前 Web 应用的根目录
Set<String> getResourcePaths(String path)	返回一个 Set 集合，集合中包含资源目录中子目录和文件的路径名称。参数 path 必须以正斜线“/”开始
URL getResource(String path)	返回映射到某个资源文件的 URL 对象。参数 path 必须以正斜线“/”开始，“/”表示当前 Web 应用的根目录
InputStream getResourceAsStream(String path)	返回一个读取参数指定的文件的输入流，参数 path 必须以正斜线“/”开始
String getMimeType(String file)	返回参数指定的文件的 MIME 类型

【示例】使用 ServletContext 接口读取 Web 资源。

GetResourceServlet.java 中 doGet 方法的代码如下：

```
    PrintWriter out = response.getWriter();
    ServletContext context = this.getServletContext();
    // 获取文件绝对路径
    String path = context.getRealPath("/Web-INF/classes/data.properties");
    FileInputStream in = new FileInputStream(path);
    Properties pros = new Properties();
    pros.load(in);
    out.println("username = " + pros.getProperty("username") + "<br>");
    out.println("password = " + pros.getProperty("password") + "<br>");
    out.println("driverClassName = " + pros.getProperty("driverClassName") +
"<br>");
    out.println("url = " + pros.getProperty("url") + "<br>");
```

ServletDemo 工程 src 文件夹下 data.propertis 文件的内容如下。

```
    username = admin888
    password = 12345678
    driverClassName = org.postgresql.Driver
    url = jdbc:postgresql://127.0.0.1:5432/postgres
```

2.4　Servlet 处理表单数据

表单在网页中主要负责数据采集功能。一个表单有以下 3 个基本组成部分。

（1）表单标签：包含了处理表单数据所用 CGI 程序（这里是用 Servlet 来做处理）的 URL 以及数据提交到服务器的方法。

（2）表单域：包含了文本框、密码框、隐藏域、多行文本框、复选框、单选框、下拉列表框和文件上传框等。

（3）表单按钮：包括提交按钮、复位按钮和一般按钮；用于将数据传送到服务器上的 CGI

脚本（这里是 Servlet）或者取消输入，还可以用表单按钮来控制其他定义了处理脚本的处理工作。

表单数据是指通过表单让用户填写内容，然后提交到服务器上；这些数据被称之为表单数据。Servlet 处理表单数据可以使用以下方法。

① getParameter()：可以调用 request.getParameter() 方法来获取表单参数的值。

② getParameterValues()：如果参数出现一次以上，则调用该方法，并返回多个值，如复选框。

③ getParameterNames()：如果要得到当前请求中的所有参数的完整列表，则调用该方法。

【示例】Servlet 处理表单数据。

Servlet 处理表单数据的示例代码，可参见 FormServlet.java 和 form.html（源码包见本书资源第 2 章的 ServletDemo 工程），先访问 form.html，填写表单，再提交给 FormServlet 处理。

2.5 Servlet 重定向和请求转发

请求转发只是把请求转发给服务器（通常是同一个 Web 应用中）的另一个组件（Servlet 或 JSP 等）；重定向则只是告诉客户（浏览器）去访问另一个 URL（可能是同一个 Web 站点甚至其他站点）。请求转发发生在服务器端，由服务器（如 Servlet）控制；重定向发生在客户端，由客户（通常是浏览器）控制。

请求转发过程在同一个请求中完成，只会返回一个响应。重定向过程则发生在两个不同的请求中，会返回两个不同响应。请求转发使用 RequestDispatcher 对象的 forward()或 include()方法。重定向则使用 HttpServletResponse 对象的 sendRedirect()方法。

请求转发和重定向方法都必须在响应提交（刷新响应正文输出到流中）之前执行，否则会抛出 IllegalStateException 异常。在转发或重定向之前，响应缓冲区中未提交的数据会被自动清除。

2.5.1 重定向

重定向后则无法在服务器端获取第一次请求对象时保存的信息。例如，在 Servlet 中将用户名保存到当前 request 对象中，并重定向到一个新的 URL，然后在新 URL 指向的地址中（假设是某个 Servlet）就无法获取原先保存在第一个请求中的信息。很明显，用户名是保存在第一次请求的对象中的，但并没有保存在本次（第二次）请求的对象中。

重定向后，浏览器地址栏 URL 变为新的 URL（因为浏览器确实给新的 URL 发送了一个新的请求）。

【示例】Servlet 网页重定向。

RedirectServlet.java 的主要代码如下：

```
    @WebServlet("/RedirectServlet")
    public class RedirectServlet extends HttpServlet {
     protected void doGet(HttpServletRequest request, HttpServletResponse
response)
            throws ServletException, IOException {
        response.setContentType("text/html;charset=UTF-8");
        response.getWriter().println(new java.util.Date());
```

```
        // 进行重定向
        response.sendRedirect(request.getContextPath() +
"/AnotherServlet");
      }
    }
```

AnotherServlet.java 的主要代码如下：

```
    @WebServlet("/AnotherServlet")
    public class AnotherServlet extends HttpServlet {
     protected void doGet(HttpServletRequest request, HttpServletResponse
response)
            throws ServletException, IOException {
        // 设置响应到客户端的文本类型为 HTML
        response.setContentType("text/html;charset=UTF-8");
        // 获取输出流
        PrintWriter out = response.getWriter();
        // 输出响应结果
        out.println("<p>重定向页面</p>");
        out.close();
      }
    }
```

（代码详见本书资源源码包第 2 章的 ServletDemo 工程。）

访问 RedirectServlet 的 URL 是 http://localhost:8080/ServletDemo/ RedirectServlet，输出结果如图 2-11 所示。

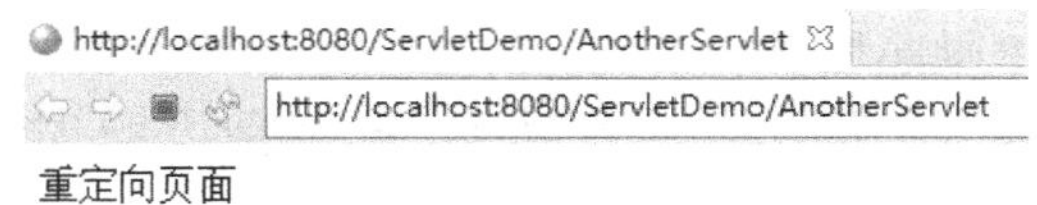

图 2-11　Servlet 网页重定向示例

sendRedirect()方法是 HttpServletResponse 对象的方法，即响应对象的方法。既然调用了响应对象的方法，就表示本次请求过程已经结束了，服务器即将向客户端返回本次请求的响应。事实上，服务器确实返回了一个状态码为“302”，首部“Location”值为新的 URL 的响应。此后，浏览器会根据“Location”首部指定的 URL，重新发起一次新的请求，转向这个目标页面，所以重定向实际上发生在两个不同的请求中。

2.5.2　请求转发

请求转发后可以在服务器端获取本次请求对象上保存的信息（例如，在 Servlet 中将用户名保存到当前 request 对象中，转发给另一组件（如 JSP）后，另一组件可以通过 request 对象取得用户名信息）。请求转发后，浏览器地址栏 URL 不会发生改变。

【示例】实现请求转发。

ForwardServlet.java 的主要代码如下：

```
    @WebServlet("/ForwardServlet")
    public class ForwardServlet extends HttpServlet {
     protected void doGet(HttpServletRequest request, HttpServletResponse
response)
            throws ServletException, IOException {
        request.setAttribute("bookname", "《Android 应用开发实践教程》");
```

```
        RequestDispatcher                    dispatcher                    =
request.getRequestDispatcher("/OtherServlet");
        dispatcher.forward(request, response);
    }
}
```

OtherServlet.java 的主要代码如下：

```
@WebServlet("/OtherServlet")
public class OtherServlet extends HttpServlet {
 protected void doGet(HttpServletRequest request, HttpServletResponse
response)
        throws ServletException, IOException {
        response.setContentType("text/html;charset=UTF-8");
        String bookname = (String) request.getAttribute("bookname");
        PrintWriter out = response.getWriter();
        out.println("<p>请求转发的结果页面</p>");
        out.println("读取的 request 对象的 bookname 属性值为：" + bookname);
    }
}
```

（代码详见本书资源源码包第 2 章的 ServletDemo 工程。）

访问 ForwardServlet 的 URL 是 http://localhost:8080/ServletDemo/ ForwardServlet，输出结果如图 2-12 所示。

图 2-12　请求转发示例

RequestDispatcher 对象是通过调用 HttpServletRequest 对象的 getRequestDispatcher()方法得到的，所以 forward()或 include()本质上属于请求对象的方法，所以请求转发始终发生在一个请求当中。

从响应速度上来说，请求转发相对较快，因为请求转发过程在同一请求中；而重定向相对较慢，因为重定向过程发生在两个不同的请求中。

2.5.3　Servlet 中请求转发时 forword()和 include()的区别

1. 定义

forward()：表示在服务器端从一个 Servlet 中将请求转发到另一个资源（Servlet、JSP 或 HTML 等），本意是让第一个组件对请求做出预处理（或者什么都不做），而让另一组件处理并返回响应。

include()：表示在响应中包含另一个资源（Servlet、JSP 或 HTML 等）的响应内容，最终被包含的页面产生的任何响应都将并入原来的 response 对象，然后一起输出到客户端。

2. 关于状态码和响应头

forward()：调用者和被调用者设置的状态码和响应头都不会被忽略。

include()：被调用者（如被包含的 Servlet）不能改变响应消息的状态码和响应头，即会忽略被调用者设置的状态码和响应头。

3．谁负责发回响应

forward()：表示转发，控制权也同时交给了另一个组件，所以最终由另一组件返回响应。
include()：表示包含，则控制权还在自己身上，所以最终还是由自己返回响应。

4．请求转发后的代码是否执行

forward()：转发后还会返回主页面继续执行，但不可以继续输出响应信息。
include()：转发后还会返回主页面继续执行，仍然可以继续输出响应信息。

5．关于 forward()，引用 Java EE 文档中的说明

（1）必须在响应被提交到客户端（刷新响应正文输出到流中）前调用 forward（即在调用 forward 之前必须清空响应缓冲区），否则会抛出 IllegalStateException 异常。

（2）在 forward 之前，响应缓冲区中未提交的数据会被自动清除。所以容器将忽略原 Servlet 所有其他输出。

6．补充说明：关于 Servlet 中的输出缓冲区

Handler 是一个消息处理类，主要用于异步消息的处理：当发出一条消息之后，首先进入一个消息队列，发送消息的函数即刻返回，而另外一部分逐个在消息队列中将消息取出，然后对消息进行处理，也就是说，发送消息和接收消息不是同步处理的。

（1）在 Servlet 中使用 ServletOutputStream 和 PrintWriter 输出响应正文时，数据首先被写入 Servlet 引擎提供的一个输出缓冲区中。直到满足以下条件之一时，Servlet 引擎才会把缓冲区中的内容真正发送到客户端。

① 输出缓冲区被填满。

② Servlet 已经写入了所有的响应内容。

③ Servlet 调用响应对象的 flushBuffer()方法，强制将缓冲区内的响应正文数据发送到客户端。

④ Servlet 调用 ServletOutputStream 或 PrintWriter 对象的 flush()方法或 close()方法。

（2）为了确保 ServletOutputStream 或 PrintWriter 输出的所有数据都能被提交给客户端，建议在所有数据输出完毕后，调用 ServletOutputStream 或 PrintWriter 的 close()方法。

（3）使用输出缓冲区后，Servlet 引擎就可以将响应状态行、各响应头和响应正文严格按照 HTTP 消息的位置顺序进行调整后再输出到客户端。

（4）如果在提交响应到客户端时，输出缓冲区中已经装入了所有的响应内容，Servlet 引擎将计算响应正文部分的大小并自动设置 Content-Length 头字段。

（5）缓冲区自动刷新（清出）功能。[注意是刷新（flush），而不是清除（flushBuffer）]

① 如果设置为自动刷新，则在缓冲区满或者使用 flush()方法显式清出时，都会向客户端输出信息。

② 如果设置为不自动刷新，则必须明确使用 flush()方法清出数据，否则如果缓冲区满了，将会产生 IOException 异常。

（6）使用缓冲区能够减少数据传输的次数，提高程序的运行效率。但也有可能产生响应延迟的问题，因为在缓冲区满或使用 flush()显式清出之前，数据并不会真正发送到客户端。

2.6 Servlet 数据库访问

2.6.1 JDBC 基础

Java 数据库连接(Java Database Connectivity，JDBC)是一种用于执行 SQL 语句的 Java API，可以为多种关系数据库提供统一访问，它由一组用 Java 语言编写的类和接口组成。JDBC 提供了一种基准，据此可以构建更高级的工具和接口，使数据库开发人员能够更为便利地编写数据库应用程序。

本书假定读者已经学习过 Java 数据库编程，具备一定的 JDBC 基础，懂得 JDBC 应用程序的工作方式，所以关于 JDBC 编程的相关知识在此不再赘述。

2.6.2 创建测试数据

在 MySQL 中创建 book 数据库，并创建 book 数据表，表结构如下：

```
CREATE TABLE 'book' (
  'bookId' int(4) NOT NULL,
  'bookName' varchar(50) NOT NULL,
  PRIMARY KEY ('bookId')
) ENGINE=InnoDB DEFAULT CHARSET=utf8;
```

在 book 数据表中插入以下几条测试数据：

```
INSERT INTO 'book' VALUES ('9801', 'Android应用开发实践教程');
INSERT INTO 'book' VALUES ('9802', 'Web应用开发');
INSERT INTO 'book' VALUES ('9803', 'iOS程序设计');
```

（详见源码包 ch2 文件夹下的 book.sql 文件。）

2.6.3 访问数据库

这里访问的是 MySQL 数据库，所以需要把 mysql-connector-java-5.1.39.jar 文件复制至 Web-INF 的 lib 目录下。

【示例】Servlet 数据库访问。

在 ServletDemo 工程的 com.mialab.servlet_demo 包下创建 DataAccessServlet.java，DataAccessServlet.java 主要代码如下：

```
@WebServlet("/DataAccessServlet")
public class DataAccessServlet extends HttpServlet {
 // JDBC 驱动名及数据库 URL
 static final String JDBC_DRIVER = "com.mysql.jdbc.Driver";
 static final String DB_URL = "jdbc:mysql://localhost:3306/book";
 // 数据库的用户名与密码，根据自己的需要来设置
 static final String USER = "root";
 static final String PASS = "1";
 protected void doGet(HttpServletRequest request, HttpServletResponse
response)
        throws ServletException, IOException {
     Connection conn = null;
     Statement stmt = null;
     // 设置响应内容类型
```

```
        response.setContentType("text/html;charset=UTF-8");
        PrintWriter out = response.getWriter();
        out.println("<h2>Servlet 数据库访问</h2>");
        try {
            // 注册 JDBC 驱动器
            Class.forName("com.mysql.jdbc.Driver");
            // 打开一个连接
            conn = DriverManager.getConnection(DB_URL, USER, PASS);
            // 执行 SQL 查询
            stmt = conn.createStatement();
            String sql;
            sql = "select bookId,bookName from book";
            ResultSet rs = stmt.executeQuery(sql);
            // 展开结果集数据库
            while (rs.next()) {
                // 通过字段检索
                int bookId = rs.getInt("bookId");
                String bookName = rs.getString("bookName");
                // 输出数据
                out.println("bookId: " + bookId);
                out.println(", bookName: " + bookName);
                out.println("<br />");
            }
            // 完成后关闭
            rs.close();
            stmt.close();
            conn.close();
        } catch (SQLException se) {
            se.printStackTrace();    // 处理 JDBC 错误
        } catch (Exception e) {
            e.printStackTrace();    // 处理 Class.forName 错误
        } finally {
            // 最后是用于关闭资源的块
        }
    }
}
```

访问 DataAccessServlet 的 URL 是 http://localhost:8080/ServletDemo/DataAccessServlet，浏览器窗口输出结果如图 2-13 所示。

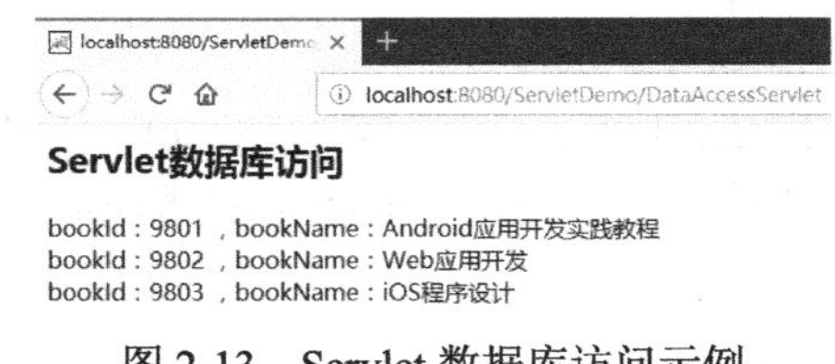

图 2-13　Servlet 数据库访问示例

（代码详见 ch2 文件夹下的 ServletDemo 工程。）

2.7　Servlet 异常处理

当一个 Servlet 抛出一个异常时，Web 容器在使用了 exception-type 元素的 Web.xml 中搜索与抛出异常类型相匹配的配置。可以在 Web.xml 中使用 error-page 元素来指定对特定异常或 HTTP 状态码出相应的 Servlet 调用。

假设有一个 ErrorHandler 的 Servlet 在任何已定义的异常或错误出现时被调用，以下将是在 Web.xml 中创建的项。

```
<!-- Servlet 定义 -->
<servlet>
      <servlet-name>ErrorHandler</servlet-name>
      <servlet-class>ErrorHandler</servlet-class>
</servlet>
<!-- Servlet 映射 -->
<servlet-mapping>
      <servlet-name>ErrorHandler</servlet-name>
      <url-pattern>/ErrorHandler</url-pattern>
</servlet-mapping>
<!-- error-code 相关的错误页面 -->
<error-page>
   <error-code>404</error-code>
   <location>/ErrorHandler</location>
</error-page>
<error-page>
   <error-code>403</error-code>
   <location>/ErrorHandler</location>
</error-page>
<!-- exception-type 相关的错误页面 -->
<error-page>
   <exception-type>
      javax.servlet.ServletException
   </exception-type >
   <location>/ErrorHandler</location>
</error-page>
<error-page>
   <exception-type>java.io.IOException</exception-type >
   <location>/ErrorHandler</location>
</error-page>
```

关于上面的 Web.xml 中的异常处有以下几点需要说明。

① ErrorHandler 与其他的 Servlet 的定义方式一样，且在 Web.xml 中进行配置。

② 如果有错误状态代码出现，不管为 404（Not Found，即未找到）或 403（Forbidden，即禁止），都会调用 ErrorHandler。

③ 如果 Web 应用程序抛出 ServletException 或 IOException，Web 容器会调用 ErrorHandler。

④ 可以定义不同的错误处理程序来处理不同类型的错误或异常。

如果要对所有的异常有一个通用的错误处理程序，那么应该定义下面的 error-page，而不是为每个异常定义单独的 error-page 元素：

```
<error-page>
   <exception-type>java.lang.Throwable</exception-type >
   <location>/ErrorHandler</location>
</error-page>
```

以下是错误/异常处理的 Servlet 可以访问的请求属性列表，用来分析错误/异常的性质。

（1）javax.servlet.error.status_code：该属性给出状态码，状态码可被存储，并可在存储为

java.lang.Integer 数据类型后被分析。

（2）javax.servlet.error.exception_type：该属性给出异常类型的信息，异常类型可被存储，并可在存储为 java.lang.Class 数据类型后被分析。

（3）javax.servlet.error.message：该属性给出确切错误消息的信息，信息可被存储，并可在存储为 java.lang.String 数据类型后被分析。

（4）javax.servlet.error.request_uri：该属性给出有关 URL 调用 Servlet 的信息，信息可被存储，并可在存储为 java.lang.String 数据类型后被分析。

（5）javax.servlet.error.exception：该属性给出异常产生的信息，信息可被存储，并可在存储为 java.lang.Throwable 数据类型后被分析。

（6）javax.servlet.error.servlet_name：该属性给出 Servlet 的名称，名称可被存储，并可在存储为 java.lang.String 数据类型后被分析。

【示例】Servlet 异常处理。

在 ServletDemo 工程的 com.mialab.servlet_demo 包下创建 ErrorHandlerServlet.java 和 ExceptionServlet.java。ErrorHandlerServlet.java 主要代码如下：

```
    public class ErrorHandlerServlet extends HttpServlet {
     protected void doGet(HttpServletRequest request, HttpServletResponse
response)
            throws ServletException, IOException {
        Throwable throwable =
            (Throwable)
request.getAttribute("javax.servlet.error.exception");
        Integer statusCode =
            (Integer)
request.getAttribute("javax.servlet.error.status_code");
        String servletName =
            (String)
request.getAttribute("javax.servlet.error.servlet_name");
        if (servletName == null) {
            servletName = "Unknown";
        }
        String requestUri = (String) request.getAttribute("javax.servlet.
error.request_uri");
        if (requestUri == null) {
            requestUri = "Unknown";
        }
        // 设置响应内容类型
        response.setContentType("text/html;charset=UTF-8");
        PrintWriter out = response.getWriter();
        String title = "Servlet 处理 Error/Exception";
        String docType = "<!DOCTYPE html>\n";
        out.println(docType + "<html>\n" + "<head><title>" + title +
"</title></head>\n" +
            "<body bgcolor=\"#f0f0f0\">\n");
        out.println("<h3>Servlet 异常/错误处理</h3>");
        if (throwable == null && statusCode == null) {
            out.println("<h3>错误信息丢失</h2>");
            out.println("请返回 <a href=\"" +
```

```
                response.encodeURL("http://localhost:8080/") + "\">主页</a>。
");
        } else if (statusCode != null) {
            out.println("错误代码 : " + statusCode + "<br><br>");
            out.println("Servlet Name : " + servletName + "</br></br>");
            out.println("异常类型 : " + throwable.getClass().getName() +
"</br></br>");
            out.println("请求 URI: " + requestUri + "<br><br>");
            out.println("异常信息: " + throwable.getMessage());
        }
        out.println("</body>");
        out.println("</html>");
    }
}
```

ExceptionServlet.java 主要代码如下：

```
@WebServlet("/ExceptionServlet")
public class ExceptionServlet extends HttpServlet {
 protected void doGet(HttpServletRequest request, HttpServletResponse
response)
        throws ServletException, IOException {
    int x = 126/0;      //除数为0
 }
}
```

在 Web.xml 中须添加如下设置代码：

```
<!-- 异常/错误处理 -->
<servlet>
 <servlet-name>ErrorHandler</servlet-name>
 <servlet-class>com.mialab.servlet_demo.ErrorHandlerServlet</servlet-cl
ass>
</servlet>
<servlet-mapping>
 <servlet-name>ErrorHandler</servlet-name>
 <url-pattern>/ErrorHandler</url-pattern>
</servlet-mapping>
<error-page>
 <error-code>404</error-code>
 <location>/ErrorHandler</location>
</error-page>
<error-page>
 <exception-type>java.lang.Throwable</exception-type>
 <location>/ErrorHandler</location>
</error-page>
```

访问 ExceptionServlet 的 URL 是 http://localhost:8080/ServletDemo/ExceptionServlet，浏览器输出结果如图 2-14 所示。但浏览器往往要做设置，默认设置很可能不认可开发者的定制。要说明的是，我们使用外部的火狐浏览器来做测试，需要进入火狐浏览器的“选项→隐私与安全”标签页，勾选“允许 Firefox 向 Mozilla 发送错误报告”复选框，如图 2-15 所示。

（代码详见本书资源 ch2 文件夹中的 ServletDemo 工程。）

图 2-14　Servlet 异常处理　　　　图 2-15　允许 Firefox 向 Mozilla 发送错误报告

2.8　异步 Servlet

在 Servlet 3.0 之前，Servlet 采用 Thread-Per-Request 的方式处理请求，即每一次 HTTP 请求都由某一个线程从头到尾负责处理。如果一个请求需要进行 IO 操作，如访问数据库、调用第三方服务接口等，那么其所对应的线程将同步地等待 IO 操作完成，而 IO 操作是非常慢的，所以此时的线程并不能及时地释放回线程池以供后续使用，在并发量越来越大的情况下，这将带来严重的性能问题。即便是像 Spring、Struts 这样的高层框架也脱离不了这样的桎梏，因为它们都是建立在 Servlet 之上的。为了解决这样的问题，Servlet 3.0 引入了异步处理，然后在 Servlet 3.1 中又引入了非阻塞 IO 来进一步增强异步处理的性能。

在 Servlet 3.0 中，可以从 HttpServletRequest 对象中获得一个 AsyncContext 对象，该对象构成了异步处理的上下文，Request 和 Response 对象都可从中获取。AsyncContext 可以从当前线程传给其他的线程，在新的线程中完成对请求的处理并返回结果给客户端，初始线程便可以返回给容器线程池以处理更多的请求。如此，通过将请求从一个线程传给另一个线程处理的过程便构成了 Servlet 3.0 中的异步处理。

编写异步 Servlet 的步骤如下。

（1）调用 ServletRequest 中的 startAsync 方法。startAsync 方法返回一个 AsyncContext。

（2）调用 AsyncContext 的 setTimeout()，传递容器等待任务完成的超时时间的毫秒数。

（3）调用 asyncContext.start，传递一个 Runnable 来执行一个长时间运行的任务。

（4）调用 Runnable 的 asyncContext.complete 或 asyncContext.dispatch 来完成任务。

【示例】异步 Servlet。

在 ServletDemo 工程的 com.mialab.servlet_demo 包下创建 SimpleAsyncServlet.java。SimpleAsyncServlet 主要代码如下：

```
    @WebServlet(name = "/SimpleAsyncServlet", urlPatterns = { "/simpleAsync" },
     asyncSupported = true)
    public class SimpleAsyncServlet extends HttpServlet {
     protected void doGet(HttpServletRequest request, HttpServletResponse
response)
            throws ServletException, IOException {
        final AsyncContext asyncContext = request.startAsync();
        System.out.println(Thread.currentThread().getName());
        request.setAttribute("boyName", "李寻欢");
        asyncContext.setTimeout(6000);
        asyncContext.start(new Runnable() {
```

```
            @Override
            public void run() {
                try {
                    int millis = ThreadLocalRandom.current().nextInt(3000);
                    String currentThread = Thread.currentThread().getName();
                    System.out.println(currentThread + " sleep for " + millis
+
                        " milliseconds.");
                    Thread.sleep(millis);
                } catch (InterruptedException e) {
                    e.printStackTrace();
                }
                request.setAttribute("girlName", "我不是潘金莲");
                asyncContext.dispatch("/boygirl.jsp");
            }
        });
    }
}
```

在 ServletDemo 工程的 WebContent 文件夹下创建 boygirl.jsp。boygirl.jsp 的代码如下：

```
<%@ page language="java" contentType="text/html; charset=UTF-8"
    pageEncoding="UTF-8"%>
<!DOCTYPE html PUBLIC "-//W3C//DTD HTML 4.01 Transitional//EN"
"http://www.w3.org/TR/html4/loose.dtd">
<html>
<head>
<meta http-equiv="Content-Type" content="text/html; charset=UTF-8">
<title>访问 Request 域</title>
</head>
<body>
<p>Hello</p><hr>
<p>girlName: ${girlName}</p>
<p>boyName: ${boyName}</p>
</body>
</html>
```

访问 SimpleAsyncServlet 的 URL 是 http://localhost:8080/ServletDemo/SimpleAsync，浏览器输出的结果如图 2-16 所示。

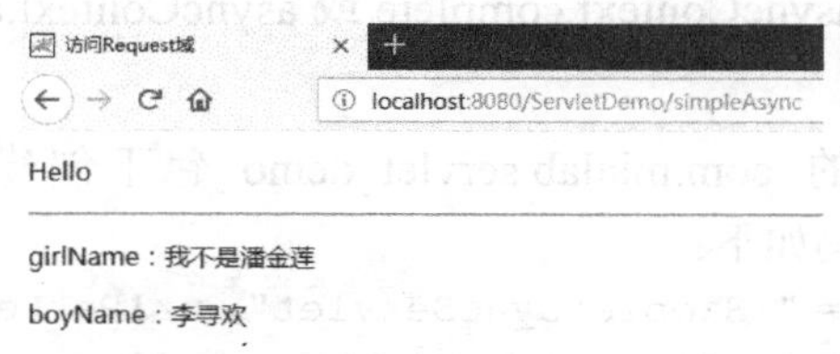

图 2-16　异步 Servlet 示例

（代码详见本书资源 ch2 文件夹下的 ServletDemo 工程。）

2.9　本章小结

Servlet 运行在服务器端，由 Servlet 容器所管理。Servlet 容器也称 Servlet 引擎，是 Web 服务器或应用服务器的一部分。Java Servlet 通常情况下与使用 CGI 实现的程序可以达到异曲同工的效果。与传统的CGI和许多其他类似CGI的技术相比，Java Servlet具有更高的效率，更容易被使用，功能更强大，具有更好的可移植性，也更节省投资。

Servlet可以由 javax.servlet和javax.servlet.http包创建，它是Java企业版的标准组成部分，Java企业版是支持大型开发项目的Java类库的扩展版本。

Servlet 负责执行的主要任务如下。

（1）读取客户端（浏览器）发送的显式的数据。这包括网页上的HTML表单，也可以是来自applet或自定义的HTTP客户端程序的表单。

（2）读取客户端（浏览器）发送的隐式的HTTP请求数据。这包括Cookies、媒体类型和浏览器能理解的压缩格式等。

（3）处理数据并生成结果。这个过程可能需要访问数据库，执行RMI或CORBA调用，调用Web服务，或者直接计算得出对应的响应。

（4）发送显式的数据（即文档）到客户端（浏览器）。该文档的格式可以是多种多样的，包括文本文件（HTML或XML）、二进制文件（GIF图像）、Excel等。

（5）发送隐式的HTTP响应到客户端（浏览器）。这包括告诉浏览器或其他客户端被返回的文档类型（如HTML），设置Cookies和缓存参数，以及其他类似的任务。

习题 2

1．如何用记事本写一个Servlet，并在Tomcat服务器上部署运行？

2．Servlet的体系结构是怎样的？

3．Servlet的生命周期是怎样的？试编程加以说明。

4．Servlet如何处理表单数据？试编程加以说明。

5．什么是Servlet的重定向和请求转发？试编程加以说明。

6．什么是Servlet的请求转发？试编程加以说明。

7．Servlet是如何访问数据库并实现数据库记录的增、删、改、查的？试编程加以说明。

8．Servlet异常处理是怎样的？试编程加以说明。

9．什么是异步Servlet?

10．Servlet处理HTTP请求的流程是什么？

11．Servlet如何产生HTTP回应？

12．RequestDispatcher接口的forward()方法与include()方法的区别是什么？

13．如何初始化Web应用程序？试编程加以说明。

14．如何用Context起始参数建立JDBC数据库连接，并进行数据库查询？试编程加以说明。

15．如何存取Servlet起始参数？试编程加以说明。

第3章 JSP语言基础

本章导读

JSP 是 Servlet 的扩展，其目的是简化建立和管理动态页面的工作。JSP 和 Servlet 在本质上是同一种技术，Servlet 是 JSP 的早期版本，JSP 是 Servlet 的另外一种表现形式。本章主要内容有：（1）JSP 基本语法（包括 JSP 运行机制、JSP 指令以及 JSP 脚本等）；（2）JSP 动作元素；（3）JSP 内置对象；（4）JSP 举例。

3.1 JSP基本语法

3.1.1 JSP 简介

Java 服务器页面（Java Server Pages，JSP）是一种动态页面开发技术，其根本是一个简化的 Servlet 设计，它是由 Sun 公司（现已被甲骨文收购）倡导、许多公司参与建立的一种动态网页技术标准。JSP 技术是在传统的网页 HTML 文件中插入 Java 程序段（Scriptlet）和 JSP 标记（Tag），从而形成 JSP 文件。

JSP 的主要目的是将表示逻辑从 Servlet 中分离出来。Java Servlet 是 JSP 的技术基础，而且大型的 Web 应用程序的开发需要 Java Servlet 和 JSP 配合才能完成。用 JSP 开发的 Web 应用是跨平台的，既能在 Linux 下运行，又能在其他操作系统上运行。

JSP 生命周期的运行过程可概括如下。

（1）客户端第一次发出请求访问 JSP 文件。

（2）JSP 容器先将 JSP 文件转换成一个 Java 源文件（Java Servlet 源程序），在转换过程中，如果发现 JSP 文件中存在任何语法错误，则中断转换过程，并返回出错信息。

（3）产生 Servlet 源文件之后，Servlet 容器就可以将它编译成 Servlet 类文件，然后由“类载入器”（Class Loader）将 Servlet 类文件载入 Servlet 容器的 JVM。

（4）成功载入 Servlet 类文件之后，Servlet 容器将会建立一个 Servlet 实体，然后运行 jspInit() 方法，进入 JSP 初始化程序。

（5）每当用户送出 JSP 运行请求时，Servlet 容器将会调用_jspService()方法，并且传入 request 与 response 对象。注意，当 JSP 被编译成 Servlet 时，所有 HTML 标签、Script Tag 与 Expression Tag 的内容都会包含在_jspService()方法中。（可参考后面的 JSP 脚本。）

（6）如果 Servlet 容器想要移除某个 JSP 所产生的 Servlet 实体，则将调用其 jspDestroy() 方法，并清除相关的系统资源。

3.1.2 JSP 运行机制

1. JSP 编译

当浏览器请求 JSP 页面时，JSP 引擎会先去检查是否需要编译这个文件。如果这个文件没有被编译过，或者在上次编译后被更改过，则编译这个 JSP 文件。

编译的过程包括以下 3 个步骤。

① 解析 JSP 文件。

② 将 JSP 文件转为 Servlet。

③ 编译 Servlet。

2. JSP 初始化

容器载入 JSP 文件后，会在为请求提供任何服务前调用 jspInit()方法。如果需要执行自定义的 JSP 初始化任务，则重写 jspInit()方法即可。

```
public void jspInit(){  // 初始化代码 }
```

3. JSP 执行

这一阶段描述了 JSP 生命周期中一切与请求相关的交互行为，直到被销毁。当 JSP 网页完成初始化后，JSP 引擎将会调用_jspService()方法。_jspService()方法需要一个 HttpServletRequest 对象和一个 HttpServletResponse 对象作为它的参数。

```
void _jspService(HttpServletRequestrequest,HttpServletResponseresponse){
 // 服务端处理代码
}
```

4. JSP 清理

JSP 生命周期的销毁阶段描述了当一个 JSP 网页从容器中被移除时所发生的一切。jspDestroy()方法在 JSP 中等价于 Servlet 中的销毁方法。当需要执行任意清理工作时覆盖 jspDestroy()方法，如释放数据库连接或者关闭文件夹时等。

```
public void jspDestroy() {  // 清理代码 }
```

3.1.3 第一个 JSP 程序

hello.jsp 的作用是输出 "Hello World"， 主要代码如下：

```
<html>
   <body>
         <%  out.println("Hello World! ");  %>
   </body>
</html>
```

在浏览器中进行访问，如图 3-1 所示。

在这里，hello.jsp 部署路径是 E:\tomcat8\webapps\hello2\hello.jsp。

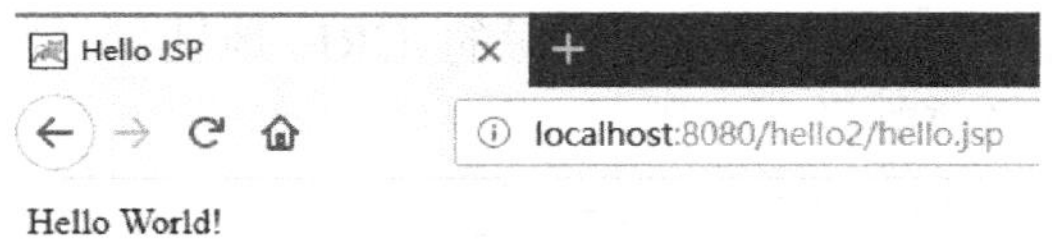

图 3-1 hello.jsp

3.1.4 JSP 指令

JSP 指令用来设置整个 JSP 页面相关的属性，如网页的编码方式和脚本语言。JSP 指令是被服务器解释并执行的，不会产生任何内容输出到网页中。也就是说，指令标识对于客户端浏览器是不可见的。

语法格式如下：<%@ 指令名 attribute1="value1" attribute2="value2"……%>。

指令可以有很多个属性，它们以键值对的形式存在，并用逗号隔开。例如：

```
<%@ page language="java" contentType="text/html; charset=UTF-8"
    pageEncoding="UTF-8"%>
```

指令名：用于指定指令名称，在 JSP 中包含 page、include 和 taglib 3 条指令。

属性：在一个指令中，可以设置多个属性。

注意：指令标识<%@和%>是完整的标记，不能添加空格。

1. page 指令

page 指令为容器提供当前页面的使用说明。一个 JSP 页面可以包含多个 page 指令。

page 指令的语法格式：<%@ page attribute1="value1" attribute2="value2"…… %>。

表 3-1 列出了与 page 指令相关的属性。

表 3-1 与 page 指令相关的属性

属性名称	描 述
language	定义本页面的脚本语言类型，默认是 Java
import	指定导入的 Java 软件包或类名列表
pageEncoding	定义本页面的字符编码
contentType	指定 MIME 类型和 JSP 页面响应时的编码方式，默认为 text/htm；charset=ISO8859-1
session	指定 JSP 页面中是否使用 session 对象，默认为 true
isErrorPage	指定当前页面是否可以作为另一个 JSP 页面的错误处理页面
errorPage	指定当 JSP 页面发生异常时需要转向的错误处理页面
buffer	定义隐式对象 out 的缓冲大小，以 KB 为单位
autoFlush	默认值为 true，表示当输出缓冲满时会自写入输出流。而值为 false，表示仅当调用隐式对象的 flush 方法时，才会写入输出流。因此，若缓冲溢出，则会抛出异常
isELIgnored	指定 JSP 页面是否忽略 EL 表达式

2. include 指令

JSP 可以通过 include 指令来包含其他文件。被包含的文件可以是 JSP 文件、HTML 文件或文本文件。包含的文件就好像是该 JSP 文件的一部分，会被同时编译执行。

include 指令的语法格式如下：<%@ include file="文件相对 url 地址" %> 。

include 指令中的文件名实际上是一个相对的 URL 地址。 如果没有给文件关联一个路径，则 JSP 编译器默认在当前路径下寻找。

【示例】应用 include 指令包含网站 Banner。

在 Eclipse 中创建 Dynamic Web Project，名为 JSPDemo，在其 WebContent 文件夹中创建 home.jsp 和 banner.jsp 文件。home.jsp 的<body>部分代码如下：

```
<font color="red"><h2>电影天堂</h2></font>
<%@ include file = "banner.jsp" %>
```

banner.jsp 的<body>部分代码如下：

```
<b>最新影片 | 经典影片 | 国内电影 | 欧美电影 | 欧美电视 | 最新综艺</b>
```

3. taglib 指令

JSP API 允许用户自定义标签，一个自定义标签库就是自定义标签的集合。taglib 指令引入一个自定义标签集合的定义，包括库路径、自定义标签。

taglib 指令的语法：<%@ taglib uri="uri" prefix="prefixOfTag" %> 。

uri 属性确定了标签库的位置，prefix 属性了指定标签库的前缀。

3.1.5 JSP 脚本

JSP 脚本元素是指嵌套在<%和%>之中的一条或多条 Java 程序代码。JSP 脚本元素可以将 Java 代码嵌入到 HTML 页面中。JSP 脚本元素主要包含如下 3 种类型：JSP 程序段（JSP Scriptlets）、JSP 表达式和 JSP 声明。

1. JSP 表达式

JSP 表达式用于向页面中输出信息，其语法格式如下：<%= 表达式 %>

JSP 表达式可以是任何 Java 语言的完整表达式，该表达式的最终运算结果将被转换为字符串。注意：<%与=之间不可以有空格。

【示例】使用 JSP 表达式显示当前时间。

在 JSPDemo 工程的 WebContent 文件夹中创建 datetime.jsp，在<body>部分加入以下代码。

```
<%= new Date().toLocaleString() %>
```

还须在 datetime.jsp 文件开头加入以下内容：

```
<%@ page import = "java.util.*" %>
```

（详见本书资源源码包 ch3 目录下 JSPDemo 工程的 datetime.jsp。）

2. JSP 程序段

JSP 表达式只能单行出现，而且仅仅把其中的运算结果输出到客户端。当需要在 JSP 程序中实现一些复杂操作或控制时，JSP 表达式是不能满足要求的，这时候需要 JSP 程序段（JSP Scriptlets）。JSP 程序段就是插入 JSP 程序的 Java 代码段。在网页任何地方都可以插入 JSP 程序段，在程序段中可以插入任何数量的 Java 代码。

其语法格式如下：<% Java 代码 %>。

在 JSP Scriptlets 中声明的变量是 JSP 页面的局部变量，调用 JSP Scriptlets 时，会为局部变量分配内存空间，调用结束后，释放局部变量占有的内存空间。

3. JSP 声明

JSP 的声明语句用于声明变量和方法，其语法格式如下：<%! 声明变量或方法的代码 %>

在上述语法格式中，被声明的 Java 代码将被编译到 Servlet 的_jspService()方法之外，即在 JSP 声明语句中定义的变量或方法将作为类的属性而存在。<%!和%>中定义的属性是成员属性，相当于类的属性；<%!和%>中定义的方法相当于全局的方法，也相当于类中的方法。注意：<%与!之间不可以有空格。

<%和%>这对标签中的内容是在此 JSP 被编译为 Servlet 的时候，放在_jspService()方法中的，这个方法是服务器向客户端输出内容的地方，它本身就是一个方法。如果在其中定义方法，相当于是在类的方法里嵌套定义方法，这在 Java 中是不允许的。

这里把<%! %>称为 Declare Tag。利用 Declare Tag 声明的变量将会出现在 Servlet 类定义中。在 Declare Tag 内可以声明的有：Instance Variable、Instance Method、Class Variable、Class

Method，甚至 Inner Classes。

在 Declare Tag 内可以覆写 jspInit()与 jspDestroy()方法，但是不能覆写_jspService()方法。

【示例】JSP 页面访问统计。

在 JSPDemo 工程的 WebContent 文件夹中创建 visit.jsp，在<body>部分加入以下代码。

```
<%! int count = 0;  //被用户共享的count
    synchronized void visitCount() {
    count++;
}%>
<% String date = new java.util.Date().toLocaleString(); %>
<%
    visitCount();
    out.print("<p><b>" + "您是第" + count + "个访问本网站的用户。" + "<b></p>");
    out.print("<p><b>" + "访问时间是：" + date + "<b></p>");
%>
```

3.1.6 JSP 注释

JSP 注释的语法格式如下：<%-- 注释信息 --%>

服务器在将 JSP 页面编译成 Servlet 程序时，会忽略 JSP 页面中被注释的内容，不会将注释信息发送到客户端。

3.2 JSP 动作元素

JSP 标准动作（Standard Action）元素是 JSP 规范所定义的一种 XML 标签，用来运行特定动作。使用 JSP 标准动作元素的优点是可以减少 JSP 所包含的 Java 程序代码。

1. <jsp:include>动作元素

<jsp:include>用于在页面运行时引入一个静态或动态的页面，也称为动态包含。当容器把 JSP 页面翻译成 Java 文件时，并不会把 JSP 页面中动作指令 include 指定的文件与原 JSP 页面合并成一个新页面，而是告诉 Java 解释器，这个文件在 JSP 运行时才被处理。

如果包含的文件是普通的文本文件，就将文件的内容发送给客户端，由客户端负责显示；如果包含的文件是 JSP 文件，则 JSP 容器执行这个文件，然后将执行结果发送到客户端，由客户端负责显示这些结果。

<jsp:include>动作元素的语法格式如下：<jsp:include page="relativeURL" flush="true | false" />。

这里，page 属性用于指定被引入资源的相对路径；flush 属性用于指定是否将当前页面的输出内容刷新到客户端。默认情况下，flush 属性的值为 false。

【示例】使用<jsp:include>。

在 JSPDemo 工程的 WebContent 文件夹中创建 include.jsp 和 sum.jsp 页面，include.jsp 页面的<body>部分代码如下。

```
<h2>求和</h2>
<jsp:include page="sum.jsp" flush="true">
```

```
    <jsp:param value="100" name="max_n" />
    <jsp:param value="10" name="min_n" />
  </jsp:include>
```

sum.jsp 页面的<body>部分代码如下。

```
  <%
    Thread.sleep(5000);
    int max_n = Integer.parseInt(request.getParameter("max_n"));
    int min_n = Integer.parseInt(request.getParameter("min_n"));
    int sum = 0;
    for (int i = min_n; i <= max_n; i++)
        sum += i;
    out.print("<p>从" + min_n + "到" + max_n + "的和为: " + sum);
  %>
```

2. <jsp:forward> 动作元素

<jsp:forward>动作元素将当前请求转发到其他 Web 资源（JSP 页面、Servlet 等），在执行请求转发之后的当前页面将不再执行，而是执行该元素指定的目标页面。

<jsp: forward >动作元素的语法格式如下：`<jsp: forward page="relativeURL" />`。

<jsp: forward >的功能和 Servlet 的 RequestDispatcher 对象的 forward 方法类似，调用者和被调用者共享同一个 request 对象。

【示例】使用<jsp:forward>。

代码详见本书资源源码包 ch3 目录下 JSPDemo 工程的 forward.jsp 和 forward_book.jsp。

3.3　JSP 内置对象

JSP 隐式对象是 JSP 容器为每个页面提供的 Java 对象，开发者可以直接使用它们而不用显式声明。JSP 隐式对象也被称为预定义变量。JSP 所支持的九大隐式对象如表 3-2 所示。

表 3-2　JSP 隐式对象

对象名称	类型	描述
request	javax.servlet.http. HttpServletRequest	请求对象，提供客户端 HTTP 请求数据的方法
response	javax.servlet.http. HttpServletResponse	服务器向客户端的回应信息
out	javax.servlet.jsp.JspWriter	输出对象，用于把结果输出至网页上
session	javax.servlet.http. HttpSession	会话对象，用来保存服务器与每个客户端会话过程中的信息
application	javax.servlet. ServletContext	包含 Web 应用程序配置信息的 ServletContext 对象
config	javax.servlet. ServletConfig	包含 Servlet（JSP 编译后所产生的）相关信息的 ServletConfig 对象
pageContext	javax.servlet.jsp. PageContext	页面上下文对象，PageContext 类的实例，提供对 JSP 页面所有对象以及命名空间的访问
page	javax.servlet.jsp.HttpJspPage	类似于 Java 类中的 this 关键字，指当前页面转换后的 Servlet 类的实例
Exception	java.lang.Throwable	Exception 类的对象，代表发生错误的 JSP 页面中对应的异常对象

1. request 对象

request 对象是 javax.servlet.http.HttpServletRequest 类的实例。每当客户端请求一个 JSP 页面时，JSP 引擎就会制造一个新的 request 对象来代表这个请求。request 对象的作用域是一次 request 请求。

request 对象主要用于接收客户端通过 HTTP 传送给服务器的数据。

request 对象拥有 HttpServletRequest 接口的所有方法。

2．response 对象

response 对象是 javax.servlet.http.HttpServletResponse 类的实例，表示服务器对客户端的响应。当服务器创建 request 对象时会同时创建用于响应这个客户端的 response 对象。

response 对象也定义了处理 HTTP 头模块的接口。通过这个对象，开发者们可以添加新的 Cookies、时间戳、HTTP 状态码等。

response 对象拥有 HttpServletResponse 接口的所有方法。

3．out 对象

out 对象是 javax.servlet.jsp.JspWriter 类的实例，用来在 response 对象中写入内容。

out 对象与 HttpServletResponse 接口的 getWriter()方法获得的 PrintWriter 对象功能相同，并都由 java.io.Writer 继承而来。

最初的 JspWriter 类对象根据页面是否有缓存来进行不同的实例化操作。可以在 page 指令中使用 buffered='false'属性来轻松关闭缓存。

JspWriter 类包含了大部分 java.io.PrintWriter 类中的方法。不过，JspWriter 新增了一些专为处理缓存而设计的方法。此外，JspWriter 类会抛出 IOExceptions 异常，而 PrintWriter 不会。

4．session 对象

session 对象是 javax.servlet.http.HttpSession 类的实例，和 Java Servlets 中的 session 对象有一样的行为。session 对象用来跟踪在各个客户端请求间的会话。

【示例】登录验证。

代码详见本书资源源码包 ch3 目录下 JSPDemo 工程的 login.jsp、loginValidate.jsp、welcome.jsp 和 logout.jsp。在 login.jsp 页面中输入用户名和密码，表单提交给 loginValidate.jsp 验证。

5．application 对象

application 对象即应用程序上下文对象，表示当前应用程序运行环境，用以获取应用程序上下文环境中的信息。application 对象在容器启动时实例化，在容器关闭时销毁，作用域为整个 Web 容器的生命周期。

application 对象实现了 javax.servlet.ServletContext 接口，具有 ServletContext 接口的所有功能，application 对象常用方法如下：

① void setAttribute(String name, Object object)：以名/值对的方式存储 application 域属性。

② Object getAttribute(String name)：根据属性名获取属性值。

③ void removeAttribute(String name)：根据属性名从 application 域中移除属性。

6．pageContext 对象

pageContext 对象是 javax.servlet.jsp.PageContext 类的实例，用来代表整个 JSP 页面。这个对象主要用来访问页面信息，同时过滤掉大部分实现细节。

这个对象存储了 request 对象和 response 对象的引用。application 对象、config 对象、session 对象以及 out 对象也可以通过访问这个对象的属性来导出。

pageContext 对象也包含了传给 JSP 页面的指令信息，包括缓存信息、ErrorPage URL、页面 scope 等。pageContext 类定义了一些字段，这些字段包括 PAGE_SCOPE、REQUEST_SCOPE、SESSION_SCOPE、APPLICATION_SCOPE。它也提供了 40 余种方法，有一半继承自 javax.servlet.jsp.JspContext 类。其中一个重要的方法就是 removeArribute()，它可接收一个或两个参数。例如，pageContext.removeArribute("attrName")可移除 4 个 scope 中的相关属性，但是下面的方法只移除特定 scope 中的相关属性。

pageContext.removeAttribute("attrName", PAGE_SCOPE);

pageContext 对象获取内置对象的方法及描述如表 3-3 所示。

表 3-3 pageContext 对象获取内置对象的方法及描述

方 法	描 述
ServletRequest getRequest()	获取当前 JSP 页面的请求对象
ServletResponse getResponse()	获取当前 JSP 页面的响应对象
HttpSession getSession()	获取和当前 JSP 页面有联系的会话对象
ServletConfig getServletConfig()	获取当前 JSP 页面的 ServletConfig 对象
ServletContext getServletContext()	获取当前 JSP 页面的运行环境中的 application 对象
Object getPage()	获取当前 JSP 页面的 Servlet 实体 page 对象
Exception getException()	获取当前 JSP 页面的异常 Exception 对象
JspWriter getOut()	获取当前 JSP 页面的输出流 out 对象

pageContext 对象存取域属性的方法及描述如表 3-4 所示。

表 3-4 pageContext 对象存取域属性的方法及描述

方 法	描 述
void setAttribute(String name, Object value, int scope)	以名/值对的方式存储 scope 范围域中的属性
Object getAttribute(String name, int scope)	获取范围为 scope、名为 name 的属性对象
void removeAttribute(String name, int scope)	从 scope 范围内移除名为 name 的属性
Enumeration<String> getAttributeNamesInScope(int scope)	从 scope 范围中获取所有属性的属性

pageContext.java 部分源码如下：

```
    public abstract class PageContext extends JspContext {
      public static final int PAGE_SCOPE = 1;
      public static final int REQUEST_SCOPE = 2;
      public static final int SESSION_SCOPE = 3;
      public static final int APPLICATION_SCOPE = 4;
      public static final String PAGE = "javax.servlet.jsp.jspPage";
      public static final String PAGECONTEXT = "javax.servlet.jsp.jspPage
Context";
      public static final String REQUEST = "javax.servlet.jsp.jspRequest";
      public static final String RESPONSE = "javax.servlet.jsp.jspResponse";
      public static final String CONFIG = "javax.servlet.jsp.jspConfig";
      public static final String SESSION = "javax.servlet.jsp.jspSession";
      public static final String OUT = "javax.servlet.jsp.jspOut";
      public static final String APPLICATION = "javax.servlet.jsp.jspApp
lication";
      public static final String EXCEPTION = "javax.servlet.jsp.jspException";
      public abstract HttpSession getSession();
      public abstract ServletRequest getRequest();
```

```
    public abstract ServletResponse getResponse();
    …
  }
```

【示例】使用 pageContext 对象。

在 JSPDemo 工程的 WebContent 文件夹中创建 pageContext.jsp，其<body>部分的代码如下。

```
  <%   //获取 request 对象
    HttpServletRequest req = (HttpServletRequest) pageContext.getRequest();
    //设置 page 范围内的属性
    pageContext.setAttribute("boyName", "东北虎", pageContext.PAGE_SCOPE);
    //设置 request 范围内的属性
    req.setAttribute("girlName", "我不是潘金莲");
    //设置 session 范围内的属性
    pageContext.setAttribute("user", "西门吹雪", pageContext.SESSION_SCOPE);
    //设置 application 范围内的属性
    pageContext.setAttribute("driverClassName", "com.mysql.jdbc.Driver",
        pageContext.APPLICATION_SCOPE);
    //获得的 page 范围属性
    String boyName = (String) pageContext.getAttribute("boyName",
        pageContext.PAGE_SCOPE);
    //获得的 request 范围属性
    String girlName = (String) pageContext.getAttribute("girlName",
        pageContext.REQUEST_SCOPE);
    //获得的 session 范围属性
    String user = (String) pageContext.getAttribute("user",
        pageContext.SESSION_SCOPE);
    //获得的 application 范围属性
    String driverClassName = (String) pageContext.getAttribute("driverClass
Name",
        pageContext.APPLICATION_SCOPE);
  %>
  <%="page 范围(属性 boyName 的值): " + boyName %><br>
  <%="request 范围(属性 girlName 的值): " + girlName %><br>
  <%="session 范围(属性 user 的值): " + user %><br>
  <%="application 范围(属性 driverClassName 的值): " + driverClassName %><br>
```

7. config 对象

config 对象是 javax.servlet.ServletConfig 类的实例，直接包装了 Servlet 的 ServletConfig 类的对象。这个对象允许开发者访问 Servlet 或者 JSP 引擎的初始化参数，如文件路径等。

以下是 config 对象的使用方法，不是很重要，所以不常用：

```
  config.getServletName();
```

它返回包含在<servlet-name>元素中的 servlet 名称，注意，<servlet-name>元素在 Web-INF\Web.xml 文件中定义。

8. page 对象

这个对象就是页面实例的引用。它可以被看作整个 JSP 页面的代表。

page 对象就是 this 对象的同义词。

9. exception 对象

exception 对象包装了从先前页面中抛出的异常信息。它通常被用来产生对出错条件的适当响应。

【示例】JSP 异常处理。

在 JSPDemo 工程的 WebContent 文件夹中创建 exception.jsp 和 error.jsp，exception.jsp 是产生异常的页面，error.jsp 是处理异常的页面。

exception.jsp 的 Page 指令中须设置 errorPage="error.jsp"，其<body>部分代码如下。

```
<%   int[] array = { 2, 4, 6 };
 System.out.println(array[3]);
%>
```

error.jsp 的 Page 指令中须设置 isErrorPage="true"，其<body>部分代码如下。

```
<h3>处理异常页面</h3>
<%out.println(exception);%>
```

使用外部的火狐浏览器访问 exception.jsp 页面，输出结果如图 3-2 所示。需要进入火狐浏览器的“选项→隐私与安全”标签页，勾选“允许 Firefox 向 Mozilla 发送错误报告”复选框。

处理异常页面

java.lang.ArrayIndexOutOfBoundsException: 3

图 3-2　JSP 异常处理

3.4　JSP 综合示例

这里实现一个猜数小游戏，其运行效果如图 3-3 所示。

（a）猜数小游戏①　（b）猜数小游戏②　（c）猜数小游戏③　（d）猜数小游戏④

图 3-3　猜数游戏

代码详见本书资源源码包 ch3 目录下 NumberGuess 工程 WebContent 目录下的 numguess.jsp 文件和 src 目录下 com.mialab.num 包中的 NumberGuessBean.java 文件。

3.5　本章小结

本章主要讲解了 JSP 的运行原理、JSP 语法、JSP 指令、JSP 动作元素和 JSP 隐式对象。

JSP 实现了 HTML 语法中的 Java 扩展（以 <% %> 形式实现扩展）。JSP 与 Servlet 一样，是在服务器端执行的。通常，返回给客户端的就是一个 HTML 文本，因此，客户端只要有浏览器就能浏览。JSP 将网页逻辑与网页设计的显示分离，支持可重用的基于组件的设计，使基于 Web 的应用程序的开发变得迅速和容易。

习题 3

1．JSP 拥有哪些内置对象，其用途与存取范围又是怎样的？

2．依照先后顺序，JSP 生命周期分为哪些阶段？

3．在 JSP 内使用 Declare Tag（<%! %>）与 Script Tag（<% %>）声明的变量有何不同？

4．JSP 的 include 指令元素和<jsp:include>动作元素有何异同？

5．Page 指令包含的常用属性有哪些？如何使用？

6．关于 JSP 生命周期的描述，下列哪些为真（选择两个答案）？

A．JSP 先解释成 Servlet 源文件，再编译成 Servlet 类文件（*.class）

B．每当用户端运行 JSP 时，jspInit()方法都会运行一次

C．每当用户端运行 JSP 时，jspService()方法都会运行一次

D．每当用户端运行 JSP 时，jspDestroy()方法都会运行一次

7．编写一个 JSP 程序，实现用户登录，当用户输入的用户或密码错误时，将页面重定向到错误提示页，并在该页面显示 30 秒后，自动返回到用户登录页面。

8．编写 JSP 页面，实现简单网站计数器的功能。

9．编写 JSP 页面，实现 JSP 异常处理机制。

10．在 Web.xml 文件中使用<error-page>元素为整个 Web 应用程序设置错误处理页面。假设处理状态码为 404 的页面是 404.jsp，处理状态码为 500 的页面是 500.jsp。

第 4 章 会话管理

本章导读

HTTP 是一种无状态协议，每次客户端访问 Web 页面时，客户端打开一个单独的浏览器窗口连接到 Web 服务器，由于服务器不会自动保存之前客户端请求的相关信息，所以无法识别一个 HTTP 请求是否为第一次访问。这意味着需要有相应的技术来维持 Web 客户端和服务器之间的会话，这就是会话跟踪。会话跟踪（或会话管理）成为 Web 应用开发中一个不可避免的主题。本章主要介绍会话跟踪技术的几种解决方案：（1）Cookie 技术；（2）Session 技术；（3）URL 重写技术；（4）隐藏表单域技术。

4.1 Cookies

4.1.1 Cookie 剖析

Cookie 是存储在客户端计算机上的文本文件，其中保留了各种跟踪信息。Java Servlet 显然支持 HTTP Cookie。识别返回用户包括 3 个步骤：服务器脚本向浏览器发送一组 Cookie；浏览器将这些信息存储在本地计算机上，以备将来使用；当下一次浏览器向 Web 服务器发送任何请求时，浏览器会把这些 Cookie 信息发送到服务器，服务器将使用这些信息来识别用户。

Cookie 通常设置在 HTTP 头信息中，如设置 Cookie 的 Servlet 会发送如下的头信息：

```
    Set-Cookie: name=sky; expires=Friday, 09-Feb-08 22:08:38 GMT; path=/;
domain=mialab.com
```

可以看到，Set-Cookie 头包含了一个名/值对、一个 GMT 日期、一个路径和一个域。名称和值会被 URL 编码。expires 字段是一个指令，告诉浏览器在给定的时间和日期之后“忘记”该 Cookie。

如果浏览器被配置为存储 Cookie，则它将会保留此信息直到截止日期。如果用户的浏览器指向任何匹配该 Cookie 的路径和域的页面，则它会重新发送 Cookie 到服务器。浏览器的头信息如下所示：

```
GET / HTTP/1.0
Connection: Keep-Alive
User-Agent: Mozilla/4.6 (X11; I; Linux 2.2.6-15apmac ppc)
…
Cookie: name=sky
```

此时，Servlet 就能够通过请求方法 request.getCookies()访问 Cookie，该方法将返回一个 Cookie 对象的数组。

4.1.2 在 Servlet 中操作 Cookie

1．通过 Servlet 设置 Cookie

通过 Servlet 设置 Cookie 包括以下 3 个步骤。

（1）创建一个 Cookie 对象：可以调用带有 cookie 名称和 cookie 值的 Cookie 构造函数，

cookie 名称和 cookie 值都是字符串。例如：

```
Cookie cookie = new Cookie("key","value");
```

（2）设置最大生存周期：使用 setMaxAge 方法来指定 cookie 能够保持的有效的时间（以秒为单位）。例如，设置一个最长有效期为 24 小时的 cookie：

```
cookie.setMaxAge(60*60*24);
```

（3）发送 Cookie 到 HTTP 响应头：使用 response.addCookie 来添加 HTTP 响应头中的 Cookie。例如：

```
response.addCookie(cookie);
```

2．通过 Servlet 读取 Cookie

要读取 Cookie，需要通过调用 HttpServletRequest 的 getCookies()方法创建一个 javax.servlet.http.Cookie 对象的数组，再循环遍历数组，并使用 getName()和 getValue()方法来访问每个 cookie 和关联的值。

3．通过 Servlet 删除 Cookie

删除 Cookie 是非常简单的，只要按照以下 3 个步骤进行操作即可。

（1）读取一个现有的 cookie，并把它存储在 Cookie 对象中。

（2）使用 setMaxAge()方法设置 cookie 的年龄（存活期）为零，来删除现有的 cookie。

（3）把这个 cookie 添加到响应头。

如下为删除一个名为 userName 的 cookie 代码：

```
Cookie cookie = new Cookie("userName","");
cookie.setMaxAge(0);
response.addCookie(cookie);
```

4.1.3　Cookie API

Cookie 类在 javax.servlet.http 包中，Cookie 类的常用方法如表 4-1 所示，以下也是在 Servlet 中操作 Cookie 时可使用的有用的方法列表。

表 4-1　Cookie 类的常用方法

方　法	功能描述
String getName()	用于返回 Cookie 的名称
String getValue()	返回 Cookie 的值
void setValue(String newValue)	用于为 Cookie 设置一个新的值
void setMaxAge(int expiry)	该方法设置 Cookie 过期的时间（以秒为单位）。如果不这样设置，Cookie 只会在当前 session 中持续有效
int getMaxAge()	该方法返回 Cookie 的最大生存周期（以秒为单位），默认情况下，-1 表示 Cookie 将持续下去，直到浏览器关闭
void setPath(String uri)	用于设置该 Cookie 项的有效目录路径
String getPath()	用于返回该 Cookie 项的有效目录路径
void setDomain(String pattern)	用于设置该 Cookie 项的有效域
String getDomain()	用于返回该 Cookie 项的有效域

4.1.4　使用 Cookie 示例

1．在 IE 浏览器中设置 Cookie

在使用 Cookie 时，要保证浏览器接收 Cookie。IE 浏览器中设置 Cookie 的方法如下：选择 IE 浏览器的“工具”菜单中的“Internet 选项”选项，选择“隐私”选项卡，在“高级隐私设置”对话框中勾选“总是允许会话 cookie”复选框，“第一方 Cookie”和“第三方 Cookie”都选择“接受”。

这里使用的 IE 浏览器版本是 Internet Explorer 11。

2. 在 Firefox 中设置 Cookie

打开 Firefox，选择火狐浏览器的“打开”菜单中的“选项”选项，选择“隐私与安全”标签页，在“隐私与安全”标签页中勾选“接受来自网站的 Cookie”复选框，“接受第三方 Cookie”选择“始终”；“保存，直到”选择“它们过期”。

这里使用的火狐浏览器版本是 Firefox Quantum 59.0.2(64 位)。

3. 使用 Cookie 示例

【示例】显示用户上次访问时间。

在 Eclipse 中创建 Dynamic Web Project，名为 CookieDemo。在 CookieDemo 工程的 src 目录下创建 Package “com.mialab.cookie”，在 com.mialab.cookie 包中创建一个 Servlet，名为 CookieDemoServlet.java。CookieDemoServlet.java 的主要代码如下。

```
@WebServlet("/CookieDemoServlet")
public class CookieDemoServlet extends HttpServlet {
 protected void doGet(HttpServletRequest request, HttpServletResponse
response)
        throws ServletException, IOException {
    response.setContentType("text/html;charset=UTF-8");
    PrintWriter out = response.getWriter();
    SimpleDateFormat sdf = new SimpleDateFormat("yyyy-MM-dd HH:mm:ss");
    String nowTime = sdf.format(new Date());
    String lastAccessTime = "";
    int vistedCount = 0;
    Cookie[] cookies = request.getCookies(); //获取客户端浏览器保存的所有
Cookie
    if (cookies != null)
        for (Cookie cookie : cookies) {
            // 判断是否为记录最近访问时间的 Cookie
            if ("lastAccessTime".equals(cookie.getName())) {
                lastAccessTime = cookie.getValue();
            }
            // 判断是否为记录访问次数的 Cookie
            if ("vistedCount".equals(cookie.getName())) {
                vistedCount = Integer.valueOf(cookie.getValue());
            }
        }
    // 若曾经访问过，则输出上次的访问时间
    if (!"".equals(lastAccessTime))
        out.println("您上一次的访问时间是: "+lastAccessTime);
    out.println("您是第"+(vistedCount+1)+"次访问本网站");//输出访问次数
    // 以本次访问时间重建同名新 Cookie
    Cookie lastAccessObj = new Cookie("lastAccessTime", nowTime);
    // 设置最大存活时间：一年
    lastAccessObj.setMaxAge(365 * 24 * 60 * 60);
    // 以新访问次数重建同名新 Cookie
```

```
        Cookie visitCountObj = new Cookie("vistedCount",
                String.valueOf(vistedCount + 1));
        // 设置最大存活时间：一年
        visitCountObj.setMaxAge(365 * 24 * 60 * 60);
        response.addCookie(lastAccessObj);   //将上述新建 Cookie 响应到客户端
        response.addCookie(visitCountObj);
    }
}
```

访问的效果如图 4-1 所示。

图 4-1　CookieDemoServlet 的第 1 次访问和第 2 次访问

（代码详见教学资源源码包 ch4 目录下的 CookieDemo 工程。）

4.2 HttpSession 对象

4.2.1 Session 简介

为了让 Web 应用程序能够记住用户送出的不同请求，Servlet 规范内定义了一个 HttpSession 接口，允许 Servlet 容器针对每个用户建立一个 HTTP 会话（即 HttpSession 对象），每个 HTTP 会话将被赋予唯一的“会话编号”（Session ID）。

HttpSession 对象会在用户第一次访问服务器时由容器创建（只有访问 JSP、Servlet 等动态资源时才会创建，访问 HTML 等静态资源并不会创建），当用户调用其失效方法（invalidate()方法）或超过其最大不活动（超时，timeout）时间时会失效。

注意，所有保存在 HttpSession 中的数据不会被发送到客户端，不同于其他会话管理技术，Servlet 容器为每个 HttpSession 生成唯一的标识，并将该标识发送给浏览器，或创建一个名为 JSESSIONID 的 Cookie，或者在 URL 后附加一个名为 jsessionid 的参数。在后续的请求中，浏览器会将标识提交给服务端，这样服务器就可以识别该请求是由哪个用户发起的。Servlet 容器会自动选择一种方式传递会话标识，无须开发人员介入。

HttpServletRequest 接口提供了获取 HttpSession 对象的方法。

（1）HttpServletRequest 接口的 getSession()方法：取得目前的 HTTP 会话。如果没有，Servlet 容器将会建立一个新的 HTTP 会话，并且赋予新的 Session ID。

（2）HttpServletRequest 接口的 getSession(boolean create)方法：允许传入 true 或 false 参数。如果调用 getSession(true)，那么执行方式与调用 getSession()完全相同。如果调用 getSession(false)，而用户之前的请求并未建立过 HTTP 会话，那么此方法将返回 null（并不会建立新的 HTTP 会话）。

4.2.2 HttpSession API

HttpSession 接口提供了存取会话域属性和管理会话生命周期的方法，如表 4-1 所示。

表 4-2　HttpSession 接口中的常用方法

方　法	功能描述
getAttribute(String key)	通过 key 获取对象值
void setAttribute(String key, Object value)	使用指定的名称绑定一个对象到该 HTTPSession 会话
public void removeAttribute(String name)	该方法将从 HTTPSession 会话移除指定名称的对象
void invalidate()	指示该 HTTPSession 会话无效
int getMaxInactiveInterval()	该方法返回 Servlet 容器在客户端访问时保持 HTTPSession 会话打开的最大时间间隔，以秒为单位
void setMaxInactiveInterval(int interval)	设置 HTTP 会话的默认超时间隔
String getId()	获取 HTTPSession 对象标识 Session ID
long getLastAccessedTime()	该方法返回客户端最后一次发送与该 HTTPSession 会话相关的请求的时间，以毫秒为单位

可同时采用下列两种方式来终止 HTTP 会话。

（1）在所有数据处理动作结束后，调用 HTTP 会话的 invalidate()方法。

（2）在 Web.xml 内设定 HTTP 会话的“超时”时间。例如：

```
<session-config>
 <session-timeout>30</session-timeout>
</session-config>
```

在上面的配置信息中，设置的时间值是以分钟为单位的。

4.2.3　使用 HttpSession 示例

【示例】管理 HTTP 会话信息，实现购物车功能。

在 Eclipse 中创建 Dynamic Web Project，名为 SessionDemo。在 SessionDemo 工程的 src 目录下创建 Package“com.mialab.session”，在 com.mialab.session 包中创建 3 个 Servlet，分别为 ProcessOneServlet.java、ProcessTwoServlet.java 和 ConfirmServlet.java，在 WebContent 目录下创建 2 个 JSP 页面：bookList.jsp 和 address.jsp。

在 bookList.jsp 页面中选择欲购置的图书，将表单数据提交给 ProcessOneServlet 处理，如图 4-2 所示。ProcessOneServlet 将欲购置的图书数据放置在 HttpSession 中，并将请求重定向到 address.jsp 页面。在 address.jsp 页面中输入客户名和邮递地址，并提交给 ProcessTwoServlet 处理，如图 4-3 所示。ProcessTwoServlet 仍将客户信息和邮递地址数据放置在 HttpSession 中，并将请求重定向到 ConfirmServlet。ConfirmServlet 将购物信息从 HttpSession 中取出，并加以显示，从而实现购物车功能，如图 4-4 所示。

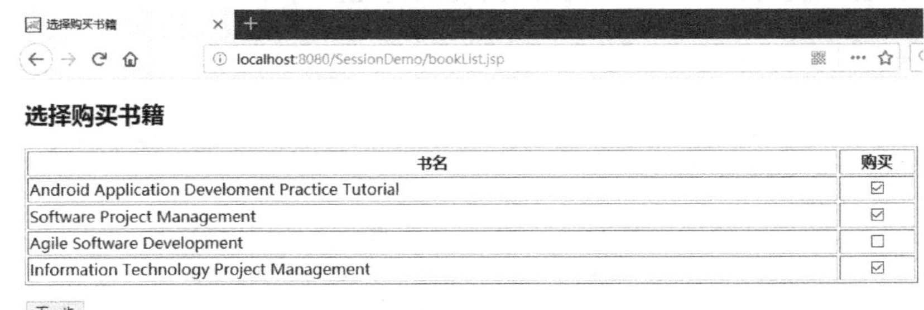

图 4-2　bookList.jsp

图 4-3 address.jsp

图 4-4 实现购物车功能

ProcessOneServlet.java 主要代码如下：

```
    @WebServlet(name = "ProcessOneServlet", urlPatterns = { "/ProcessOne" })
    public class ProcessOneServlet extends HttpServlet {
     protected void doPost(HttpServletRequest request, HttpServletResponse
response)
              throws ServletException, IOException {
         response.setContentType("text/html;charset=UTF-8");
         PrintWriter out = response.getWriter();
         HttpSession session = request.getSession(true);
         String[] selectedBooks = request.getParameterValues("buy");
         session.setAttribute("choosed", selectedBooks);
         response.sendRedirect("./address.jsp");
     }
    }
```

ProcessTwoServlet.java 主要代码如下：

```
    @WebServlet(name = "ProcessTwoServlet", urlPatterns = { "/ProcessTwo" })
    public class ProcessTwoServlet extends HttpServlet {
     protected void doPost(HttpServletRequest request, HttpServletResponse
response)
              throws ServletException, IOException {
         response.setContentType("text/html;charset=UTF-8");
         PrintWriter out = response.getWriter();
         HttpSession session = request.getSession(true);
         String customer =
              new
String(request.getParameter("customer").getBytes("iso-8859-1"));
         String address =
              new
String(request.getParameter("address").getBytes("iso-8859-1"));
         session.setAttribute("A_customer", customer);
         session.setAttribute("A_address", address);
         response.sendRedirect("Confirm");
```

```
        }
    }
```

ConfirmServlet.java 主要代码如下：

```
    @WebServlet(name = "/ConfirmServlet", urlPatterns = { "/Confirm" })
    public class ConfirmServlet extends HttpServlet {
     protected void doGet(HttpServletRequest request, HttpServletResponse
response)
            throws ServletException, IOException {
        response.setContentType("text/html;charset=UTF-8");
        PrintWriter out = response.getWriter();
        HttpSession session = request.getSession(true);
        String[] books = (String[]) session.getAttribute("choosed");
        String customer = (String) session.getAttribute("A_customer");
        String address = (String) session.getAttribute("A_address");
        //以下代码实现数据（购物信息）显示
     }
    }
```

4.3 URL 重写

4.3.1 为什么需要 URL 重写

在某些情况下，用户为了安全性可能会关闭 Web 浏览器的 Cookie 功能。这时候如果从用户端浏览器送出 HTTP 请求，并不会包含 Session ID，因此，Servlet 容器无法利用同一个 HTTP 会话来保存不同的 HTTP 请求内容。

例如，关闭火狐浏览器的 Cookie 功能，再来测试前面的 SessionDemo 工程，最终将输出："无法从 HTTP 会话内取出所需信息！"，如图 4-5 所示。

图 4-5　关闭 Firefox 的 Cookie 功能

为了避免用户关闭Cookie后导致Web应用程序无法如期运行，必须采取另一种方法来"强迫用户端浏览器送出相同的 Session ID"，其实现方式是"改写用户端浏览器所送出的 URL 请求，在 URL 请求后面插入一个 Session ID"。这样，Servlet 容器就可以根据用户端传送 HTTP 请求时夹带的 Session ID 来取得相同的 HTTP 会话。

该如何改写用户端浏览器送出的 URL 呢？一般来说，用户输入数据的 HTML 网页都是固定的，并不能随意更改。所以，要实现 URL 重写的唯一方法是动态产生 HTML 网页，并且在<form>标签的 action 属性内夹带一个 Session ID。

4.3.2 encodeURL()和 encodeRedirectURL()

URL 重写通过 HttpServletResponse 的 encodeURL()方法和 encodeRedirectURL()方法实现，其中 encodeRedirectURL()方法主要对使用 sendRedirect ()方法的 URL 进行重写。

URL 重写方法根据请求信息中是否包含 Set-Cookie 请求头来决定是否进行 URL 重写，若包含该请求头，则会将 URL 原样输出；若不包含，则会将会话标识重写到 URL 中。

例如，encodeURL()方法的运行操作可分为以下几个步骤。

（1）检查用户端浏览器是否关闭 Cookie 功能。

（2）如果 Cookie 功能被关闭，encodeURL()会改写传入该方法的 URL，方式是在 URL 后面加上目前的 Session ID（假设 Servlet 容器已建立一个 HTTP 会话）。

（3）如果 Cookie 功能并未被关闭，或是目前并未建立 HTTP 会话，则 encodeURL()会保留传入该方法的 URL，不做任何改动。

4.3.3 使用 URL 重写示例

HttpSession 接口提供了存取会话。

【示例】使用 URL 重写管理 HTTP 会话，实现购物车功能。

在 Eclipse 中创建 Dynamic Web Project，名为 SessionDemo。在 SessionDemo 工程的 src 目录下创建 Package “com.mialab.session”，在 com.mialab.session 包中创建 3 个 Servlet，分别为 ProcessOneServlet.java、ProcessTwoServlet.java 和 ConfirmServlet.java，在 WebContent 目录下创建 2 个 JSP 页面：bookList.jsp 和 address.jsp。

4.4 隐藏表单域

一个 Web 服务器可以发送一个隐藏的 HTML 表单字段，以及一个唯一的 SESSION ID，如下所示：

```
<input type="hidden" name="session_id" value="123456789">
```

当表单被提交时，指定的名称和值会被自动包含在表单数据中。每次当用户端浏览器发送回请求时，session_id 值可以用于保持不同的用户端浏览器的跟踪。这也是一种保持 SESSION 跟踪的有效方式。但是单击常规的超文本链接（<A HREF...>）不会导致表单提交，因此，在这里隐藏的表单字段并不支持常规的会话跟踪。

4.5 本章小结

本章主要讲解了会话跟踪技术的 4 种解决方案，主要有 Cookie 技术、Session 技术、URL 重写技术和隐藏表单域技术。

Cookie 是由服务器端生成的，用户端是浏览器，用户端浏览器会将 Cookie 的 key/value 保存到某个目录下的文本文件内，下次请求同一网站时就发送该 Cookie 给服务器（前提是浏览器设置为启用 Cookie）。浏览器访问网站时会携带这些 Cookie 信息，达到鉴定身份的目的。Cookie 名称和值可以由服务器端开发自定义，对于 JSP 而言，也可以直接写入 JSESSIONID，这样服务器可以知道该用户是否是合法用户以及是否需要重新登录等，服务器可以设置或读取 Cookies 中包含的信息，以此维护用户和服务器会话中的状态。

Session 是通过 Cookie 技术实现的，依赖于名为 JSESSIONID 的 Cookie，它将信息保存在服务器端。Session 中能够存储复杂的 Java 对象。Session 机制是一种服务器端的机制，服务器使用一种类似于散列表的结构（也可能就是使用散列表）来保存信息。

URL 重写是一种很好的维持 HTTP 会话的方式，它在浏览器不支持 Cookie 时能够很好地工作，但是它的缺点是会动态生成每个 URL 来为页面分配一个 HTTP Session ID。为了在整个交互过程中始终保持状态，就必须在每个客户端可能请求的路径后面都包含这个 Session ID。URL 重写技术是实现动态网站会话跟踪的重要保障。

隐藏表单域技术就是在表单域中添加一个隐藏字段，以便在表单提交时能够把 Session ID 传递回服务器。利用表单的隐藏表单域，可以在完全脱离浏览器对 Cookie 的使用限制以及在用户无法从页面显示看到隐藏标识的情况下，将标识随请求一起传送给服务器处理，从而实现会话的跟踪。这种技术现在已较少应用，可以用 URL 重写技术替代。

习题 4

1．Cookie 技术与 Session 技术的主要区别是什么？
2．会话对象的生命周期是怎样的？
3．如何终止 HTTP 会话？
4．会话 Cookie 和持久 Cookie 的区别是什么？
5．何时使用 URL 重写技术？又如何使用？请编程加以说明。
6．在火狐浏览器中如何关闭 Cookie 功能？又如何开启 Cookie？
7．编程实现：使用 Session 技术实现购物车功能。
8．关于 URL 重写技术，下列哪些为真？（选择两个答案）
A．需使用 ServletResponse 接口的 encodeURL()方法
B．适用于用户端浏览器已使用 Cookie 的情况
C．适用于用户端浏览器已关闭 Cookie 的情况
D．改写用户端浏览器送出的 Servlet URL，方式是在 URL 后夹带 Session ID 相关信息

第5章 EL和JSTL

本章导读

表达式语言（Expression Language，EL）是 JSP 2.0 的重要特性。EL 设计成可以轻松地编写免脚本的 JSP 页面，也就是说，页面不使用任何 JSP 声明、表达式或者 Scriptlets。JSP 标准标签库（JSP Standard Tag Library，JSTL）主要提供给 Java Web 开发人员一个标准通用的标签函数库，标签库同时支持 EL 用于获取数据，Web 开发人员能够用此标签库取代直接在页面中嵌入 Java 程序的做法，以提高程序的可读性和易维护性。本章主要内容有：（1）EL 简介、EL 的运算符和优先级；（2）EL 隐式对象；（3）定义和使用 EL 函数；（4）JSTL 简介；（5）JSTL 安装与配置；（6）JSTL 的核心标签库。

5.1 JSP 表达式语言

5.1.1 EL 简介

EL 可以简化 JSP 页面的书写，JSP 用户可以用它来访问应用程序数据。EL 最初是在 JSTL 1.0 中定义的，从 JSTL 1.1 开始，Sun 公司将 EL 从 JSTL 规范中分离出来，正式独立为 JSP 2.0 标准规范之一。EL 在容器默认配置下处于启用状态。

EL 表达式以“${”开头，并以“}”结束。其语法格式如下：`${expression}` 。

EL 表达式中的常量包括布尔常量、整型常量、字符串常量和 NULL 常量。

1．EL 中的标识符

EL 的标识符可以由任意的大小写字母、数字和下画线组成。但 EL 的标识符：①不能以数字开头；②不能是 EL 中的保留字；③不能是隐式对象；④不能包含单引号、双引号、减号和正斜杠等特殊字符。

2．变量与保留关键字

EL 表达式中的变量须从 JSP 四大作用域范围中依序进行查找。例如，${student}，表达式将按照 page、request、session 和 application 范围的顺序依次查找名 student 的属性，如果中途找到，就直接回传，不再继续找下去；但是当全部的范围都没有找到时，就回传 null。以下是 EL 中的保留关键字，不能用做标识符。

```
and、or、not、eq、ne、lt、gt、le、ge、true、false、null、instanceof、empty、div、mod
```

3．[ ] 和 . 运算符

通过 EL 提供的“[ ]”和“.”运算符可以访问数据。

（1）“.”运算符：EL 表达式可使用点操作符来访问对象的属性。例如，${object.propertyname} 。

（2）“[ ]”运算符：与点操作符类似，也用于访问对象的属性，属性名须用双引号括起来。例如，${object["propertyname"]}。但是，如果属性中包含了特殊字符，如“.”或“-”等非字母或数字的符号，就一定要用“[ ]”运算符。

（3）“[ ]”运算符可以访问 Map 对象的 key 关键字的值，如${map["key"]}。

4．EL 的错误处理

对于 JSP 页面的错误处理，EL 不提供警告，只提供默认值和错误，默认值是空字符串，错误是抛出一个异常。EL 对以下几种常见错误的处理方式如下。

（1）若在 EL 中访问一个不存在的变量，则表达式输出空字符串，而不是输出 null。

（2）若在 EL 中访问一个不存在对象的属性，则表达式输出空字符串，而不会抛出 NullPointerException 异常。

（3）若在 EL 中访问一个存在对象的不存在属性，则表达式会抛出 PropertyNotFound Exception 异常。

5.1.2 EL 的运算符和优先级

1．算术运算符

EL 中使用的算术运算符如表 5-2 所示。在使用这些运算符时，须注意以下几点。

（1）“-”运算符既可以作为减号，也可以作为符号。

（2）“/”或“div” 运算符在进行除法运算时，商为小数。

表 5-1 EL 算术运算符

算术运算符	说 明	举 例	结 果
+	加	${128+69}	197
-	减	${128-69}	59
*	乘	${18*2}	36
/（或 div）	除	${10/4}或${10 div 4}	2.5
%（或 mod）	求余	${10/4}或${10 div 4}	2

2．关系运算符

EL 的关系运算符如表 5-2 所示。

表 5-2 EL 关系运算符

关系运算符	说 明	举 例	结 果
= =或 eq	等于	${18= =18}或${18 eq 18}	true
!=或 ne	不等于	${18!=18}或${18 ne 18}	false
<或 lt	小于	${18<18}或${18 lt 18}	false
>或 gt	大于	${18>18}或${18 gt 18}	false
<=或 le	小于等于	${18<=18}或${18 le 18}	true
>=或 ge	大于等于	${18>=18}或${18 ge 18}	true

3．逻辑运算符

EL 的逻辑运算符如表 5-3 所示。

表 5-3 EL 逻辑运算符

逻辑运算符	说 明	举 例	结 果
&&或 and	逻辑与	${A&&B}或${A and B}	true/false
||或 or	逻辑或	${A\|\|B}或${A or B}	true/false
!或 not	逻辑非	${!A }或${not A}	true/false

4．empty 运算符

EL 的 empty 运算符用于判断某个对象是否为 null 或空（空字符串""、空的数组或者空的集合），结果为布尔类型，其基本语法格式如下：${empty var}。

例如，定义两个 request 范围内的变量 userA 和 userB，分别设置值为 null 和""。

```
<%request.setAttribute("userA","");%>
<%request.setAttribute("userB",null);%>
```

再使用 empty 运算符判断 userA 和 userB 是否为空。

```
${empty userA}              //返回值为true
${empty userB}              //返回值为true
```

5．条件运算符

EL 的条件运算符和 Java 语言中的用法完全一致。其语法格式如下：

```
${条件表达式?表达式 1:表达式 2}
```

上面的 EL 表达式的值：如果条件表达式为真，则返回表达式 1 的值，否则返回表达式 2 的值。

通常情况下，条件运算符可以用 JSTL 中的条件标签<c:if>或<c:choose>替代。

6．运算符优先级

EL 运算符的优先级如表 5-4 所示，优先级从上到下、从左到右依次降低。

表 5-4 EL 运算符的优先级

运 算 符	优 先 级
[]	1
()	2
-（负号）、not、!、empty	3
*、/、div、%、mod	4
+（加号）、-（减号）	5
<、>、<=、>=、lt、gt、le、 ge	6
= =、!=、eq、ne	7
&&、and	8
||、or	9
?:	10

5.1.3 EL 隐式对象

为了能够获得 Web 应用程序中的相关数据，EL 提供了 11 个隐式对象，这些对象类似于 JSP 的内置对象，也是直接通过对对象名进行操作的在 EL 的隐式对象中，除 pageContext 是 JavaBean 对象，对应于 javax.servlet.jsp.PageContext 类型外，其他的隐式对象都对应于 java.util.Map 类型。这些隐式对象可分为页面上下文对象、访问作用域范围的隐式对象和访问环境信息的隐式对象 3 种，如表 5-5 所示。

表 5-5　EL 隐式对象

分　类	对　象	描　述
页面上下文对象	pageContext	相当于 JSP 页面中的 pageContext 对象，用于获取 ServletContext、request、response 和 session 等其他 JSP 内置对象
访问作用域范围的隐式对象	pageScope	这是一个 Map，其中包含了全页面范围内的所有属性。属性名称就是 Map 的 key
	requestScope	这是一个 Map，其中包含了当前 HttpServletRequest 对象中的所有属性，并用属性名称作为 key
	sessionScope	这是一个包含了 HttpSession 对象中所有属性的 Map，并用属性名称作为 key
	applicationScope	这是一个包含了 ServletContext 对象中所有属性的 Map，并用属性名称作为 key
访问环境信息的隐式对象	param	这是一个包含所有请求参数，并用参数名作为 key 的 Map。每个 key 的值就是指定名称的第一个参数值
	paramValues	这是一个包含所有请求参数，并用参数名作为 key 的 Map。每个 key 的值就是一个字符串数组，其中包含了指定参数名称的所有参数值。就算该参数只有一个值，它也仍然会返回一个带有一个元素的数组
	header	这是一个包含请求标题，并用标题名作为 key 的 Map。每个 key 的值就是指定标题名称的第一个标题。换句话说，如果一个标题的值不止一个，则只返回第一个值。要想获得多个值的标题，则要用 headerValues 代替
	headerValues	这是一个包含请求标题，并用标题名作为 key 的 Map。每个 key 的值就是一个字符串数组，其中包含了指定标题名称的所有参数值。就算该标题只有一个值，它也仍然会返回一个带有一个元素的数组
	initParam	这是一个包含所有环境初始化参数，并用参数名称作为 key 的 Map
	cookie	这是一个包含了当前请求对象中所有 Cookie 对象的 Map。Cookie 名称就是 key 名称，并且每个 key 都映射到一个 Cookie 对象

【示例】pageContext 隐式对象的使用。

在 pageContext.jsp 的<body>部分插入以下代码：

```
获取请求 URI：${pageContext.request.requestURI }<br>
获取请求 URL：${pageContext.request.requestURL }<br>
获取响应的内容类型：${pageContext.response.contentType }<br>
获取服务器信息：${pageContext.servletContext.serverInfo }<br>
获取 Servlet 注册名：${pageContext.servletConfig.servletName }<br>
获取 session 创建时间：${pageContext.session.creationTime }<br>
获取 session 会话最长空闲时间:${pageContext.session.maxInactiveInterval }<br>
获取 session 会话的 id：${pageContext.session.id }
```

运行 pageContext.jsp 页面，会有以下输出：

```
获取请求 URI：/ELDemo/pageContext.jsp
获取请求 URL：http://localhost:8080/ELDemo/pageContext.jsp
获取响应的内容类型：text/html;charset=UTF-8
获取服务器信息：Apache Tomcat/8.0.45
获取 Servlet 注册名：jsp
获取 session 创建时间：1523803900700
获取 session 会话最长空闲时间：1800
获取 session 会话的 id：07E4433E5953AFD29B8C0A39C2B6771D
```

【示例】作用域范围隐式对象的使用。

在 scope.jsp 的<body>部分插入以下代码：

```
<% pageContext.setAttribute("cityName", "Soochow"); %>
<% request.setAttribute("university", "华东师范大学"); %>
<% session.setAttribute("bookName", "《Android应用开发实践教程》"); %>
<% application.setAttribute("cityName", "New York"); %>
<p>pageScope隐式对象中key为"cityName"的值：${pageScope.cityName }</p>
<p>requestScope 隐式对象中 key 为"university"的值：${requestScope.
university }</p>
<p>session隐式对象中key为"bookName"的值：${sessionScope.bookName }</p>
<p>applicationScope隐式对象中key为"cityName"的值：
   ${applicationScope.cityName }</p>
<p>表达式\${cityName}的值：${pageScope.cityName}</p>
```

运行scope.jsp页面，会有以下输出：

```
pageScope隐式对象中key为"cityName"的值：Soochow
requestScope隐式对象中key为"university"的值：华东师范大学
session隐式对象中key为"bookName"的值：《Android应用开发实践教程》
applicationScope隐式对象中key为"cityName"的值：New York
表达式${cityName}的值：Soochow
```

【示例】param和paramValues隐式对象的使用。

在param.jsp的<body>部分插入以下代码：

```
<h2>个人信息</h2>
<form action="${pageContext.request.contextPath}/param_result.jsp">
<p>姓名:<input type="text" name="user" /></p>
<p>爱好:<input type="checkbox" name="hobby" value="swim">游泳
<input type="checkbox" name="hobby" value="pingpang">乒乓球
<input type="checkbox" name="hobby" value="music">音乐
<input type="checkbox" name="hobby" value="Go">围棋</p>
<p><input type="submit" value="提交"></p>
</form>
```

在param_result.jsp的<body>部分插入以下代码：

```
<p>姓名:${param.user}</p>
<p>爱好:${paramValues.hobby[0]} ${paramValues.hobby[1]} 
${paramValues.hobby[2]}</p>
```

运行param.jsp页面，结果如图5-1所示，提交表单数据会跳转至param_result.jsp页面，如图5-2所示。

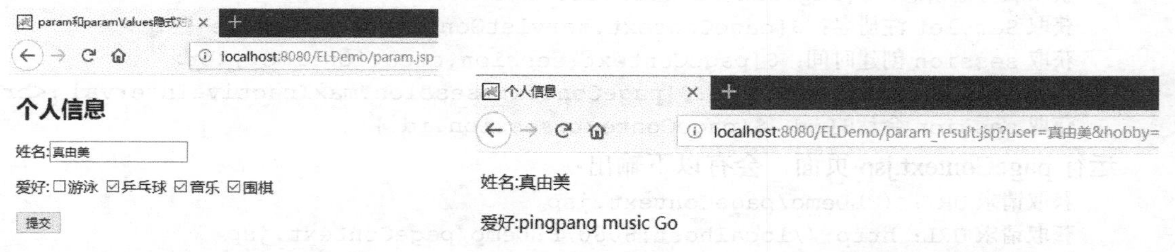

图5-1 param.jsp页面　　　　图5-2 param_result.jsp页面

【示例】Cookie隐式对象的使用。

在cookie.jsp的<body>部分插入以下代码：

```
<h3>Cookie隐式对象的使用</h3>
<p>获取名为JSESSIONID的Cookie对象：${cookie.JSESSIONID }</p>
<p>获取名为JSESSIONID的Cookie对象的名称：${cookie.JSESSIONID.name }</p>
<p>获取名为JSESSIONID的Cookie对象的值：${cookie.JSESSIONID.value }</p>
<hr>
```

```
<%
 Cookie cookie2 = new Cookie("username", "admin");
 response.addCookie(cookie2);
%>
<p>获取名为 username 的 Cookie 对象：${cookie.username }</p>
<p>获取名为 username 的 Cookie 对象的名称：${cookie.username.name }</p>
<p>获取名为 username 的 Cookie 对象的值：${cookie.username.value }</p>
<hr>
<%
 Cookie[] cookies = request.getCookies();
 if (cookies != null)
      for (Cookie c : cookies) {
           out.println(c);
           out.println(c.getName());
           out.println(c.getValue());
           out.println("<br>");
      }
%>
```

运行 cookie.jsp 页面，结果如图 5-3 所示，需要注意的是：浏览器的设置要接受 Cookie。譬如笔者使用的 Firefox（火狐）浏览器作测试的，须单击“选项”菜单，找到“隐私与安全”标签页，在“Cookie 和网站数据”设置中选择“接受来自网站的 Cookie 和网站数据”，关闭火狐浏览器，再重新打开，接受 Cookie 即可生效。

图 5-3　Cookie 隐式对象的使用

【示例】header 隐式对象的使用。

在 cookie.jsp 的<body>部分插入以下代码：

```
<h3>header 隐式对象的使用</h3>
<p>获取请求头 Host 的值：${header.host }</p>
<p>获取请求头的 connection 属性（是否需要持久连接）：${header.connection }</p>
```

运行 cookie.jsp 页面，结果如图 5-4 所示。

图 5-4　header 隐式对象的使用

【示例】initParam 隐式对象的使用。

Web.xml 的<context-param>（context 起始参数）设置如下：

```
<context-param>
    <param-name>driverClassName</param-name>
    <param-value>com.mysql.jdbc.Driver</param-value>
</context-param>
```

在 initParam.jsp 的<body>部分插入以下代码：

```
<h3>initParam 隐式对象的使用</h3>
<p>driverClassName 的值：${initParam.driverClassName }</p>
```

运行 initParam.jsp 页面，结果如图 5-5 所示。

initParam隐式对象的使用

driverClassName的值：com.mysql.jdbc.Driver

图 5-5 initParam 隐式对象的使用

（代码详见本书教学资源包 code 文件夹下 ch5 目录中的 ELDemo 工程。）

5.1.4 定义和使用 EL 函数

EL 自定义函数就是提供一种语法并允许在 EL 中调用某个 Java 类的静态方法。

EL 自定义函数的开发与应用包括以下 3 个步骤。

（1）编写一个 Java 类，并在该类中编写公用的静态方法，用于实现自定义 EL 函数的具体功能。

（2）编写标签库描述文件（扩展名为.tld），对函数进行声明，保存到 Web 应用的 Web-INF 文件夹下。

（3）在 JSP 页面中引用标签库，并调用定义的 EL 函数，实现相应的功能。

【示例】定义和使用 EL 函数。

1．编写 EL 自定义函数映射的 Java 类及类中的静态方法

customEL.java 主要代码如下：

```
public final class CustomEL {
 public static String cFilter(String message) {
      if (message == null)
         return (null);
      char content[] = new char[message.length()];
      message.getChars(0, message.length(), content, 0);
      StringBuilder result = new StringBuilder(content.length + 50);
      for (int i = 0; i < content.length; i++) {
         switch (content[i]) {
         case '<':
            result.append("&lt;");
            break;
         case '>':
            result.append("&gt;");
            break;
         …
```

```
        default:
            result.append(content[i]);
        }
    }
    return (result.toString());
}
}
```

2. 编写标签库描述文件

编写标签库描述文件 mytaglib.tld，将其保存到 Web-INF 文件夹下。

mytaglib.tld 主要内容如下：

```
<tlib-version>1.0</tlib-version>
<short-name>customEL</short-name>
<uri>http://com.mialab.customEL/cFilter</uri>
<function>
 <description>HTML 过滤</description>
 <name>cfilter</name>
 <function-class>com.mialab.customEL.CustomEL</function-class>
 <function-signature>java.lang.String   cFilter(   java.lang.String   )
</function-signature>
</function>
```

3. 测试

customEL.jsp 页面主要内容如下：

```
<%@ page language="java" contentType="text/html; charset=UTF-8"
 pageEncoding="UTF-8"%>
<%@taglib prefix="mlab" uri="http://com.mialab.customEL/cFilter"%>
......
<body>
 <h3>白日依山尽，黄河入海流</h3>
 ${mlab:cfilter("<h3>白日依山尽，黄河入海流</h3>") }
</body>
</html>
```

运行 customEL.jsp 页面，结果如图 5-6 所示。

白日依山尽，黄河入海流

<h3>白日依山尽，黄河入海流</h3>

图 5-6 EL 自定义函数示例

（代码详见本书教学资源包 code 文件夹下 ch5 目录中的 customEL 工程。）

5.2 JSP 标准标签库

5.2.1 JSTL 简介

JSTL 是一个 JSP 标签集合，它封装了 JSP 应用的通用核心功能。

JSTL 支持通用的、结构化的任务，如迭代、条件判断、XML 文档操作、国际化标签、SQL 标签。除了这些，它还提供了一个框架来使用集成 JSTL 的自定义标签。

根据 JSTL 标签所提供的功能，可以将其分为以下 5 个类别。

（1）核心标签。

（2）格式化标签。

（3）SQL 标签。

（4）XML 标签。

（5）JSTL 函数。

JSTL 格式化标签用来格式化并输出文本、日期、时间、数字。引用格式化标签库的语法如下：

```
<%@ taglib prefix="fmt" uri="http://java.sun.com/jsp/jstl/fmt" %>
```

JSTL SQL 标签库提供了与关系型数据库（Oracle、MySQL、SQL Server 等）进行交互的标签。引用 SQL 标签库的语法如下：

```
<%@ taglib prefix="sql" uri="http://java.sun.com/jsp/jstl/sql" %>
```

JSTL XML 标签库提供了创建和操作 XML 文档的标签。引用 XML 标签库的语法如下：

```
<%@ taglib prefix="x" uri="http://java.sun.com/jsp/jstl/xml" %>
```

JSTL 包含一系列标准函数，大部分是通用的字符串处理函数。引用 JSTL 函数库的语法如下：

```
<%@taglib prefix="fn" uri="http://java.sun.com/jsp/jstl/functions" %>
```

5.2.2 JSTL 安装与配置

JSTL 的官方下载网址是 http://tomcat.apache.org/taglibs/ standard/，这里选择 JSTL 1.2，将下载的 4 个 JAR 包复制到 JSTLDemo 工程中的 WebContent\WEB-INF\lib 目录下，如图 5-7 所示。（事先在 Eclipse 中建立 Dynamic Web Project，名为 JSTLDemo。）

在 JSP 页面使用 JSTL 时，使用 taglib 指令指定需要使用的函数库前缀和 URI，例如，使用核心标签库：<%@ taglib prefix="c" uri="http://java.sun.com/jsp/jstl/core" %>。

倘若编译 JSP 页面时有问题，可分别选择 4 个 JAR 包，右击，在弹出的快捷菜单中选择“Build Path→Add to Build Path”选项。

```
JSTLDemo
  src
  JRE System Library [jdk1.8.0_92]
  Apache Tomcat v8.0 [Apache Tomcat v8.0]
  Web App Libraries
  build
  WebContent
    META-INF
    WEB-INF
      lib
        taglibs-standard-compat-1.2.5.jar
        taglibs-standard-impl-1.2.5.jar
        taglibs-standard-jstlel-1.2.5.jar
        taglibs-standard-spec-1.2.5.jar
      Web.xml
    catch_test.jsp
    choose_test.jsp
    forEach_test.jsp
    hello.jsp
    if_test.jsp
    import_test.jsp
    main.jsp
    navigation.jsp
    redirect_test.jsp
    remove_test.jsp
    url_test.jsp
```

图 5-7 JSTLDemo 工程

5.2.3 核心标签库

JSTL 核心标签库包含 Web 应用中最常使用的标签，是 JSTL 中比较重要的标签库。核心标签库的标签按功能可分为 4 类：表达式标签、条件标签、迭代标签、URL 相关标签，如表 5-6 所示。引用核心标签库的语法如下：

<%@ taglib prefix="c" uri="http://java.sun.com/jsp/jstl/core" %>

表 5-6 JSTL 的核心标签库

标签分类	标　签	描　述
表达式标签	<c:out>	用于在 JSP 中显示数据，就像<%= ... >
	<c:set>	用于设置各种范围域的属性

续表

标签分类	标　签	描　述
表达式标签	<c:remove>	用于删除各种范围域属性
	<c:catch>	用来处理产生错误的异常状况，并且将错误信息存储起来
条件标签	<c:if>	与在一般程序中使用的 if 一样
	<c:choose>	本身只当做<c:when>和<c:otherwise>的父标签
	<c:when>	<c:choose>的子标签，用来判断条件是否成立
	<c:otherwise>	<c:choose>的子标签，接在<c:when>标签后，当<c:when>标签判断为 false 时被执行
迭代标签	<c:forEach>	基础迭代标签，接受多种集合类型
	<c:forTokens>	根据指定的分隔符来分隔内容并迭代输出
URL 相关标签	<c:import>	用于在 JSP 页面中导入站内或其他网站的静态或动态文件
	<c:url>	使用正确的 URL 重写规则以构造一个 URL
	<c:redirect>	重定向至一个新的 URL.
	<c:param>	用来给包含或重定向的页面传递参数

【示例】<c:out>输出标签、<c:set>变量设置标签和<c:remove>变量移除标签。

hello.jsp 的<body>部分主要代码如下：

```
<c:set var="mystr" value="Hello，我们开始学习 JSTL！"></c:set>
<c:out value="${mystr}"></c:out>
```

<c:out>标签用于将表达式的值输出到 JSP 页面中。运行 hello.jsp 页面，结果如图 5-8 所示。

remove_test.jsp 的<body>部分主要代码如下：

```
<c:set var="user" value="一剑冲天" scope="session" />
session 中的 user：<c:out value="${user}" />
<br><hr>
<c:remove var="user" scope="session" />
从 session 中移除 user 再访问：<c:out value="${user}" />
```

<c:set>标签用于在指定范围（page、request、session 或 application）中定义保存某个值的变量，或为指定的对象设置属性值。使用该标签可以在页面中定义变量。<c:remove>标签用于移除指定的 JSP 范围内的变量。运行 remove_test.jsp 页面，结果如图 5-9 所示。

图 5-8　<c:out>标签　　　　图 5-9　<c:set>和<c:remove>标签

【示例】< c:catch>捕获异常标签。

catch_test.jsp 的<body>部分主要代码如下：

```
<c:catch var="exception1"><%=169/0%></c:catch>
<c:out value="${exception1}" /><br>
<c:out value="${exception1.message}" />
```

<c:catch>标签用于捕获嵌套在标签体中的内容抛出的异常。

运行 catch_test.jsp 页面，结果如图 5-10 所示。

图 5-10　<c:catch>标签

【示例】<c:if>、<c:choose>、<c:when>和<c:otherwise>等条件标签。

if_test.jsp 的<body>部分主要代码如下：

```
<c:set var="user" value="一剑冲天" scope="session" />
<c:if test="${not empty sessionScope.user}">
 您好，${sessionScope.user}
</c:if>
```

<c:if>标签用于进行条件判断。运行 if_test.jsp 页面，结果如图 5-11 所示。

choose_test.jsp 的<body>部分主要代码如下：

```
<c:set var="user" value="admin2" scope="session" />
<c:set var="actor1" value="admin1" scope="session" />
<c:set var="actor2" value="admin2" scope="session" />
<c:choose>
 <c:when test="${sessionScope.user eq sessionScope.actor1}">欢迎您，一级管理员！
 </c:when>
 <c:when test="${sessionScope.user eq sessionScope.actor2}">欢迎您，二级管理员！
 </c:when>
 <c:otherwise>对不起，您只是游客，无权访问该资源！</c:otherwise>
</c:choose>
```

<c:choose>标签用于指定多个条件选择，必须与<c:when>和<c:otherwise>标签一起使用。这类似于 if…else if…else 的复杂条件判断结构。运行 choose_test.jsp 页面，结果如图 5-12 所示。

图5-11 <c:if>标签

图5-12 choose_test.jsp

【示例】<c:forEach>迭代标签

forEach_test.jsp 的 <body>部分。主要代码如下：

```
<%
   List<String> bookList = new ArrayList<String>();
 bookList.add("《Android 应用开发实践教程》");
 bookList.add("《Web 应用开发》");
 bookList.add("《软件项目管理》");
 bookList.add("《iOS 程序设计》");
 session.setAttribute("bookList", bookList);
%>
<c:forEach items="${sessionScope.bookList}" var="book" varStatus="bookvst">
<p>图书序号：${bookvst.index+1}，图书名称：${book}</p>
</c:forEach>
```

<c:forEach>标签用于遍历集合。items 指定将要迭代的集合对象，var 用于指定将当前迭代到的元素保存到 page 域中的属性名称，varStatus 表示当前被迭代到对象的状态信息。varStatus 的 index 属性表示当前迭代成员的索引值，varStatus 的 count 属性表示当前已迭代成员的数量。运行 forEach_test.jsp 页面，结果如图 5-13 所示。

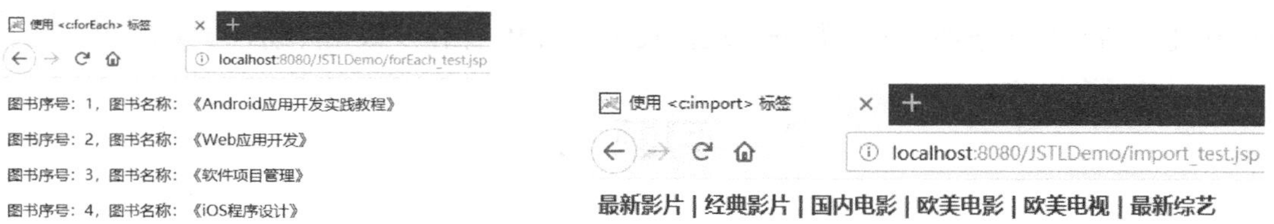

图 5-13　使用<c:forEach>标签　　　　图 5-14　使用<c:import>标签

【示例】<c:import >标签。

import_test.jsp 的 <body>部分主要代码如下：

```
<c:import url="navigation.jsp" />
```

navigation.jsp 的 <body>部分主要代码如下：

```
<font color="blue"><b>最新影片 | 经典影片 | 国内电影... | 最新综艺</b></font>
```

<c:import>标签用于在 JSP 页面中导入一个 URL 地址指向的资源内容，可以是一个静态或动态文件，也可以是当前应用或同一服务器下的其他应用中的资源。

运行 import_test.jsp 页面，结果如图 5-14 所示。

【示例】<c:url>、<c:set>和<c:remove>变量移除标签。

url_test.jsp 的 <body>部分主要代码如下：

```
<c:url var="path" value="navigation.jsp" scope="page" />
<a href="${pageScope.path}">电影天堂首页</a>
```

<c:url>标签用于在 JSP 页面中构造一个 URL 地址。运行 url_test.jsp 页面，结果如图 5-15 所示。单击“电影天堂首页”超链接，可跳转到 navigation.jsp 页面。

redirect_test.jsp 的 <body>部分主要代码如下：

```
<c:redirect url="main.jsp">
 <c:param name="user" value="诗仙李白" />
</c:redirect>
```

main.jsp 的 <body>部分主要代码如下：

```
<p>原来你是${param.user}</p>
```

<c:redirect >标签用于执行 response.sendRedirect()方法的功能，将当前访问请求重定向到其他资源。<c:param>标签只用于为其他标签提供参数信息，它与<c:import>、<c:redirect >和<c:url>标签组合可以实现动态定制参数，从而使标签完成更复杂的程序应用。

运行 redirect_test.jsp 页面，结果如图 5-16 所示。

图 5-15　使用<c:url>标签　　　　图 5-16　<c: redirect >和<c:param>标签

5.3　本章小结

本章主要介绍了 EL（表达式语言）和 JSTL。EL 提供了在 JSP 中简化表达式的方法，可使 JSP 的代码更加简化。EL 表达式的基本形式为${var}，所有的表达式都以“${”符号开头，以“}”符号结尾。如果在 JSP 文件的模板文件中使用 EL 表达式，那么表达式的值会输出到网页上。JSP 标准标签库是一个实现 Web 应用程序中常见的通用功能的定制标记库集，这些功能包括迭代和条件判断、数据管理格式化、XML 操作及数据库访问。

习题 5

1. 编写一个 JSP 程序，实现通过 EL 获取并显示用户注册信息的功能。要求包括用户名、密码、性别（采用单选按钮）、爱好（采用复选框）等信息。

2. EL 表达式的功能和特点是什么？

3. EL 中“.”操作符和“[ ]”操作符使用时有何区别？

4. 在一个 Servlet 中创建一个对象集合类，如 List<Book>，将此对象集合类存入 Request 对象属性，请求转发到 bookList.jsp；在 bookList.jsp 中遍历并使用 EL 表达式输出 Book 对象的属性值。

5. JSTL 由哪些标签库组成？各自的作用是什么？

6. 在 Web 应用中如何引入 JSTL？

7. 编写 JSP 程序，使用<c:choose>、<c:when>和<c:otherwise>标签根据当前是星期几而显示不同的提示信息。

8. 编写 JSP 程序，实现用户注册功能，要求注册协议通过文本文件导入。

第6章 过滤器与监听器

本章导读

过滤器（Filter）不能处理用户请求，也不能对客户端生成响应。Filter 主要用于对 HttpServletRequest 进行预处理，也可以对 HttpServletResponse 进行后处理，是典型的处理链。监听器（Listener）可以监听由于 Web 应用中状态改变而引起的 Servlet 容器产生的相应事件，并对这些事件做出反应和处理。本章主要内容有：（1）Servlet 过滤器；（2）Filter 工作原理、Filter 核心接口、Filter 生命周期、Filter 配置；（3）Filter 应用；（4）Servlet 监听器；（5）Servlet 上下文监听、HTTP 会话监听、Servlet 请求监听。

6.1 Servlet 过滤器

Filter 是拦截 Request 请求的对象，在用户的请求访问资源前处理 ServletRequest 以及 ServletResponse，它可用于日志记录、加解密、Session 检查、验证用户访问权限、对请求进行重新编码和压缩响应信息等。通过 Filter 可以拦截处理某个资源或者某些资源。Filter 的配置可以通过 Annotation 或者部署描述来完成。当一个资源或者某些资源需要被多个 Filter 所使用到，且它的触发顺序很重要时，只能通过部署描述来配置。

6.1.1 Filter 工作原理

当客户端发出 Web 资源的请求时，Web 服务器根据应用程序配置文件设置的过滤规则进行检查。若客户请求满足过滤规则，则对客户请求/响应进行拦截，对请求头和请求数据进行检查或改动，并依次通过过滤器链，最后把请求/响应交给请求的 Web 资源处理。

请求信息在过滤器链中可以被修改，也可以根据条件让请求不发往资源处理器，并直接向客户机发回一个响应。当资源处理器完成了对资源的处理后，响应信息将逐级逆向返回。同样，在这个过程中，用户可以修改响应信息，从而完成一定的任务。

Filter 主要用于对用户请求进行预处理，也可以对 HttpServletResponse 进行后处理，是典型的处理链。

6.1.2 Filter 核心接口

1．javax.servlet.Filter 接口

每一个过滤器对象都要直接或间接地实现 Filter 接口，在该接口中定义了如下 3 个方法。

（1）void init(FilterConfig fConfig)：用于完成 Filter 的初始化。

（2）void destroy( )：用于 Filter 销毁前，完成某些资源的回收。

（3）void doFilter(ServletRequest request,ServletResponse response,FilterChain chain)：实现

过滤功能，该方法是对每个请求及响应增加的额外处理。

2．javax.servlet.FilterConfig 接口

javax.servlet.FilterConfig 接口由容器实现，容器将其实例作为参数传入 Filter 对象的初始化方法 init()中，来获取过滤器的初始化参数和 Servlet 的相关信息。在 FilterConfig 接口中定义了如下 4 个方法。

（1）String getFilterName( )：用于获取过滤器的名称。

（2）ServletContext get ServletContext( )：获取 Servlet 上下文。

（3）String getInitParameter(String name)：获取过滤器的初始化参数值。

（4）Enumeration getInitParameterNames( )：获取过滤器所有初始化参数名的枚举集合。

3．javax.servlet.FilterChain 接口

javax.servlet.FilterChain 接口由容器实现，容器将其实例作为参数传入过滤器对象的 doFilter()方法中。过滤器对象使用 FilterChain 对象调用过滤器链中的下一个过滤器，如果该过滤器是链中最后一个过滤器，那么将调用目标资源。

FilterChain 接口只有一个方法：

```
void doFilter(ServletRequest request,ServletResponse response)
```

该方法将使过滤器链中的下一个过滤器被调用，如果调用该方法的过滤器是链中最后一个过滤器，那么目标资源被调用。

6.1.3 Filter 生命周期

过滤器的生命周期分为 4 个阶段：加载和实例化→初始化→执行 doFilter()方法→销毁。

6.1.4 Filter 配置

（1）使用@WebFilter 标注配置 Filter，该名称为 TestFilter，描述了两个初始化参数，且适用于所有资源。其代码如下：

```
@WebFilter(filterName = "TestFilter", urlPatterns = { "/*" }, initParams = {
    @WebInitParam(name = "loginPage", value = "login.jsp"),
    @WebInitParam(name = "loginServlet", value = "LoginProcessServlet") })
```

（2）如果在 Web.xml 中进行配置，则其代码如下：

```
<filter>
    <filter-name>TestFilter</filter-name>
    <filter-class>com.mialab.
filterdemo.filter.TestFilter</filter-class>
    <init-param>
      <param-name> loginPage </param-name>
      <param-value> login.jsp </param-value>
    </init-param>
    <init-param>
      <param-name> loginServlet </param-name>
      <param-value> LoginProcessServlet </param-value>
    </init-param>
</filter>
<filter-mapping>
```

```
        <filter-name>TestFilter</filter-name>
        <url-pattern>/*</url-pattern>
    </filter-mapping>
```

6.1.5 Filter 应用

1. 日志 Filter

“日志 Filter”示例实现功能：把 Request 请求的 URL 记录到日志文件 mylog.txt 中。

在 Eclipse 中创建“Dynamic Web Project”，工程名为 FilterDemo。FilterDemo 工程目录如图 6-1 所示，右击 test.jsp，选择“Run As→Run on Server”选项。此时，FilterDemo 工程会自动部署到 Tomcat 服务器的 Webapps 目录中。依次访问 test.jsp、a.jsp、HelloServlet、TestServlet，得到 mylog.txt 日志记录，如图 6-2 所示。

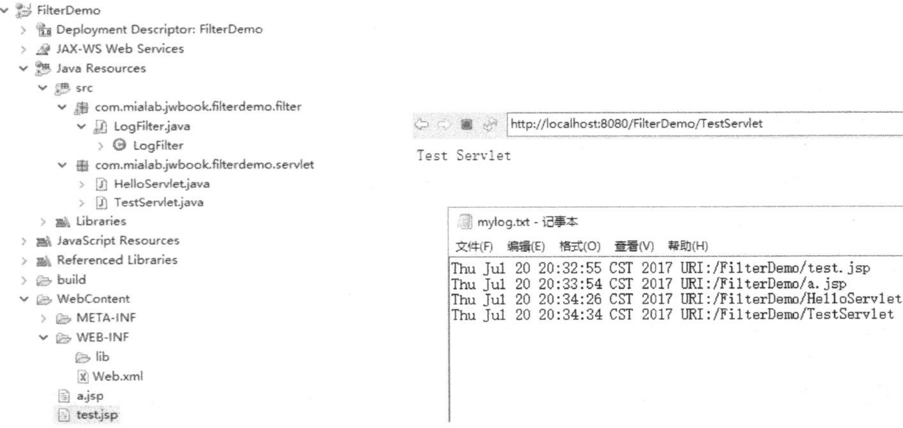

图 6-1 FilterDemo 工程目录　　　图 6-2 mylog.txt 日志记录

LogFilter.java 主要代码如下：

```
public class LogFilter implements Filter {
 private PrintWriter logger;
 private String prefix;
 public LogFilter() { }
 public void init(FilterConfig fConfig) throws ServletException {
     prefix = fConfig.getInitParameter("prefix");
     String logFileName = fConfig.getInitParameter("logFileName");
     String appPath = fConfig.getServletContext().getRealPath("/");
     System.out.println("logFileName:" + logFileName);
     try {
         logger = new PrintWriter(new File(appPath, logFileName));
     } catch (FileNotFoundException e) {
         e.printStackTrace();
         throw new ServletException(e.getMessage());
     }
 }
 public void doFilter(ServletRequest request, ServletResponse response,
         FilterChain filterChain) throws IOException, ServletException {
     System.out.println("LogFilter.doFilter");
     HttpServletRequest httpServletRequest = (HttpServletRequest) request;
 logger.println(newDate()+""+prefix+httpServletRequest.getRequestURI());
```

```
        logger.flush();
        filterChain.doFilter(request, response);
    }
    public void destroy() {
        System.out.println("destroying filter");
        if (logger != null) {
            logger.close();
        }
    }
}
```

Web.xml 中配置（或者注册）过滤器的代码如下：

```
<filter>
  <filter-name>LogFilter</filter-name>
  <filter-class>com.mialab.jwbook.filterdemo.filter.LogFilter</filter-class>
  <init-param>
      <param-name>logFileName</param-name>
      <param-value>mylog.txt</param-value>
  </init-param>
  <init-param>
      <param-name>prefix</param-name>
      <param-value>URI:</param-value>
  </init-param>
</filter>
<filter-mapping>
  <filter-name>LogFilter</filter-name>
  <url-pattern>/*</url-pattern>
</filter-mapping>
```

（代码详见本书教学资源包 code 文件夹下 ch6 目录中的 FilterDemo 工程。）

2．控制用户访问权限 Filter

“控制用户访问权限 Filter”示例实现功能：欲访问信息页面 info.jsp，通过过滤器对会话对象 Session 进行检查，以控制用户访问权限。这里是直接访问 info.jsp，因会话中无用户名，故限制其访问，并跳转至 login.jsp。

在 Eclipse 中创建“Dynamic Web Project”，工程名为 FilterDemo2。FilterDemo2 工程目录如图 6-3 所示，右击 info.jsp，选择“Run As→Run on Server”选项。此时，FilterDemo2 工程会自动部署到 Tomcat 服务器的 Webapps 目录中。因会话中无用户名，故跳转至 login.jsp，如图 6-4 所示。

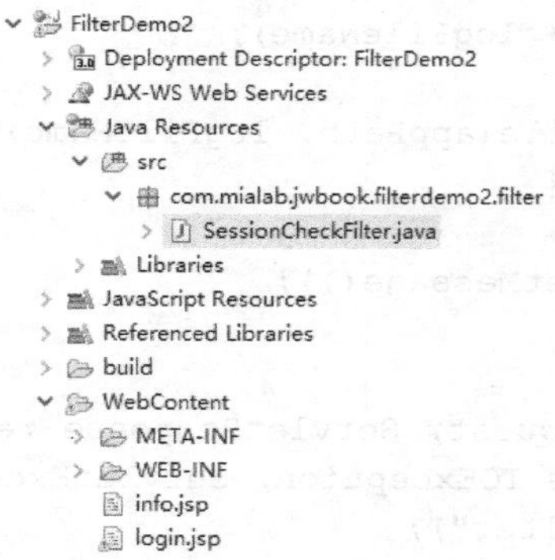

图 6-3　FilterDemo2 工程目录

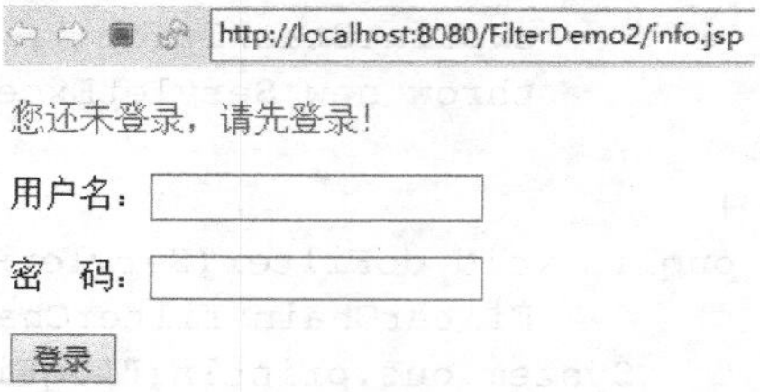

图 6-4　访问 info.jsp

SessionCheckFilter.java 主要代码如下：

```
    @WebFilter(filterName = "SessionCheckFilter", urlPatterns = { "/*" },
initParams = {
          @WebInitParam(name = "loginPage", value = "login.jsp"),
          @WebInitParam(name = "loginServlet", value = "LoginProcessServlet") })
    public class SessionCheckFilter implements Filter {
     private FilterConfig config;        // 用于获取初始化参数
     public SessionCheckFilter() { }     // 默认构造方法
     public void init(FilterConfig fConfig) throws ServletException {
          this.config = fConfig;
     }
     public void doFilter(ServletRequest request, ServletResponse response,
               FilterChain chain) throws IOException, ServletException {
          String loginPage = config.getInitParameter("loginPage"); // 获取初始
化参数
          String loginServlet = config.getInitParameter("loginServlet");
          // 获取会话对象
          HttpSession session = ((HttpServletRequest) request).getSession();
          // 获取请求资源路径（不包含请求参数）
          String requestPath = ((HttpServletRequest) request).getServletPath();
          if  (session.getAttribute("user")  != null  || requestPath.endsWith
(loginPage)  //【1】
                    || requestPath.endsWith(loginServlet)) {   //【2】
               *如果用户会话域属性 user 存在，并且请求资源为登录页面和
                    登录处理的 Servlet，则“放行”请求*1
               chain.doFilter(request, response);
          } else {
               // 对请求进行拦截，返回登录页面
               request.setAttribute("tip", "您还未登录，请先登录！");
               request.getRequestDispatcher(loginPage).forward(request,
response);
          }
     }
     public void destroy() {this.config = null;}
    }
```

【1】和【2】表明，对于用户登录页面 login.jsp 和处理登录操作的 LoginProcessServlet，不能进行访问限制。

login.jsp 主要代码如下：

```
    <body>
    <p><font color="red">${tip}</font></p>
    <form action="LoginProcessServlet" method="post">
     <p>用户名：<input type="text" name="username"></p>
     <p>密 码：<input type="password" name="userpass"></p>
     <p><input type="submit" value="登录"></p>
    </form>
    </body>
```

（代码详见本书教学资源包 code 文件夹下 ch6 目录中的 FilterDemo2 工程。）

6.2 Servlet 监听器

6.2.1 Servlet 监听器概述

当 Web 应用在 Web 容器中运行时，Web 应用内部会不断发生各种事件，如 Web 应用被启动、Web 应用被停止，用户 session 开始、用户 session 结束、用户请求到达等。通常来说，这些 Web 事件对开发者是透明的。Servlet API 提供了大量监听器接口来帮助开发者实现对 Web 应用内特定事件的监听，常用的 Web 事件监听器接口可分为如下 3 类：监听 Servlet context events、监听 HTTP session events、监听 Servlet request events。

当在 Eclipse 中创建 Listener 时，便会打开如图 6-5 所示的窗口，选择所创建监听器应实现的接口（一个或多个）。

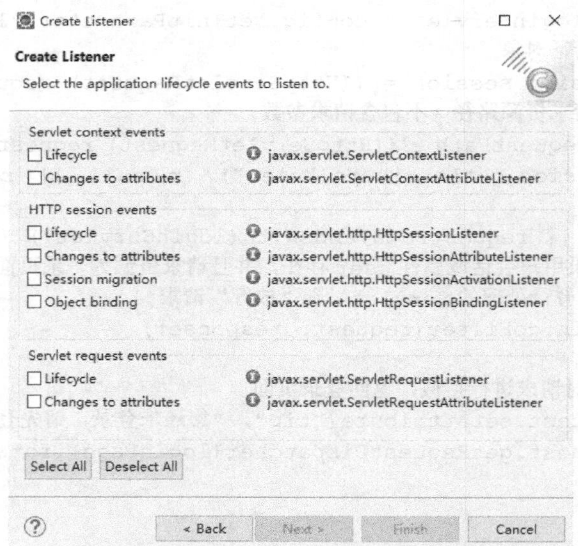

图 6-5 创建监听器，选择欲实现的 Listener 接口

一般来说，监听器的实现通过以下两个步骤来完成。

（1）定义监听器实现类，实现监听器接口的所有方法。

（2）通过 Annotation 或在 Web.xml 文件中注册（或配置）Listener。

6.2.2 Servlet 上下文监听

1. ServletContextListener

ServletContextListener 接口用于监听 Web 应用的 ServletContext 对象的创建和销毁事件。每个 Web 应用对应一个 ServletContext 对象，在 Web 容器启动时创建，在容器关闭时销毁。

当 Web 应用程序中声明了一个实现 ServletContextListener 接口的事件监听器后，Web 容器在创建或销毁此对象时就会产生一个 ServletContextEvent 事件对象，然后再执行监听器中的相应事件处理方法，并将 ServletContextEvent 事件对象传递给这些方法。

在 ServletContextListener 接口中定义了如下两个事件处理方法。

（1）void contextInitialized(ServletContextEvent event)：当 ServletContext 对象被创建时，

Web 容器将调用此方法。该方法接收 ServletContextEvent 事件对象。

（2）void contextDestroyed(ServletContextEvent event)：当 ServletContext 对象被销毁时，Web 容器调用此方法，同时向其传递 ServletContextEvent 事件对象。

java.servlet.ServletContextEvent 是一个 java.util.EventObject 的子类，它定义了访问 ServletContext 的 getServletContext 方法，通过此方法能够轻松地获取到 ServletContext。

2. ServletContextAttributeListener

当一个 ServletContext 范围内的属性被添加、删除或替换时，ServletContextAttributeListener 接口的实现类会接收到消息。这个接口定义了如下 3 个方法。

```
void  attributeAdded(ServletContextAttributeEvent event)
void  attributeRemoved(ServletContextAttributeEvent event)
void  attributeReplaced(ServletContextAttributeEvent event)
```

3. "Web 应用访问计数器"示例

在 Eclipse 中创建"Dynamic Web Project"，工程名为 ListenerDemo。ListenerDemo 工程目录如图 6-6 所示，右击 CountServlet.java，选择"Run As→Run on Server"选项，此时，ListenerDemo 工程会自动部署到 Tomcat 服务器的 Webapps 目录中。count.txt 中初始值为 28，是上次 Web 应用终止时写入的值。访问 CountServlet，结果如图 6-7 所示。

图 6-6　ListenerDemo 工程目录　　　　图 6-7　访问 CountServlet

AppListener.java 主要代码如下：

```
@WebListener
public class AppListener implements ServletContextListener {
  // Web 应用初始化时，容器调用此方法
  public void contextInitialized(ServletContextEvent sce) {
      ServletContext context = sce.getServletContext();  //获取 Servlet
Context 对象
      // 输出应用初始化日志信息
      context.log(context.getServletContextName() + "应用开始初始化。");
      try {
          // 从文件中读取计数器的数值
          BufferedReader reader = new BufferedReader(
              new InputStreamReader(context.getResourceAsStream("/count.
txt")));
          String strcount = reader.readLine();
```

```
            if (strcount == null || "".equals(strcount))
                strcount = "0";
            int count = Integer.parseInt(strcount);
            reader.close();
            context.setAttribute("count", count); //把计数器对象保存到 Web 应用
范围内
        } catch (IOException e) {
            e.printStackTrace();
        }
    }

    //Web 应用停止时，容器调用此方法
    public void contextDestroyed(ServletContextEvent sce) {
        ServletContext context = sce.getServletContext();  // 获取 Servlet
Context 对象
        // 输出应用停止日志信息
        context.log(context.getServletContextName() + "应用停止。");
        // 从 Web 应用范围内获得计数器对象
        Integer counter = (Integer) context.getAttribute("count");
        if (counter != null) {
            try {
                // 把计数器的数值写到发布根目录下的 count.txt 文件中
                String filepath = context.getRealPath("/") + "/count.txt";
                PrintWriter pw = new PrintWriter(filepath);
                pw.println(counter.intValue());
                pw.close();
            } catch (IOException e) {
                e.printStackTrace();
            }
        }
    }
}
```

在 Web 应用终止时，把保存在应用域属性中的计数器数值永久性地保存到 count.txt 文件中。每次 Web 应用启动时再从 count.txt 文件中读取计数器的数值，并将其存入应用域属性。

CountServlet.java 主要代码如下：

```
@WebServlet("/CountServlet")
public class CountServlet extends HttpServlet {
 private static final long serialVersionUID = 1L;
 public CountServlet() {
     super();
 }
 protected void doGet(HttpServletRequest request, HttpServletResponse
response)
         throws ServletException, IOException {
     // 设置响应到客户端的 MIME 类型和字符编码方式
     response.setContentType("text/html;charset=UTF-8");
     // 获取 ServletContext 对象
     ServletContext context = super.getServletContext();
     // 从 ServletContext 对象获取 count 属性中存放的计数值
     Integer count = (Integer) context.getAttribute("count");
     if (count == null) {
         count = 1;
```

```
        } else {
            count = count + 1;
        }
        // 将更新后的数值存放到 ServletContext 对象的 count 属性中
        context.setAttribute("count", count);
        // 获取输出流
        PrintWriter out = response.getWriter();
        // 输出计数信息
        out.println("<p>本网站目前访问人数是： " + count + "</p>");
    }
}
```

每次访问 CountServlet 时，从 ServletContext 对象获取 count 属性中存放的计数值。如果 count 有值，则 count 值+1；如果 count 值为 null，则说明是第一次访问 CountServlet，count 值为 1。除此之外，还须将更新后的 count 数值存放到 ServletContext 对象的 count 属性中。

（代码详见本书教学资源包 code 文件夹下 ch6 目录中的 ListenerDemo 工程。）

6.2.3 HTTP 会话监听

1. HttpSessionListener

当一个 HttpSession 创建或销毁时，容器都会通知所有的 HttpSessionListener。HttpSessionListener 接口有如下两个方法：

```
void sessionCreated(HttpSessionEvent event)
void sessionDestroyed(HttpSessionEvent event)
```

这两个方法都可以接收到一个继承于 java.util.EventObject 的 HttpSessionEvent 对象。可以通过调用 HttpSessionEvent 对象的 getSession 方法来获取当前的 HttpSession。

getSession 方法如下：

```
HttpSession  getSession()
```

以下两种情况下就会发生 sessionDestoryed（会话销毁）事件：

（1）执行 HttpSession 对象的 invalidate() 方法时，将发生会话销毁。例如，在后面“统计在线用户信息”示例中的 usersOnline.jsp 页面中加上链接页面 logout.jsp，代码如下：

```
<a href = "logout.jsp">安全退出</a>
```

在 logout.jsp 中加入以下代码，在跳转到 logout.jsp 页面时，就会触发 sessionDestroyed（会话销毁）事件。

```
<%  session.invalidate();  %>
```

（2）如果用户长时间没有访问服务器，超过了会话最大超时时间，服务器就会自动销毁超时的会话。会话超时时间可以在 Web.xml 中进行设置，时间单位默认是分钟，并且只能是整数，如果是零或负数，那么会话就永远不会超时。譬如，Web.xml 中有以下代码：

```
<session-config>
<session-timeout>30</session-timeout>
</session-config>
```

2. HttpSessionAttributeListener

HttpSessionAttributeListener 接口和 ServletContextAttributeListener 接口类似，它响应的是 HttpSession 范围内属性的添加、删除和替换。

HttpSessionAttributeListener 接口有以下方法：

```
void  attributeAdded(HttpSessionBindingEvent event)
```

```
void attributeRemoved(HttpSessionBindingEvent event)
void attributeReplaced(HttpSessionBindingEvent event)
```

3．HttpSessionActivationListener

当 HttpSession 对象持久化到一个存储设备时，如果绑定到该 HttpSession 对象的对象实例对应的类实现了 HttpSessionActivationListener 接口，则 Web 容器将调用该接口的 sessionWillPassivate 方法。如果 HttpSession 对象从存储设备中恢复到内存中，Web 容器会调用该接口的 sessionDidActivate 方法。这两个方法都接收一个 HttpSessionEvent 类型的参数。

```
void sessionWillPassivate(HttpSessionEvent event)
void sessionDidActivate(HttpSessionEvent event)
```

在分布式环境下，会用多个容器来进行负载均衡，有可能需要将 session 保存起来，在容器之间传递。例如，当一个容器内存不足时，会把很少用到的对象转存到其他容器上。此时，容器就会通知所有 HttpSessionActivationListener 接口的实现类。

4．HttpSessionBindingListener

当有属性绑定或者解绑到 HttpSession 上时，HttpSessionBindingListener 会被调用。HttpSessionBindingListener 接口中共定义了两个方法——valueBound()和 valueUnbound()，分别对应数据绑定和取消绑定两个事件。

```
void valueBound(HttpSessionBindingEvent event)
void valueUnbound(HttpSessionBindingEvent event)
```

HttpSessionBindingListener 使用方法与 HttpSessionListener 完全不同。所谓对 session 进行数据绑定，就是调用 session.setAttribute()把 HttpSessionBindingListener 保存到 session 中。譬如，如下的代码表明名为 Person 的实体 Bean 实现了 HttpSessionBindingListener 接口，也就变成了 HttpSessionBindingListener 接口对象（或者监听器对象），当把其放进 session 中或从 session 中移除时，便会触发相应的事件。

```
public class Person implements HttpSessionBindingListener {
    private String name;
    private double salary;
    public String getName() { return name; }
    public void setName(String name) { this.name = name; }
    public double getSalary() { return salary; }
    public void seSalary (double salary) { this. salary = salary; }
    @Override
    public void valueBound(HttpSessionBindingEvent event) {
        String attributeName = event.getName();
        System.out.println(attributeName + " valueBound");
    }
    @Override
    public void valueUnbound(HttpSessionBindingEvent event) {
        String attributeName = event.getName();
        System.out.println(attributeName + " valueUnbound");
    }
}
```

HttpSessionListener 和 HttpSessionBindingListener 之间的区别在于：HttpSessionListener 只需要设置到 Web.xml 中就可以监听整个应用中的所有 session；HttpSessionBindingListener 必须实例化后放入某一个 session，才可以进行监听。

valueUnbound 的触发条件通常有以下几种情况。

（1）执行 session.invalidate() 时。

（2）session 超时，自动销毁时。

（3）执行 session.setAttribute("属性名", "其他对象"); //这里是替代操作时。

（4）执行 session.removeAttribute("属性名"); //将 listener 对象从 session 中移除时。

我们可以看到，从监听范围上比较，HttpSessionListener 设置一次就可以监听所有 session，而 HttpSessionBindingListener 只与一个 session 绑定或解除绑定，是一对一的。

5. "统计在线用户信息"示例

在 Eclipse 中创建"Dynamic Web Project"，工程名为 ListenerDemo2。ListenerDemo2 工程目录如图 6-8 所示，右击 usersOnline.jsp，选择"Run As→Run on Server"选项，此时，ListenerDemo2 工程会自动部署到 Tomcat 服务器的 Webapps 目录中。从不同浏览器访问 usersOnline.jsp，结果如图 6-9 所示。

图 6-8 ListenerDemo2 工程目录　　图 6-9 从不同浏览器访问 usersOnline.jsp

AppSessionListener.java 主要代码如下：

```
@WebListener
public class AppSessionListener implements HttpSessionListener {
  // 当用户与服务器之间开始 session 时触发该方法
  public void sessionCreated(HttpSessionEvent se) {
      HttpSession session = se.getSession();
      ServletContext application = session.getServletContext();
      String sessionId = session.getId();  // 获取 session ID
      if (session.isNew()) {  // 如果是一次新的会话
          String user = (String) session.getAttribute("user");
          // 未登录用户当作游客处理
          user = (user == null) ? "未登录游客" : user;
          Map<String, String> users =
              (Map<String, String>) application.getAttribute("online");
          if (users == null) {
              users = new Hashtable<String, String>();
          }
          users.put(sessionId, user);  // 将用户在线信息放入 Map
          application.setAttribute("online", users);
      }
  }

  // 当用户与服务器之间 session 断开时触发该方法
  public void sessionDestroyed(HttpSessionEvent se) {
```

```
        HttpSession session = se.getSession();
        ServletContext application = session.getServletContext();
        String sessionId = session.getId();
        Map<String, String> users =
            (Map<String, String>) application.getAttribute("online");
        if (users != null) {
            users.remove(sessionId);  // 删除该用户的在线信息
        }
        application.setAttribute("online", users);
        System.out.println(users.size());
    }
}
```

usersOnline.jsp 主要代码如下：

```
<body>
 <center><h2>在线用户信息</h2></center>   <hr>
 <table border=1 align="center" width=50%>
     <%
         Map<String, String> users =
             (Map<String, String>) application.getAttribute("online");
         for (String sessionId : users.keySet()) {
     %>
     <tr>
         <td><%=sessionId%></td>
         <td><%=users.get(sessionId)%></td>
     </tr>
     <% } %>
 </table>
</body>
```

（代码详见本书教学资源包 code 文件夹下 ch6 目录中的 ListenerDemo2 工程。）

6.2.4 Servlet 请求监听

1.ServletRequestListener

ServletRequestListener 监听器会对 ServletRequest 的创建和销毁事件进行响应。ServletRequestListener 接口有以下两个方法：

```
void  requestInitialized(ServletRequestEvent event)
void  requestDestroyed(ServletRequestEvent event)
```

当一个 ServletRequest 被创建时，requestInitialized 方法会被调用。当 ServletRequest 销毁时，requestDestroyed 方法会被调用。这两个方法都会接收到一个 ServletRequestEvent 对象，可以通过使用这个对象的 getServletRequest 方法来获取 ServletRequest 对象。另外，ServletRequestEvent 接口也提供了一个 getServletContext 方法来获取 ServletContext。

ServletRequestEvent 接口的两个方法如下所示：

```
ServletRequest  getServletRequest()
ServletContext  getServletContext()
```

2. ServletRequestAttributeListener

当一个 ServletRequest 范围内的属性被添加、删除或替换时，ServletRequestAttribute Listener 接口会被调用。ServletRequestAttributeListener 接口提供了如下 3 个方法。

```
void  attributeAdded(ServletRequestAttributeEvent event)
```

```
void  attributeRemoved(ServletRequestAttribute Event event)
void  attributeReplaced(ServletRequestAttribute Event event)
```

通过 ServletRequestAttributeEvent 类提供的 getName 和 getValue 方法可以访问到属性的名称和值：java.lang.String getName() 和 java.lang.Object getValue() 。

3．“计算 Servlet Request Lifecycle 时间”示例

在 Eclipse 中创建“Dynamic Web Project”，工程名为 ListenerDemo3。ListenerDemo3 工程目录如图 6-10 所示，右击 testRequest.jsp，选择“Run As→Run on Server”选项，此时，ListenerDemo3 工程会自动部署到 Tomcat 服务器的 Webapps 目录中。访问 testRequest.jsp，结果如图 6-11 所示。

图 6-10　ListenerDemo3 工程目录

图 6-11　访问 testRequest.jsp

RequestStateListener.java 主要代码如下：

```
@WebListener
public class RequestStateListener implements ServletRequestListener {
 public void requestInitialized(ServletRequestEvent sre)  {
     System.out.println("call requestInitialized() method! ");
     ServletRequest servletRequest = sre.getServletRequest();
       servletRequest.setAttribute("start", System.nanoTime());
   }

   public void requestDestroyed(ServletRequestEvent sre)  {
     System.out.println("call requestDestroyed() method! ");
     ServletRequest servletRequest = sre.getServletRequest();
       Long start = (Long) servletRequest.getAttribute("start");
       Long end = System.nanoTime();
       HttpServletRequest httpServletRequest =
               (HttpServletRequest) servletRequest;
       String uri = httpServletRequest.getRequestURI();
       System.out.println("time taken to execute " + uri +
               ":" + ((end - start) / 1000) + " microseconds !");
   }
}
```

System.currentTimeMillis()返回的是从 1970.1.1 UTC（世界标准时间）零点开始到现在的时间，精确到毫秒。相对于 System.currentTimeMillis()而言，System.nanoTime() 提供相对精确的计时，它返回最准确的可用系统计时器的当前值，以毫微秒为单位。

（代码详见本书教学资源包 code 文件夹下 ch6 目录中的 ListenerDemo3 工程。）

6.3 本章小结

本章主要介绍了 Servlet 过滤器和监听器。

Servlet 过滤器可以动态地拦截请求和响应，以变换或使用包含在请求或响应中的信息。可以将一个或多个 Servlet 过滤器附加到一个 Servlet 或一组 Servlet 中。Servlet 过滤器也可以附加到 JSP 文件和 HTML 页面中。调用 Servlet 前应调用所有附加的 Servlet 过滤器。

Servlet 过滤器是可用于 Servlet 编程的 Java 类，可以实现以下目标。

（1）在客户端的请求访问后端资源之前，拦截这些请求。

（2）在服务器的响应发送回客户端之前，处理这些响应。

Servlet 事件监听器就是一个实现了特定接口的 Java 程序，专门用于监听 Web 应用程序中 ServletContext、HttpSession 和 ServletRequest 等域对象的创建和销毁过程，监听这些域对象属性的修改以及感知绑定到 HttpSession 域中某个对象的状态。

习题 6

1．过滤器的生命周期是什么？过滤器是否为单向的过滤过程？请编程加以说明。

2．监听器的作用是什么？有哪些常用的监听器？

3．如何使用过滤器来控制用户访问权限？请编程加以说明。

4．如何使用监听器来统计在线用户信息？请编程加以说明。

第 7 章 AJAX 技术

本章导读

异步的 JavaScript 和 XML（Asynchronous JavaScript and XML，AJAX）是一种对传统的 Web 应用模式加以扩展的技术，使得“不刷新页面向服务器发起请求”成为可能。AJAX 最大的优点是在不重新加载整个页面的情况下，可以与服务器交换数据并更新部分网页内容。本章主要内容有：（1）实现 AJAX 应用的一般步骤；（2）使用 XMLHttpRequest 对象；（3）AJAX 示例。

7.1 实现 AJAX 应用的一般步骤

AJAX 不需要任何浏览器插件，但需要用户允许 JavaScript 在浏览器上执行。

AJAX 应用通过在客户端浏览器和服务器之间引入一个媒介“AJAX Engine”来发送异步请求，客户端可以在响应未到达之前继续当前页面的其他操作，AJAX Engine 则继续监听服务器的响应状态，在服务器完成响应后，获取响应结果并更新当前页面内容。

一般来说，一个 AJAX 应用的实现需要经过以下几个步骤。

（1）在页面中定义 AJAX 请求的触发事件。

（2）创建 XMLHttpRequest 对象。

（3）确定请求地址和请求参数。

（4）调用 XMLHttpRequest 对象的 open()方法建立对服务器的调用。

（5）通过 XMLHttpRequest 对象的 onreadystatechange 属性指定响应事件处理函数。

（6）在函数中根据响应状态进行数据读取和数据处理工作。

（7）调用 XMLHttpRequest 对象的 send()方法向服务器发出请求。

AJAX 的工作原理如图 7-1 所示。AJAX Engine 通过 XMLHttpRequest 对象向服务器发送 HTTP 请求，服务器处理完后返回响应结果（可能是各种类型的数据，如字符串、XML 和 JSON 等），AJAX Engine 根据响应文档类型对数据进行解析后再配合 HTML 和 CSS 渲染，将结果显示到客户端页面。

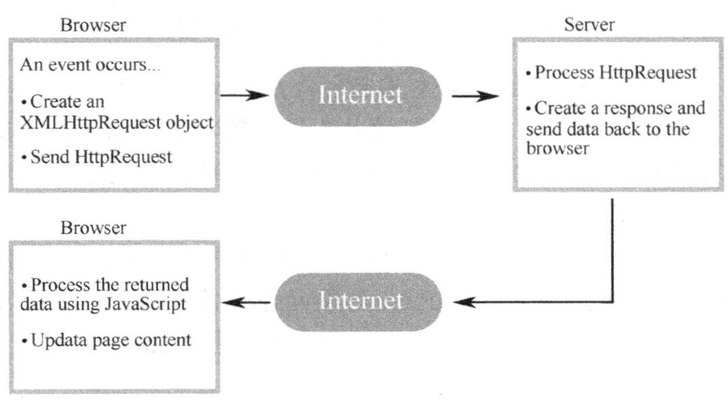

图 7-1 AJAX 的工作原理

7.2 使用 XMLHttpRequest 对象

XMLHttpRequest 是 AJAX 技术得以实现的一个重要的 JavaScript 对象，是 AJAX 的核心。

7.2.1 创建 XMLHttpRequest 对象

所有现代浏览器（IE 7+、Firefox、Chrome、Safari 以及 Opera）均内建了 XMLHttpRequest 对象。创建 XMLHttpRequest 对象的语法如下：

```
variable = new XMLHttpRequest();
```

旧版本的 Internet Explorer（IE5 和 IE6）则使用 ActiveX 对象：

```
variable = new ActiveXObject("Microsoft.XMLHTTP");
```

针对所有浏览器（包括 IE5 和 IE6）检查其是否支持 XMLHttpRequest 对象，如果支持，则创建 XMLHttpRequest 对象。如果不支持，则创建 ActiveXObject。

```
var xmlhttp;
if (window.XMLHttpRequest)
{
   // IE 7+、Firefox、Chrome、Opera、Safari 浏览器执行代码
   xmlhttp=new XMLHttpRequest();
} else {
   // IE6、IE5 浏览器执行代码
   xmlhttp=new ActiveXObject("Microsoft.XMLHTTP");
}
```

7.2.2 XMLHttpRequest 对象的常用属性和事件

XMLHttpRequest 对象的常用属性和事件如表 7-1 所示。

表 7-1 XMLHttpRequest 对象的常用属性和事件

属　性	描　述
readyState	表示异步请求过程中的各种状态【1】
responseText	包含客户端接收到的 HTTP 响应的文本内容【2】
responseXML	从服务器端进程返回的 XML 文档数据对象【3】
status	从服务器端返回的响应状态码【4】
statusText	服务器返回状态码对应的文本【5】
onreadystatechange	当 readyState 属性发生变化时触发此事件，用于触发回调函数

说明【1】：当一个 XMLHttpRequest 对象被创立后，readyState 属性标示了当前对象处于什么状态，可以通过对该属性的访问，来判读此次请求的状态并做出相应的操作。

① readyState 取值 0：未初始化状态（Uninitialized），此时已经创建了一个 XMLHttpRequest 对象，但是还没有初始化。

② readyState 取值 1：准备发送状态（Open），此时已经调用了 XMLHttpRequest 对象的 open() 方法，并且 XMLHttpRequest 对象已经准备好将一个请求发送到服务器。

③ readyState 取值 2：已发送状态（Send），此时已经通过 send()方法把一个请求发送到服务器，等待响应。

④ readyState 取值 3：正在接收状态（Receiving），此时 AJAX Engine 已经接收到 HTTP

响应头部信息，但是消息体部分还没有完全接收到。

⑤ readyState 取值 4：完成响应状态（Loaded），此时 HTTP 响应已经完全接收。

说明【2】：responseText 属性包含客户端接收到的 HTTP 响应的文本内容，当 readyState 属性为 0、1 或 2 时，responseText 属性包含一个空字符串。当 readyState 属性值为 3 时，响应中包含客户端还没有完成的响应信息。当 readyState 属性值为 4 时，responseText 属性才包含完整的响应信息。

说明【3】：只有当 readyState 属性为 4，并且响应头部的 Content-Type 的 MIME 类型被指定为 XML（text/XML 或者 application/XML）时，该属性才会有值并且被解析成一个 XML 文档，否则该属性为 null。如果回传的 XML 文档结构不良或者未完成响应回传，则该属性也会为 null。

说明【4】：status 属性描述了 HTTP 状态代码，注意，仅当 readyState 属性值为 3（正在接收中）或者 4（已加载）时，才能对此属性进行访问。如果当 readyState 属性值小于 3 时，试图去读取 status 属性值，将引发一个异常。

说明【5】：statusText 属性描述了 HTTP 状态代码文本，并且仅当 readyState 属性为 3 或者 4 时才可用。当 readyState 属性为其他值时，试图存取 statusText 属性将引发一个异常。

7.2.3 XMLHttpRequest 对象的常用方法

当我们利用前面几小节介绍的绘图方法进行了开发后，可能会有一些困惑，仅仅通过那些方法有时候不能够很好地绘制出一些特殊图形，如画一个不规则的多边形等。此时，可以考虑使用 android.graphics 提供的另外一组绘制方法——路径（Path）。

1. open()方法

XMLHttpRequest 对象是通过调用 open(method,url,async,username,password)方法来进行初始化工作的。调用该方法将得到一个可以用来进行发送的对象，open 方法有 5 个参数。前 2 个参数必选，后 3 个参数可选。

method 参数，用来指定发送请求的 HTTP 方法（取值为 GET 或 POST 或 PUT 或 DELETE 或 HEAD），参数名要大写。

url 参数，用来指定所调用资源的 URL。

async 参数，用来指定请求是否为异步，默认是 true。如果需要发送一个同步请求，则需要把该参数设置为 false。

如果需要服务器验证访问用户的情况，则可以设置 username 和 password。

2. send()方法

调用 open()方法后，就可以通过调用 send()方法按照 open()方法设定的参数将请求进行发送。当 open()方法中 async 为 true 时，在 send()方法调用后立即返回，否则将会等待直到服务器返回响应为止。

send()方法使用一个可选的参数，该参数可以包含可变类型的数据。用户可以使用它并通过 POST()方法把数据发送到服务器。另外，可以显式地使用 null 参数调用 send()方法，这与不用参数调用该方法一样。

3. abort()方法

abort()方法可以暂停一个 HttpRequest 的请求发送或者 HttpResponse 的接收，并且将 XMLHttpRequest 对象设置为未初始化状态。

4. setRequestHeader()方法

setRequestHeader()方法用来设置请求的头部信息。当 readyState 属性为 1 时，可以在调用 open()方法后调用这个方法；否则将得到一个异常。setRequestHeader(header,value)方法包含两个参数：第一个是 header 键名称，第二个是键值。

5. getResponseHeader()方法

getResponseHeader()方法用于检索响应的头部值，仅当 readyState 属性是 3 或者 4 时，才可调用该方法。否则，该方法返回一个空字符串。此外，还可以通过 getAllResponseHeader()方法获取所有的 HttpResponse 的头部信息。

7.3 AJAX 示例

这里，为了方便起见，把“更改文本内容”示例、“查询项目信息”示例、“验证注册邮箱格式和唯一性”示例都放在一个工程中，此工程名为 AjaxDemo，详见本书教学资源包 code 文件夹下 ch7 目录中的 AjaxDemo 工程。

7.3.1 更改文本内容

在 Eclipse 中创建“Dynamic Web Project”，工程名为 AjaxDemo。AjaxDemo 工程目录如图 7-2 所示，“更改文本内容”示例页面是 changeText.jsp。右击 changeText.jsp，选择“Run As→Run on Server”选项。这时，AjaxDemo 工程会自动部署到 Tomcat 服务器的 Webapps 目录中。运行 changeText.jsp，结果如图 7-3 所示。

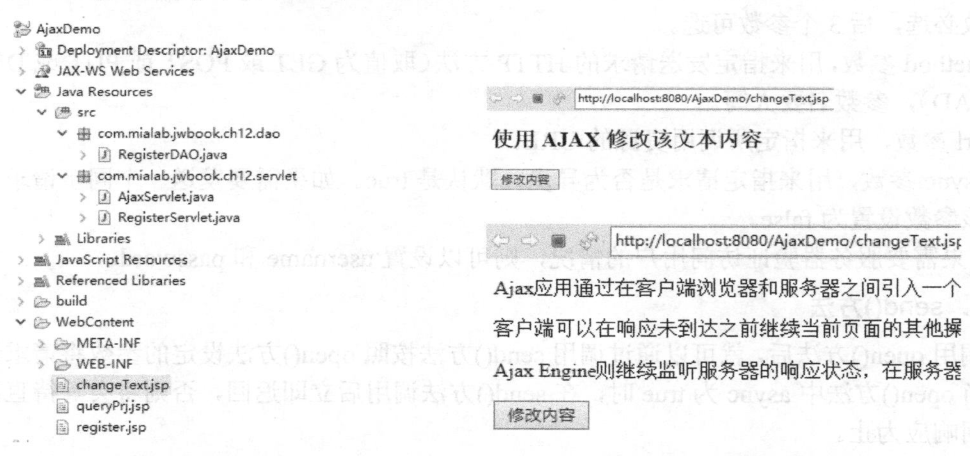

图 7-2 AjaxDemo 工程目录　　　图 7-3 运行 changeText.jsp

changeText.jsp 代码如下：

```
<%@ page language="java" contentType="text/html; charset=UTF-8"
 pageEncoding="UTF-8"%>
<!DOCTYPE html PUBLIC "-//W3C//DTD HTML 4.01 Transitional//EN"
   "http://www.w3.org/TR/html4/loose.dtd">
<html>
```

```
<head>
<meta http-equiv="Content-Type" content="text/html; charset=UTF-8">
<title>使用 AJAX 修改文本</title>
<script>
 function loadXMLDoc() {
     var xmlhttp;
     if (window.XMLHttpRequest) {
         //  IE7+、Firefox、Chrome、Opera、Safari 浏览器执行代码
         xmlhttp = new XMLHttpRequest();
     } else {
         //  IE6、IE5 浏览器执行代码
         xmlhttp = new ActiveXObject("Microsoft.XMLHTTP");
     }
     xmlhttp.onreadystatechange = function() {
         if (xmlhttp.readyState == 4 && xmlhttp.status == 200) {
             document.getElementById("myDiv").innerHTML = xmlhttp.response
Text;
         }
     }
     xmlhttp.open("GET", "/jwbook/ajax/ajax_info.txt", true);
     xmlhttp.send();
 }
</script>
</head>
<body>
<div id="myDiv"><h2>使用 AJAX 修改该文本内容</h2></div>
<button type="button" onclick="loadXMLDoc()">修改内容</button>
</body>
</html>
```

需要注意以下几点。

（1）向服务器发送 GET 请求。

```
xmlhttp.open("GET", "/jwbook/ajax/ajax_info.txt", true);
xmlhttp.send();
```

（2）常用的 HTTP 状态码及其对应的含义如下。

① 200：请求成功。

② 202：请求被接收但处理未完成。

③ 400：错误请求。

④ 404：需要访问的资源不存在。

⑤ 500：服务器内部错误。

（3）onreadystatechange 事件：每当 readyState 改变时，就会触发 onreadystatechange 事件。在 onreadystatechange 事件中，规定每次状态改变所触发事件的事件处理程序。

当 readyState 等于 4 且 HTTP 状态码为 200 时，表示响应已就绪。

```
xmlhttp.onreadystatechange = function() {
 if (xmlhttp.readyState == 4 && xmlhttp.status == 200) {
     document.getElementById("myDiv").innerHTML =
             xmlhttp.responseText;
  }
}
```

readyState 存有 XMLHttpRequest 的状态，其值从 0 到 4 发生变化。0 表示“请求未初始

化”；1 表示“服务器连接已建立”；2 表示“请求已接收”；3 表示“请求处理中”；4 表示“请求已完成，且响应已就绪”。

onreadystatechange 事件可被触发 5 次（0～4），对应着 readyState 的每个变化。

7.3.2 查询项目信息

在 AjaxDemo 工程中，“查询项目信息”示例页面是 queryPrj.jsp。右击 queryPrj.jsp，选择“Run As→Run on Server”选项。这时，AjaxDemo 工程会自动部署到 Tomcat 服务器的 Webapps 目录中。运行 queryPrj.jsp，结果如图 7-4 所示。

http://localhost:8080/AjaxDemo/queryPrj.jsp

获取项目编号对应的项目信息

项目编号:
项目名称:
项目主持人姓名:
项目主持人工号:

http://localhost:8080/AjaxDemo/queryPrj.jsp

获取项目编号对应的项目信息

项目编号: 5869
项目名称: 《IOS程序设计》教材培育项目
项目主持人姓名: 一剑冲天
项目主持人工号: TM3612

图 7-4 运行 queryPrj.jsp

queryPrj.jsp 代码如下：

```
<%@ page language="java" contentType="text/html; charset=UTF-8"
 pageEncoding="UTF-8"%>
<!DOCTYPE html PUBLIC "-//W3C//DTD HTML 4.01 Transitional//EN"
 "http://www.w3.org/TR/html4/loose.dtd">
<html>
<head>
<meta http-equiv="Content-Type" content="text/html; charset=UTF-8">
<title>AjaxDemo：查询项目信息</title>
</head>
<script type="text/javascript">
 var xhr = false;  // 定义一个全局的 XMLHttpRequest 对象
 // 创建 XMLHttpRequest 对象
 function createXHR() {
     try {
         // IE7+、Firefox、Chrome、Opera、Safari 代码
         xhr = new XMLHttpRequest();
     } catch (e) {
         try {
             // IE6, IE5 代码
             xhr = new ActiveXObject("Microsoft.XMLHTTP");
         } catch (e1) {
             xhr = false;
         }
     }
     if (!xhr)
         alert("初始化 XMLHttpRequest 对象失败！");
 }

 // 进行 AJAX 请求和响应结果处理
 function ajaxProcess(obj) {
     createXHR();    // 创建 XMLHttpRequest 对象
     var prjnumber = obj.value;   // 获取请求数据
```

```
        var url = "AjaxServlet?prjnumber=" + prjnumber;  // 设定请求地址
        xhr.open("GET", url, true);  // 建立对服务器的调用
        // 指定响应事件处理函数
        xhr.onreadystatechange = function() {
            // 当 readyState 等于 4 且服务器响应状态码为 200 时，表示响应已就绪
            if (xhr.readyState == 4 && xhr.status == 200) {
                // 对响应结果进行处理
                var responseData = xhr.responseText.split(",");
                // 将响应数据更新到页面控件中进行显示
                document.getElementById("prjname").value = responseData[0];
                document.getElementById("host").value = responseData[1];
                document.getElementById("jobnumber").value                =
responseData[2];
            }
        };
        xhr.send(null);  // 向服务器发出请求
    }
  </script>
  <body>
   <h2>获取项目编号对应的项目信息</h2> <hr>
   <table border=0 align="center" width=70%>
       <tr>
           <td>项目编号:</td>
           <td><input name="prjnumber" id="prjnumber" type="text" size=42
               onblur="ajaxProcess(this)"></td>
       </tr>
       <tr>
           <td>项目名称:</td>
           <td><input name="prjname" id="prjname" type="text" size=42></td>
       </tr>
       <tr>
           <td>项目主持人姓名: </td>
           <td><input name="host" id="host" type="text" size=42></td>
       </tr>
       <tr>
           <td>项目主持人工号: </td>
           <td><input    name="jobnumber"    id="jobnumber"    type="text"
size=42></td>
       </tr>
   </table>
  </body>
  </html>
```

onblur 是 JavaScipt 事件，在元素失去焦点时触发。

AjaxServlet.java 代码详见本书教学资源包 code 文件夹下 ch7 中的 AjaxDemo 工程。

7.3.3 验证注册邮箱格式和唯一性

在 AjaxDemo 工程中，“验证注册邮箱格式和唯一性”示例页面是 register.jsp。右击 register.jsp，选择“Run As→Run on Server”选项。这时，AjaxDemo 工程会自动部署到 Tomcat 服务器的 Webapps 目录中。运行 register.jsp，结果如图 7-5 所示。

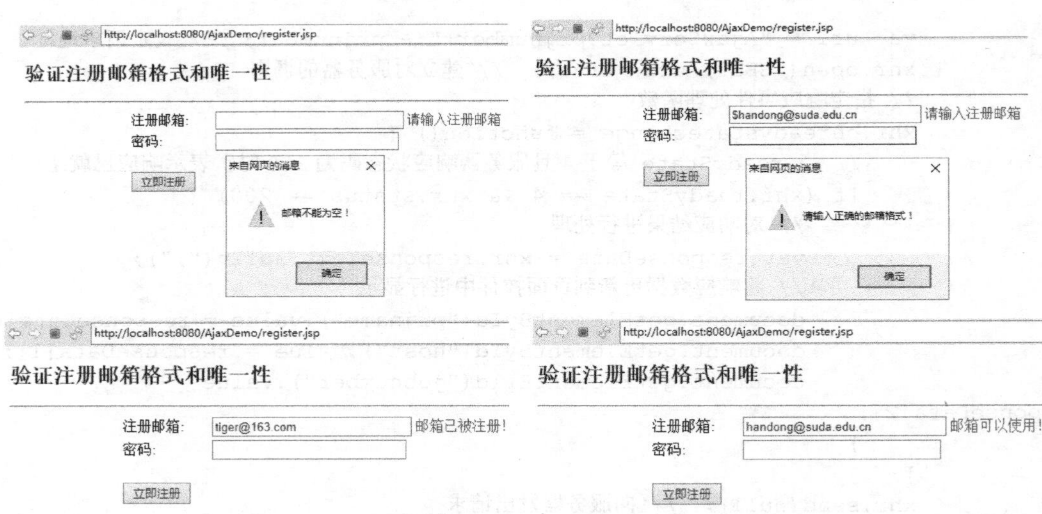

图 7-5 运行 register.jsp

register.jsp 代码如下：

```
<%@ page language="java" contentType="text/html; charset=UTF-8"
 pageEncoding="UTF-8"%>
<!DOCTYPE html PUBLIC "-//W3C//DTD HTML 4.01 Transitional//EN"
 "http://www.w3.org/TR/html4/loose.dtd">
<html>
<head>
<meta http-equiv="Content-Type" content="text/html; charset=UTF-8">
<title>注册页面</title>
<script type="text/javascript">
 var xhr = false;
 function createXHR() {
     try {
         xhr = new XMLHttpRequest();
     } catch (e) {
         try {
             xhr = new ActiveXObject("Microsoft.XMLHTTP");
         } catch (e1) {
             xhr = false;
         }
     }
     if (!xhr)
         alert("初始化 XMLHttpRequest 对象失败！");
 }

 function ajaxValidate(emailObj) {
     createXHR();
     var url = "RegisterServlet";
     var content = "type=emailAjaxValidate&email=" + emailObj.value;
     xhr.open("POST", url, true);
     xhr.onreadystatechange = function() {
         if (xhr.readyState == 4 && xhr.status == 200) {
             document.getElementById("emailValidate").innerHTML =
                 xhr.responseText;
         }
```

```
        };
        xhr.setRequestHeader("CONTENT-TYPE",
                "application/x-www-form-urlencoded");
        xhr.send(content);
    }

    function validate(emailObj) {
        var pattern = /^[A-Za-z\d]+([-_.][A-Za-z\d]+)*@([A-Za-z\d]+[.])+
            [A-Za-z]{2,5}$/;
        if (emailObj.value == "") {
            alert("邮箱不能为空！");
            email.focus();
        } else if (!pattern.test(emailObj.value)) {
            alert("请输入正确的邮箱格式！");
            email.focus();
        } else {
            ajaxValidate(emailObj);
        }
    }
   </script>
   </head>

   <body>
    <h2>验证注册邮箱格式和唯一性</h2><hr>
    <form action="" method="post">
        <table border=0 align="center" width=70%>
            <tr>
                <td>注册邮箱:</td>
                <td><input name="email" id="email" type="text" size=32
                    onblur="validate(this)"> <label style="color: red"
                    id="emailValidate">请输入注册邮箱</label></td>
            </tr>
            <tr>
                <td>密码:</td>
                <td><input name="password" type="password" id="pwd" size=32>
</td>
            </tr>
            <tr>
                <td align="center" colspan="2"><br /></td>
            </tr>
            <tr>
                <td colspan="2">
                <input name="submit" type="submit" value="立即注册">
                </td>
            </tr>
        </table>
    </form>
   </body>
   </html>
```

需要注意的是，如果是向服务器发送 POST 请求，且 send 方法要传递值为非 null 的参数（类似于 HTML 表单那样的 POST 数据），则往往需要设置正确的请求头。

假设新创建的 XMLHttpRequest 对象为 http_request，示例代码如下：

```
        var param = "user="+form1.user.value+ "&pwd="+form1.pwd.value+  "&email=
"+form1.email.value
        http_request.setRequestHeader("CONTENT-TYPE","application/x-www-form-ur
lencoded");
        http_request.send(param);
```

7.4 本章小结

AJAX 是一种在无须重新加载整个网页的情况下更新部分网页的技术。通过在后台与服务器进行少量数据交换，AJAX 可以使网页实现异步更新。这意味着可以在不重新加载整个网页的情况下，对网页的某部分进行更新。而传统的网页（不使用 AJAX）若想更新内容，就必须重载整个网页。

习题 7

1．XMLHttpRequest 对象的常用属性和事件有哪些？
2．XMLHttpRequest 对象的常用方法有哪些？
3．一个 AJAX 应用示例的实现有哪些步骤？试编程加以说明。
4．AJAX 异步请求方式和传统同步请求方式的区别有哪些？
5．使用 AJAX 技术在 JSP 网页中添加实时走动的系统时钟。
6. 在网上查阅 jQuery 的相关材料，编程实现基于 jQuery 的 AJAX 应用，并做相关说明。

第 8 章 Maven

本章导读

Maven 这个词可以翻译为“知识的积累”，也可以翻译为“专家”或“内行”。本章将介绍 Maven 这一跨平台的项目管理工具。作为 Apache 组织中的一个颇为成功的开源项目，Maven 主要服务于基于 Java 平台的项目构建、依赖管理和项目信息管理。无论是小型的开源类库项目，还是大型的企业级应用；无论是传统的瀑布式开发，还是流行的敏捷模式，Maven 都能大显身手。本章主要内容有：（1）初识 Maven；（2）Maven 安装与配置；（3）Maven 使用；（4）坐标和依赖；（5）使用 Maven 构建支持 Servlet 3.0 的 Web 应用。

8.1 初识 Maven

1. 什么是 Maven

为什么有 Maven？构建是程序员每天要做的工作，而且相当长的时间花在了此方面，而 Maven 使这一系列的工作完全自动化。我们一直在寻找避免重复的方法，这里的重复有：设计的重复、文档的重复、编码的重复、构建的重复等。而 Maven 是跨平台的，使用它能最大限度地消除构建的重复。

Maven 是一个项目管理工具，它包含了一个项目对象模型（Project Object Model）、一组标准集合、一个项目生命周期（Project Lifecycle）、一个依赖管理系统（Dependency Management System）和用来运行定义在生命周期阶段（Phase）中插件（Plugin）目标（Goal）的逻辑。当使用 Maven 的时候，可以用一个明确定义的项目对象模型来描述项目，然后 Maven 可以应用横切的逻辑，这些逻辑来自一组共享的（或者自定义的）插件。

此外，Maven 能够方便地用于管理项目报告、生成站点、管理 JAR 文件等。

2. Maven 与 Ant

（1）Maven 和 Ant 针对的是构建问题的两个不同方面。Ant 为 Java 技术开发项目提供跨平台构建任务。Maven 本身描述项目的高级方面，它从 Ant 借用了绝大多数构建任务。Ant 脚本是可以直接运行在 Maven 中的。Maven 和 Ant 最大的差别在于，Maven 的编译以及所有的脚本都有一个基础——POM。这个模型定义了项目的方方面面，各式各样的脚本在这个模型上工作，而 Ant 完全是自己定义的。

（2）Maven 对所依赖的包有明确的定义，如使用哪个包，版本是多少，一目了然。而 Ant 则通常是简单的 inclllde 所有的 JAR。导致的最终结果就是，用户根本无法确定 JBoss 中的 lib 下的 common-logging 是哪个版本的，唯一方法就是打开 META — INF 目录下的 MANIFEST.MF 进行查看。

（3）Maven 是基于中央仓库的编译，即把编译所需要的资源放在一个中央仓库里，如 JAR、TLD、POM 等。当编译的时候，Maven 会自动在仓库中找到相应的包，如果本地仓库没有，则从设定好的远程仓库中下载到本地。这一切都是自动的，而 Ant 需要自己定义。这个好处带来的结果就是，用 Maven 编译的项目在发布的时候只需要发布源码，非常小；反之，Ant 的发布则要使所有的包一起发布，显然，Maven 又胜了一筹。

（4）Ant 没有正式地约定一个一般项目的目录结构，必须明确地告诉 Ant 到哪里去找源代码，在哪里放置输出。Maven 拥有约定，因为用户遵循了约定，它已经知道用户的源代码在哪里。它把字节码放到 target/classes 里，然后在 target 中生成一个 JAR 文件。

（5）Ant 是程序化的，必须明确告诉 Ant 做什么，什么时候做，必须告诉它去编译、复制、压缩。Maven 是声明式的，用户需要做的只是创建一个 pom.xml 文件，并将源代码放到默认的目录中。Maven 会帮用户处理其他的事情。

（6）Ant 没有生命周期，用户必须定义目标和目标之间的依赖。用户必须手工为每个目标附上一个任务序列。Maven 有一个生命周期，当用户运行 mvn install 的时候被调用。这条命令告诉 Maven 执行一系列的有序的步骤，直到到达用户指定的生命周期。遍历生命周期的一个影响就是，Maven 运行了许多默认的插件目标，这些目标完成了像编译和创建一个 JAR 文件这样的工作。Maven 以插件的形式为一些一般的项目任务提供了内置的智能应用。

（7）Maven 有大量的重用脚本可以利用，如生成网站，生成 Javadoc、sourcecode reference 等。而 Ant 都需要用户自己去编写。对于 Maven，有了 pom.xml，只要从命令行窗口中运行 mvn install，就会处理资源文件，编译源代码，运行单元测试，创建一个 JAR，然后把这个 JAR 安装到本地仓库以为其他项目提供重用性，而不用做任何修改，可以运行 mvn site，然后在 target/site 目录中找到一个 index.html 文件，这个文件链接了 Javadoc 和一些关于源代码的报告。

（8）使用 Maven 还是 Ant 不是非此即彼的，Ant 在复杂的构建中还有其位置。如果目前的构建包含一些高度自定义的过程，或者已经写了一些 Ant 脚本并通过一种明确的方法完成一个明确的过程，而这个过程不适合 Maven 标准，也仍然可以在 Maven 中用这些脚本。作为一个 Maven 的核心插件，Ant 还是可用的。自定义的插件可以用 Ant 来实现，Maven 项目可以配置成在生命周期中运行 Ant 的脚本。

（9）Maven 目前不足的地方就是没有如 Ant 那样的成熟的 GUI 界面，目前使用 Maven 最好的方法是利用命令行窗口。

3. 约定优于配置

约定优于配置（Convention Over Configuration）是一个简单的概念。系统、类库、框架应该假定合理的默认值，而非要求提供不必要的配置。流行的框架如 Ruby on Rails 和 EJB3 已经开始坚持这些原则，以应对像原始的 EJB 2.1 规范那样的框架配置复杂度。

Maven 通过给项目提供明智的默认行为来融合这个概念。

在没有自定义的情况下，源代码假定在${basedir}/src/main/java 中，资源文件假定在${basedir}/src/main/resources 中，测试代码假定在${basedir}/src/test 中。

项目假定会产生一个 JAR 文件。Maven 假定用户想要把编译好的字节码放到${basedir}/target/classes 中并在${basedir}/target 中创建一个可分发的 JAR 文件。

虽然这看起来无关紧要，但是大部分基于 Ant 的构建必须为每个子项目定义这些目录。Maven 对约定优于配置的应用不仅仅是简单的目录位置，Maven 的核心插件使用了一组通用约定，以用来编译源代码、打包可分发的构件、生成 Web 站点，以及许多其他过程。

Maven 的力量来自它的“武断”，它有一个定义好的生命周期和一组知道如何构建及装配软件的通用插件。如果用户遵循这些约定，Maven 的工作仅仅是将源代码放到正确的目录中，Maven 会处理其余事情。

使用“约定优于配置”系统的一个副作用是用户可能会觉得其被强迫使用一种特殊的方法。当然，Maven 有一些核心观点不应该被怀疑，其实很多默认行为还是可配置的。例如，项目源码的资源文件的位置可以被自定义，JAR 文件的名称可以被自定义，在开发自定义插件的时候，几乎任何行为都可以被裁剪以满足特定的环境需求。如果用户不想遵循约定，Maven 也会允许用户自定义默认值来适应需求。

4．Maven 的优点

（1）平时开发项目时，一般一个项目就是一个工程。在划分模块时，都是使用 package 来进行划分的。但是，当项目有很多子模块时，即使是利用 package 来进行划分，也往往会让人无所适从。

优点一：项目非常大时，可借助 Maven 将一个项目拆分成多个工程，最好是一个模块对应一个工程，这样利于分工协作，且模块之间还是可以发送消息的。

（2）同一项目的 JAR 包复制和粘贴到 Web/INF/lib 下，问题：同样的 JAR 包重复出现在不同的工程中，既浪费空间又会使工程臃肿。

优点二：借助 Maven，可将 JAR 包仅仅保存在“仓库”中，当需要该文件时，就引用该文件接口，不需要复制文件而占用空间。

（3）如果 JAR 包都要到各个官方网站下载，则既会浪费很多时间，又可能资源不全。

优点三：借助 Maven 可以以规范的方式下载 JAR 包，因为所有的知名框架或第三方工具的 JAR 包已经按照统一的规范存放到了 Maven 的中央仓库中。

（4）一个 JAR 包依赖的其他 JAR 包可能未导入到项目而导致项目无法进行。

优点四：Maven 会自动将要加入到项目中的 JAR 包导入，还会将该 JAR 包所依赖的 JAR 包都自动导入进来。

8.2　Maven 的安装和配置

1．安装 JDK

在安装 Maven 之前，先确保已经安装了 JDK 1.6 及以上版本，并已经配置好了环境变量。

2．下载 Maven

在 http://maven.apache.org/download.cgi 地址下载 apache-maven-3.3.3-bin.zip 并解压，这里解压到 D:\apache-maven-3.3.3 目录下。

3．配置环境变量

配置 Maven 的环境变量。

先配置 M2_HOME 的环境变量，新建一个系统变量 M2_HOME，路径是 D:\apache-maven-3.3.3，如图 8-1 所示。

再配置 path 环境变量，在 path 值的末尾添加"%M2_HOME%\bin"。

打开命令行窗口，输入 mvn –version，出现如图 8-2 所示内容表示安装成功。

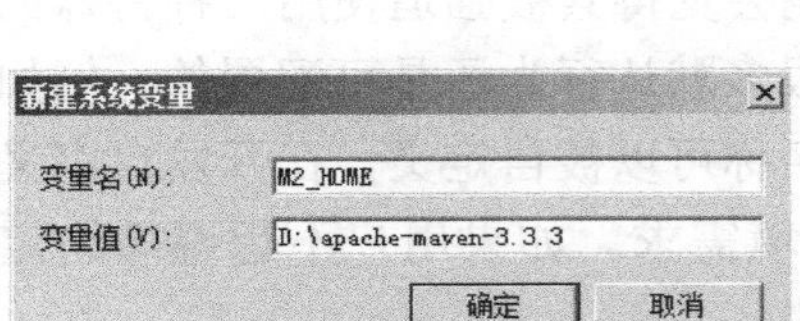

图 8-1　配置 M2_HOME 环境变量

```
管理员: C:\Windows\system32\cmd.exe
Microsoft Windows [版本 6.1.7601]
版权所有 (c) 2009 Microsoft Corporation。保留所有权利。

C:\Users\Administrator>mvn -version
Apache Maven 3.3.3 (7994120775791599e205a5524ec3e0dfe41d4a06; 2015-04-22T19:57:3
7+08:00)
Maven home: D:\apache-maven-3.3.3
Java version: 1.7.0_79, vendor: Oracle Corporation
Java home: C:\Program Files (x86)\Java\jdk1.7.0_79\jre
Default locale: zh_CN, platform encoding: GBK
OS name: "windows 7", version: "6.1", arch: "x86", family: "windows"

C:\Users\Administrator>_
```

图 8-2　输入 mvn –version

4. 给 Maven 添加本地仓库

在默认情况下，不管是在 Windows 还是 Linux 中，每个用户在自己的用户目录下都有一个路径名为.m2/repository/的仓库目录。有时候，因为某些原因（如 C 盘空间不够），用户要自定义本地仓库目录地址。这时，可以编辑文件~/.m2/settings.xml，设置 localRepository 元素的值为想要的仓库地址。这里~指用户目录，如 C:\Users\Administrator。

例如，本书的本地仓库配置为

```
<settings>
  <localRepository> D:\my_maven_repository\work\maven </localRepository>
</settings>
```

这样，该用户的本地仓库地址就被设置成了 D:\my_maven_repository\work\maven。需要注意的是，在默认情况下，~/.m2/settings.xml 文件是不存在的，用户需要从 Maven 安装目录复制$M2_HOME/conf/settings.xml 文件后再进行编辑。

一个构件只有在本地仓库中才能由其他 Maven 项目使用，那么构件如何进入到本地仓库中呢？最常见的是依赖 Maven 从远程仓库下载到本地仓库中。

还有一种常见的情况是，将本地项目的构件安装到 Maven 仓库中。例如，本地有两个项目 A 和 B，两者都无法从远程仓库获得，而同时 A 依赖于 B，为了构建 A，B 就必须先构建并安装到本地仓库中。为了安装项目，我们可以在项目中执行中 mvn clean install 命令。Install 插件的 install 目标将项目的构建输出文件安装到本地仓库中。

安装好 Maven 后，如果不执行任何 Maven 命令，本地仓库目录是不存在的。当用户输入第一条 Maven 命令之后，Maven 才会创建本地仓库，然后根据配置和需要，从远程仓库下载构件至本地仓库。

由于最原始的本地仓库是空的，Maven 必须知道至少一个可用的远程仓库，才能在执行 Maven 命令的时候下载到需要的构件。中央仓库就是这样一个默认的远程仓库，Maven 的安装文件自带了中央仓库的配置。

私服是一种特殊的远程仓库，它是架设在局域网内的仓库服务，私服代理广域网上的远程仓库，供局域网内的 Maven 用户使用。当 Maven 需要下载构件的时候，它向私服发出请求，如果私服上不存在该构件，则从外部的远程仓库下载，缓存在私服上之后，再为 Maven 的下载请求提供服务。此外，一些无法从外部仓库下载到的构件也能从本地上传到私服上供大家使用。

5. 配置用户范围 settings.xml

Maven 用户可以选择配置$M2_HOME/conf/settings.xml 或者~/.m2/settings.xml。前者是全局范围的，整台机器上的所有用户都会直接受到该配置的影响；而后者是用户范围的，只有当前用户才会受到该配置的影响。具体配置如图 8-3 所示。

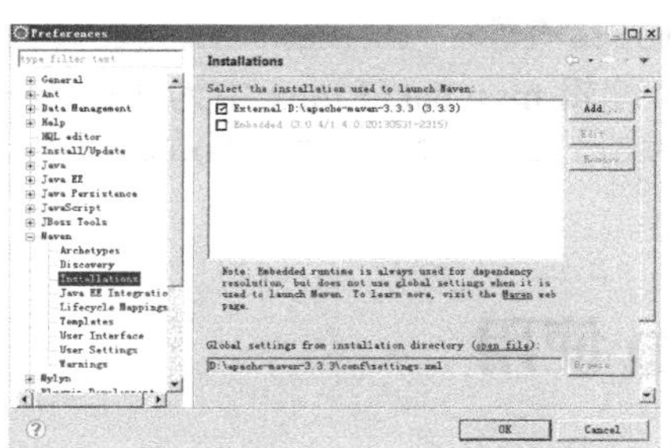

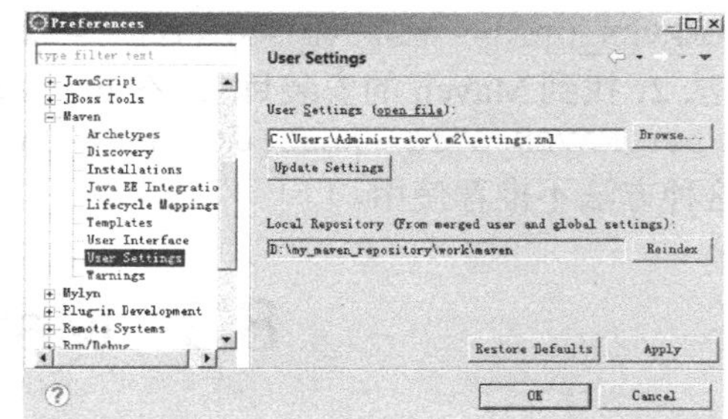

图 8-3　配置 settings.xml

6．错误处理

在 Eclipse 中使用 Maven 插件的时候，选择“Run As → Maven clean”选项时会报错，如下所示：

```
-Dmaven.multiModuleProjectDirectory system propery is not set.
Check $M2_HOME environment variable and mvn script match.
```

解决方法：选择“Window->Preferences”选项，在弹出的对话框中选择“Java->Installed JREs”节点，如图 8-4 所示。

单击“Edit”按钮，在 Default VM arguments 中进行如下设置，如图 8-5 所示。

```
-Dmaven.multiModuleProjectDirectory=$M2_HOME
```

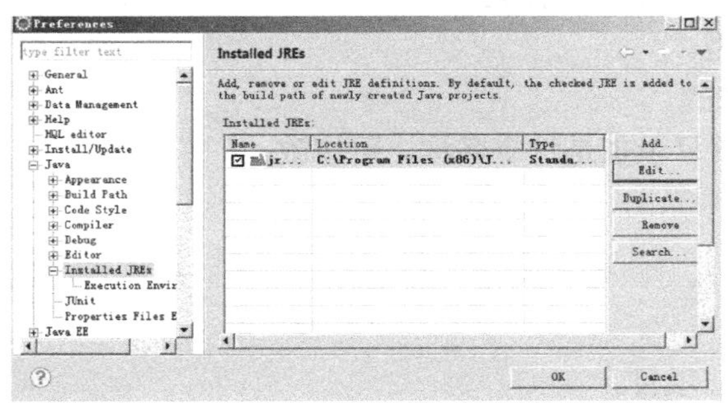

图 8-4　Installed JREs 对话框

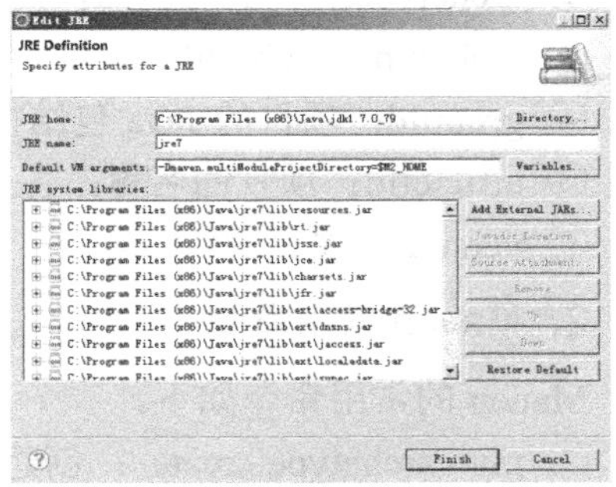

图 8-5　配置 JRE

7．设置 MAVEN_OPTS 环境变量

在前面介绍 Maven 安装目录时我们了解到，运行 mvn 命令实际上是执行了 Java 命令，既然是运行 Java，那么运行 Java 命令可用的参数当然也应该在运行 mvn 命令时可用。此时，MAVEN_OPTS 环境变量就有用了。

通常需要设置 MAVEN_OPTS 的值为：-Xms128m -Xmx512m，因为 Java 默认的最大可用内存往往不能满足 Maven 运行的需要。在项目较大时，使用 Maven 生成项目站点需要占用大量的内存，如果没有该配置，很容易得到 java.lang.OutOfMemoryError。因此，一开始就配置该变量是推荐的做法。

关于如何设置环境变量，请参考前面设置 M2_HOME 环境变量的做法，尽量不要直接修改 mvn.bat 或者 mvn 这两个 Maven 执行脚本文件。因为如果修改了脚本文件，升级 Maven 时就不得不再次修改，既麻烦又容易忘记。同理，应该尽可能不去修改任何 Maven 安装目录下的文件。

关于 Maven 中 JVM 参数的设置如下。

方法 1：在系统的环境变量中，设置 MAVEN_OPTS，用以存放 JVM 的参数，示例如下。

```
MAVEN_OPTS=-Xms256m -Xmx768m -XX:PermSize=128m -XX:MaxPermSize=256M
```

方法 2：找到 Maven 的安装目录，在 bin 目录下，编辑 mvn.bat（Linux 中为文件 mvn.sh）。

```
set MAVEN_OPTS=-Xms256m -Xmx768m -XX:PermSize=128m -XX:MaxPermSize=256M
```

（这种方法不推荐使用。）

8.3 Maven 使用

1. 生成项目

（1）新建一个 Java 项目：

```
mvn archetype:generate -DgroupId=com.mialab -DartifactId=Exercise
-DarchetypeArtifactId=maven-archetype-quickstart -Dversion=1.0
```

（2）新建一个 Web 项目：

```
mvn archetype:generate -DgroupId=com.mialab -DartifactId=PalmSuda
-DarchetypeArtifactId=maven-archetype-Webapp -Dversion=1.0
```

解释：

① archetype:create：archetype 插件的 create 目标。在 Maven 中，一个插件可以拥有多个目标。

② archetype：一个内建插件，其作用是建立项目骨架。

③ archetypeArtifactId：项目骨架的类型。

④ groupId：项目的 Java 包结构，可修改。

⑤ artifactId：项目的名称，生成的项目目录也是这个名称，可修改。

⑥ version：项目的版本。

2. Maven 常用命令

Mauen 的常用命令如下。

① mvn archetype:create：创建 Maven 项目。

② mvn compile：编译源代码。

③ mvn test-compile：编译测试代码。

④ mvn test：运行应用程序中的单元测试。

⑤ mvn site：生成项目相关信息的网站。

⑥ mvn clean：清除目标目录中的生成结果。

⑦ mvn package：依据项目生成 JAR 文件。

⑧ mvn install：在本地 Repository 中安装 JAR 包。

⑨ mvn deploy：将 JAR 包发布到远程仓库。

⑩ mvn eclipse:eclipse：生成 Eclipse 项目文件。

3. Maven 常用项目骨架

如果是 Maven 3，可简单运行：

```
mvn archetype:generate
```

（1）internal -> appfuse-basic-jsf：创建一个基于 Hibernate、Spring 和 JSF 的 Web 应用程序的原型。

（2）internal -> appfuse-basic-spring：创建一个基于 Hibernate、Spring 和 Spring MVC 的

Web 应用程序的原型。

（3）internal -> appfuse-basic-struts ：创建一个基于 Hibernate、Spring 和 Struts 2 的 Web 应用程序的原型。

（4）internal -> appfuse-modular-spring ：创建一个基于 Hibernate、Spring 和 Spring MVC 的模块化应用原型。

（5）internal -> appfuse-modular-struts ：创建一个基于 Hibernate、Spring 和 Struts 2 的模块化应用原型。

（6）internal -> maven-archetype-j2ee-simple ：一个简单的 J2EE 的 Java 应用程序。

（7）internal -> maven-archetype-marmalade-mojo ：创建一个 Maven 的插件开发项目 using marmalade。

（8）internal -> maven-archetype-mojo ：一个 Maven 的 Java 插件开发项目。

（9）internal -> maven-archetype-portlet ：一个简单的 Portlet 应用程序。

（10）internal -> maven-archetype-site-simple ：简单的网站生成项目。

（11）internal -> maven-archetype-site ：更复杂的网站项目。

（12）internal -> maven-archetype-Webapp ：一个简单的 Java Web 应用程序。

8.4 坐标和依赖

1. 坐标

Maven 的所有构件均通过坐标进行组织和管理。Maven 的坐标通过 5 个元素进行定义，其中 groupId、artifactId、version 是必需的，packaging 是可选的（默认为 JAR），classifier 是不能直接定义的。

（1）groupId：定义当前 Maven 项目所属的实际项目，和 Java 包名类似，通常与域名反向一一对应。

（2）artifactId：定义当前 Maven 项目的一个模块，在默认情况下，Maven 生成的构件，其文件名会以 artifactId 开头，如 hibernate-core-3.6.5.Final.jar。

（3）version：定义项目版本。

（4）packaging：定义项目打包方式，如 JAR、WAR、POM、ZIP，等等，默认为 JAR。

（5）classifier：定义项目的附属构件，如 hibernate-core-3.6.6.Final-sources.jar、hibernate-core-3.6.6.Final-javadoc.jar，其中 sources 和 Javadoc 就是这两个附属构件的 classifier。classifier 不能直接定义，通常由附加的插件生成。

2. 依赖

使用 Maven 可以方便地管理依赖，如下是一段在 pom.xml 文件中声明依赖的代码示例：

```
<dependencies>
    <dependency>
        <groupId>org.springframework</groupId>
        <artifactId>spring-test</artifactId>
        <version>3.2.0.RELEASE</version>
        <type>jar</type>
        <scope>test</scope>
        <systemPath>${java.home}/lib/rt.jar</systemPath>
```

```
            <optional>false</optional>
            <exclusions>
                <exclusion></exclusion>
            </exclusions>
        </dependency>
    </dependencies>
```

type：依赖类型，对应构件中定义的 packaging，可不声明，默认为 JAR。

scope：依赖范围。

optional：依赖是否可选。

exclusions：排除传递依赖。

执行不同的 Maven 命令（mvn package、mvn test、mvn install 等），会使用不同的 classpath，Maven 对应有三套 classpath：编译 classpath、测试 classpath 和运行 classpath。scope 选项的值，决定了该依赖构件会被引入到哪一个 classpath 中。

compile：编译依赖范围，默认值。此选项对编译、测试、运行 3 种 classpath 都有效，如 hibernate-core-3.6.5.Final.jar，表明在编译、测试、运行的时候都需要该依赖。

test：测试依赖范围。其只对测试有效，表明只在测试的时候需要，在编译和运行时将无法使用该类依赖，如 Junit。

provided：已提供依赖范围，编译和测试有效，运行无效。如 servlet-api，在项目运行时，Tomcat 等容器已经提供，无须 Maven 重复引入。

runtime：运行时依赖范围，测试和运行有效，编译无效。如 JDBC 驱动实现，编译时只需接口，测试或运行时才需要具体的 JDBC 驱动实现。

system：系统依赖范围。其和 provided 依赖范围一致，需要通过<systemPath>显示指定，且可以引用环境变量。

import：导入依赖范围。使用该选项时，通常需要<type>pom</type>，将目标 pom 的 dependencyManagement 配置导入合并到当前 pom 的 dependencyManagement 元素上。

8.5 本章小结

Maven 提供了一套软件项目管理的综合性方案。无论是编译、发布还是团队协作，Maven 都提供了必要的抽象，它鼓励重用，并做了除软件构建以外的许多工作。

Maven 为构建、测试、部署项目定义了一个标准的生命周期。它提供了一个框架，允许遵循 Maven 标准的所有项目方便地重用公用的构建逻辑。

Maven 的优势如下。

（1）Maven 不仅是构建工具，还是依赖管理工具和项目管理工具，提供了中央仓库，能够帮助用户自动下载构件。使用 Maven 可以进行项目高度自动化构建、依赖管理和仓库管理等。而使用 Maven 最大的好处就是可以实现依赖管理。

（2）为了解决依赖的增多、版本不一致、版本冲突、依赖臃肿等问题，Maven 通过一个坐标系统来精确地定位每一个构件。

（3）Maven 可帮助用户管理分散在各个角落的项目信息，包括项目描述、开发者列表、版本控制系统、许可证、缺陷管理系统地址等。

（4）Maven 还为全世界的 Java 开发者提供了一个免费的中央仓库，在其中几乎可以找到所有的流行开源软件。通过衍生工具，还能对其进行快速搜索。

（5）Maven 对于目录结构有要求，约定优于配置，这样，用户在项目间切换就省去了学习成本。

习题 8

1．Maven 与 Ant 的区别是什么？

2．Maven 的优点是什么？

3．如何安装配置 Maven？请简要加以说明。

4．如何理解“约定优于配置”？请加以说明。

5．何为 Maven 坐标？何为 Maven 仓库？如何编写 POM？

6．如何使用 archetype 生成 Maven 项目骨架？

第 9 章 jQuery EasyUI

本章导读

jQuery EasyUI 是一组基于 jQuery 的 UI 插件集合体，其目的是帮助 Web 开发者更轻松地打造出功能丰富且美观的用户界面。开发者不需要编写复杂的 JavaScript，也不需要对 CSS 样式有深入了解，开发者需要了解的只是一些简单的 HTML 标签。本章主要内容有：（1）jQuery 基础；（2）jQuery EasyUI；（3）jQuery EasyUI 布局；（4）jQuery EasyUI 数据网格。

9.1 jQuery 基础

jQuery 是一个快速、简洁的 JavaScript 框架，诞生于 2005 年，由 John Resign 开发。

jQuery 设计的宗旨是“Write Less，Do More”，即倡导写更少的代码，做更多的事情。它封装 JavaScript 常用的功能代码，提供了一种简便的 JavaScript 设计模式，优化了 HTML 文档操作、事件处理、动画设计和 AJAX 交互。

9.1.1 初识 jQuery

1. JavaScript 和 JavaScript 库

JavaScript 是一种属于网络的脚本语言，它的解释器被称为 JavaScript 引擎，为浏览器的一部分。JavaScript 已经被广泛用于 Web 应用开发，常用来为网页添加各式各样的动态功能，为用户提供更流畅美观的浏览效果。通常，JavaScript 是通过嵌入到 HTML 中来实现自身的功能的。近年来，随着 AJAX 概念的提出，基于 Web 的编程的交互性被提升到了一个新高度，JavaScript 也变得越来越重要。

JavaScript 高级程序设计（特别是对浏览器差异的复杂处理）通常很困难也很耗时。为了应对这些调整，许多 JavaScript 库（常被称为 JavaScript 框架）应运而生。一些较为流行的 JavaScript 框架有 Prototype（http://www.prototypejs.org/）、jQuery（http://jquery.com/）、YUI（http://developer.yahoo.com/yui/）、Dojo（http://dojotoolkit.org）、MooTools（http://mootools.net/）、Ext JS（http://www.extjs.com/）等。

jQuery 是目前最受欢迎的 JavaScript 框架。它使用 CSS 选择器来访问和操作网页上的 HTML 元素（DOM 对象）。

2. 使用 jQuery

jQuery 库的功能包括：HTML 元素选取、HTML 元素操作、CSS 操作、HTML 事件函数、JavaScript 特效和动画、HTML DOM 遍历和修改、AJAX 异步请求方式等。

访问 jQuery 官方网站（http://jquery.com/），可下载最新版本的 jQuery 库文件。

有以下两个版本的 jQuery 可供下载。

（1）Production version：用于实际的网站中，已被精简和压缩。

（2）Development version：用于测试和开发（未压缩，是可读的代码）。

以上两个版本都可以从 jquery.com 中下载。

这里下载的是发布版（网页链接上的标签为“Download the compressed, production jQuery 3.3.1”，其网页链接为 https://code.jquery.com/jquery-3.3.1.min.js）。

jQuery 库是一个 JavaScript 文件，用户可以使用 HTML 的 <script> 标签引用它。以下代码是“hello, jQuery”示例（见教学资源源码包 ch9/1 中的 hello.html）。

```
<!DOCTYPE html>
<html>
<head>
<meta charset="utf-8">
<title>hello, jQuery</title>
<script src="jQuery/jquery-3.3.1.min.js">
</script>
<script>
$(function(){
  alert("Hello, jQuery!");
});
</script>
</head>
<body>
<p>hello, jQuery</p>
</body>
</html>
```

在浏览器中预览该网页文件（ch9/1/hello.html），可以看到在当前窗口中弹出了一个提示对话框，内容为“hello, jQuery”。

在 jQuery 库中，$是 jQuery 的别名，$( )等效于 jQuery( )。jQuery( )是 jQuery 库文件的接口函数，所有 jQuery 操作都必须从该接口函数切入。

jQuery( )函数相当于页面初始化事件处理函数，当页面加载完毕后，会执行 jQuery( )函数包含的函数。$(function() {}) 是$(document).ready(function(){})的简写。

3. jQuery 对象和 DOM 对象

DOM：文档对象模型，把 HTML 分解成 DOM 树和节点。HTML 的不同元素，体现在 DOM 中就是不同节点，通过 JavaScript，可以方便地控制这些节点，动态地更改页面显示。

DOM 对象能使用 JavaScript 固有的方法，但是不能使用 jQuery 中的方法。

jQuery 对象就是通过 jQuery 包装 DOM 对象后产生的对象。jQuery 对象是 jQuery 独有的，其可以使用 jQuery 中的方法，但是不能使用 DOM 的方法。

对于一个 DOM 对象，只需要用$()把 DOM 对象包装起来，就可以获得一个 jQuery 对象，即$(DOM 对象)。例如：

```
var v=document.getElementById(“v”);  //DOM 对象
var $v=$(v);    //jQuery 对象
```

DOM 对象转换成 jQuery 对象后，就可以使用 jQuery 的方法了。

由于 jQuery 对象本身是一个集合，所以如果 jQuery 对象要转换为 DOM 对象，则必须取出其中的某一项，一般可通过索引取出。例如，可通过$("#msg")[0]得到 DOM 对象。

9.1.2 jQuery 选择器

jQuery 选择器允许对 HTML 元素组或单个元素进行操作。jQuery 选择器基于元素的 id、类、类型、属性、属性值等“查找”（或选择）HTML 元素。它基于已经存在的 CSS 选择器，除此之外，它还有一些自定义的选择器。

jQuery 中所有选择器都以美元符号开头，即$()。

1. 元素选择器

jQuery 元素选择器基于元素名选取元素。

例如，在页面中选取所有 <p> 元素：

```
$("p")
```

又如，用户单击按钮后，所有<p>元素都隐藏：

```
$(document).ready(function(){
  $("button").click(function(){
    $("p").hide();
  });
});
```

2. #id 选择器

jQuery #id 选择器通过 HTML 元素的 id 属性选取指定的元素。页面中元素的 id 应该是唯一的，所以在页面中选取唯一的元素需要通过 #id 选择器来实现。

通过 id 选取元素的语法如下：

```
$("#test")
```

当用户单击按钮后，有 id="test" 属性的元素将被隐藏：

```
$(document).ready(function(){
  $("button").click(function(){
    $("#test").hide();
  });
});
```

3. .class 选择器

jQuery 类选择器可以通过指定的 class 查找元素。

例如，选择 class="test" 的元素：

```
$(".test")
```

又如，用户单击按钮后所有带有 class="test" 属性的元素都隐藏：

```
$(document).ready(function(){
  $("button").click(function(){
    $(".test").hide();
  });
});
```

9.1.3 jQuery 事件

jQuery 事件处理方法是 jQuery 中的核心函数。事件处理程序指的是当 HTML 中发生某些事件时所调用的方法。通常会把 jQuery 代码放到<head>部分的事件处理方法中，如以下代码（见本书教学资源源码包 ch9/2 中的 event.html）：

```
<html>
<head>
<script type="text/javascript" src="jQuery/jquery-3.3.1.min.js"></script>
<script type="text/javascript">
$(document).ready(function(){
  $("button").click(function(){
    $("p").hide();
  });
});
</script>
</head>
<body>
<h2>This is a heading</h2>
<p>This is a paragraph.</p>
<p>This is another paragraph.</p>
<button>Click me</button>
</body>
</html>
```

表 9-1 列出了 jQuery 中事件方法的一些例子。如需完整的参考手册，可参考 w3school 的 jQuery 事件参考手册（http://www.w3school.com.cn/jquery/jquery_ref_events.asp）。

表 9-1 jQuerg 事件方法

Event 函数	绑定函数至
$(document). ready(function)	将函数绑定到文档的就绪事件（当文档完成加载时）
$(selector).click(function)	触发或将函数绑定到被选元素的单击事件
$(selector).dblclick(function)	触发或将函数绑定到被选元素的双击事件
$(selector).focus(function)	触发或将函数绑定到被选元素的获得焦点事件
$(selector).mouseover(function)	触发或将函数绑定到被选元素的鼠标悬停事件

9.1.4 jQuery AJAX

AJAX 即为异步 JavaScript 和 XML。简短地说，在不重载整个网页的情况下，AJAX 通过后台加载数据，并在网页上进行显示。

jQuery 提供多个与 AJAX 有关的方法。通过 jQuery AJAX 方法，可以使用 HTTP Get 和 HTTP Post 从远程服务器上请求文本、HTML、XML 或 JSON，同时，能够把这些外部数据直接载入网页的被选元素中。

jQuery get() 和 post()方法用于通过 HTTP GET 或 POST 从服务器请求数据。

1. jQuery $.get() 方法

$.get() 方法通过 HTTP GET 从服务器上请求数据。

语法：**$.get(URL,callback)。**

表示请求的 URL 参数是必需的，可选的 callback 参数是请求成功后所执行的函数名。

2. jQuery $.post() 方法

$.post() 方法通过 HTTP POST 从服务器上请求数据。

语法：**$.post(URL,data,callback);**

说明：可选的 data 参数规定了连同请求发送的数据，URL 参数和 callback 参数同 get()方法。

3. jQuery $.ajax() 方法

$.post()方法和$.get()方法较为简单，如果要处理复杂的逻辑，则需要用到 jQuery.ajax()。

jQuery.ajax()通过 HTTP 请求加载远程数据，该方法是 jQuery 底层 AJAX 实现，$.get()和$.post()可视为简单易用的高层实现。jQuery.ajax()返回其创建的 XMLHttpRequest 对象。

在最简单的情况下，$.ajax() 可以不带任何参数直接使用。

语法：jQuery.ajax([settings])。

说明：参数 settings 可选，用于配置 AJAX 请求的键值对集合。也可以通过$.ajaxSetup()设置任何选项的默认值。

以下是示例代码：

```
//GET 方式
$.ajax({
   type: "GET",
   url: "…",
 async: "true",
   dataType: "text",
   success: function(msg){ }
});
//POST 方式
$.ajax({
   type: "POST",
   url: "…",
 data: "…",
 async: "true",
   dataType: "text",
   success: function(msg){ }
});
```

以下是 jQuery 中关于 AJAX 方法一些参数的说明。

（1）url：要求为 String 类型的参数，请求地址。

（2）type：要求为 String 类型的参数，请求方式（post 或 get）默认为 get。注意，其他 HTTP 请求方法（如 put 和 delete）也可以使用，但仅部分浏览器支持。

（3）timeout：要求为 Number 类型的参数，设置请求超时时间（毫秒）。此设置将覆盖$.ajaxSetup()方法的全局设置。

（4）async：要求为 Boolean 类型的参数，默认设置为 true，所有请求均为异步请求。如果需要发送同步请求，则应将此选项设置为 false。注意，同步请求将锁住浏览器，用户其他操作必须等待请求完成后才可以执行。

（5）cache：要求为 Boolean 类型的参数，默认为 true（当 dataType 为 script 时，默认为 false），设置为 false 时将不会从浏览器缓存中加载请求信息。

（6）data：要求为 Object 或 String 类型的参数，表示发送到服务器的数据。如果已经不是字符串，则将自动转换为字符串格式。在 get 请求中，其将附加在 url 后。若要防止这种自动转换，可以查看 processData 选项。对象必须为 key/value 格式，如{foo1:"bar1",foo2:"bar2"}转换为&foo1=bar1&foo2=bar2。如果是数组，则 jQuery 将自动为不同值对应同一个名称。例如，{foo:["bar1","bar2"]}转换为&foo=bar1&foo=bar2。

（7）dataType：要求为 String 类型的参数，预期服务器返回的数据类型。如果不指定，jQuery 将自动根据 HTTP 包 MIME 信息返回 responseXML 或 responseText，并作为回调函数的参数进行传递。可用的类型如下：

① xml：返回 XML 文档，可用 jQuery 处理。

② html：返回纯文本 HTML 信息；包含的 script 标签会在插入 DOM 时执行。

③ script：返回纯文本 JavaScript 代码，不会自动缓存结果，除非设置了 cache 参数。注意，在远程请求（不在同一个域下）时，所有 post 请求都将转换为 get 请求。

④ json：返回 JSON 数据。

⑤ jsonp：JSONP 格式。使用 JSONP 形式调用函数时，如 myurl?callback=?，jQuery 将自动替换后一个“?”为正确的函数名，以执行回调函数。

⑥ text：返回纯文本字符串。

（8）beforeSend：要求为 Function 类型的参数，发送请求前可以修改 XMLHttpRequest 对象的函数，如添加自定义 HTTP 头。在 beforeSend 中，如果返回 false，则可以取消本次 AJAX 请求。XMLHttpRequest 对象是唯一的参数。

```
function(XMLHttpRequest){
 this;   //调用本次 AJAX 请求时传递的 options 参数
}
```

（9）complete：要求为 Function 类型的参数，为请求完成后调用的回调函数（请求成功或失败时均调用）。参数是 XMLHttpRequest 对象和一个描述成功请求类型的字符串。

```
function(XMLHttpRequest, textStatus){
    this;  //调用本次 AJAX 请求时传递的 options 参数
}
```

（10）success：要求为 Function 类型的参数，为请求成功后调用的回调函数。其有两个参数，参数 1 由服务器返回，并根据 dataType 参数处理数据；参数 2 为描述了状态的字符串。

```
function(data, textStatus){ //data 可能是 XMLDoc、JSONObj、HTML、TExt 等
    this;  //调用本次 AJAX 请求时传递的 options 参数
}
```

（11）error：要求为 Function 类型的参数，为请求失败时被调用的函数。该函数有 3 个参数，即 XMLHttpRequest 对象、错误信息、捕获的错误对象(可选)。AJAX 事件函数如下：

```
function(XMLHttpRequest, textStatus, errorThrown){
//通常情况下，textStatus 和 errorThrown 只有其中一个包含信息
    this;   //调用本次 AJAX 请求时传递的 options 参数
}
```

（12）contentType：要求为 String 类型的参数，当发送信息至服务器时，内容编码类型默认为“application/x-www-form-urlencoded”。该默认值适用于大多数应用场合。

（13）dataFilter：要求为 Function 类型的参数，给 AJAX 返回的是原始数据进行预处理后的函数。其提供了 data 和 type 两个参数。data 是 AJAX 返回的原始数据，type 是调用 jQuery.ajax() 时提供的 dataType 参数。函数返回的值将由 jQuery 进一步处理。

```
function(data, type){
    return data;  //返回处理后的数据
}
```

（14）global：要求为 Boolean 类型的参数，默认为 true。其表示是否触发全局 AJAX 事件，设置为 false 时将不会触发全局 AJAX 事件。ajaxStart 或 ajaxStop 可用于控制各种 AJAX 事件。

（15）ifModified：要求为 Boolean 类型的参数，默认为 false。其仅在服务器数据改变时获取新数据。服务器数据改变判断的依据是 Last-Modified 头信息。其默认值是 false，即忽略头信息。

（16）jsonp：要求为 String 类型的参数，在一个 JSONP 请求中重写回调函数的名称。该值用来替代类似“callback=?”这种 GET 或 POST 请求中 URL 参数中的“callback”部分，如{jsonp:'onJsonPLoad'}会将"onJsonPLoad=?"传给服务器。

（17）username：要求为 String 类型的参数，用于响应 HTTP 访问认证请求的用户名。

（18）password：要求为 String 类型的参数，用于响应 HTTP 访问认证请求的密码。

（19）processData：要求为 Boolean 类型的参数，默认为 true。在默认情况下，发送的数据将被转换为对象（从技术角度来讲，其并非字符串）以配合默认内容类型"application/x-www-form-urlencoded"。如果要发送 DOM 信息或者其他不希望转换的信息，则应将其设置为 false。

（20）scriptCharset：要求为 String 类型的参数，只有当请求 dataType 为“jsonp”或者“script”，并且 type 是 GET 时才会用于强制修改字符集(charset)。通常，其在本地和远程的内容编码不同时使用。

9.2 jQuery EasyUI

jQuery EasyUI 是一种基于 jQuery 的用户界面插件集合，它使 Web 开发人员能快速地在流行的 jQuery 核心和 HTML 5 上建立程序页面。在使用 EasyUI 之前必须先声明 UI 控件，有以下两种声明方法。

1. 直接在 HTML 中声明组件

```
<div class="easyui-dialog" style="width:400px;height:200px"
 data-options="title:'My Dialog',collapsible:true,iconCls:'icon-ok',
    onOpen:function(){}">
      dialog content.
</div>
```

2. 编写 JavaScript 代码来创建组件

```
<input id="cc" style="width:200px" />
$('#cc').combobox({
    url: ...,
    required: true,
    valueField: 'id',
    textField: 'text'
});
```

可以从 http://www.jeasyui.com/download/index.php 上下载所需要的 jQuery EasyUI 版本。这里下载的是 jquery-easyui-1.5.4.2.zip，解压缩到自己创建的命名为 jeasyui 文件夹中，其目录层次如图 9-1 所示。使用时，导入相应的文件即可。

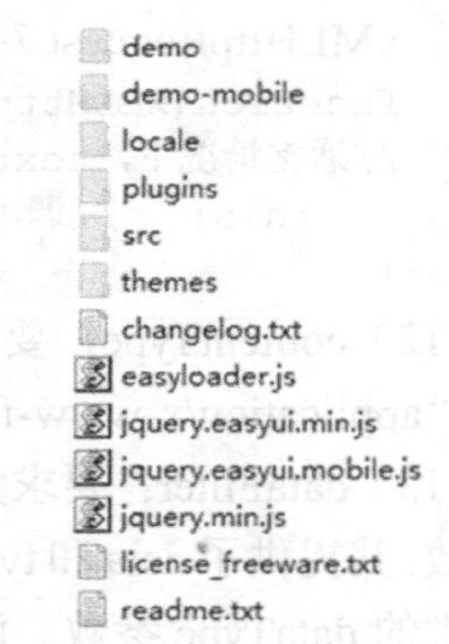

图 9-1 jquery-easyui 压缩包中的文件和文件夹

9.3 jQuery EasyUI 布局

9.3.1 创建边框布局

边框布局（Border Layout）提供了 5 个区域：east、west、north、south、center。

（1）north 区域可以用来显示网站的标题和 Logo。

（2）south 区域可以用来显示版权及一些说明信息。

（3）west 区域可以用来显示导航菜单。

（4）east 区域可以用来显示一些推广的项目。

（5）center 区域可以用来显示主要内容。

为了应用布局，可先确定一个布局容器，然后定义一些区域。布局中至少需要一个 center 区域，以下是一个布局实例（源码包见本书资源 ch9/jeasyui-layout- layout/ layout_demo.html）：

```
<html>
<head>
<title>jQuery EasyUI Demo</title>
<link rel="stylesheet" type="text/css" href="../jeasyui/themes/default/
easyui.css">
<link rel="stylesheet" type="text/css" href="../jeasyui/themes/icon.css">
<script type="text/javascript" src="../jeasyui/jquery.min.js"></script>
<script  type="text/javascript"  src="../jeasyui/jquery.easyui.min.js">
</script>
<script type="text/javascript">
 function showcontent(language){
     $('#content').html('Introduction to ' + language + ' language');
 }
 $(function(){
     showcontent('java');
 });
</script>
</head>
<body>
 <div class="easyui-layout" style="width:400px;height:200px;">
     <div region="west" split="true" title="Navigator" style="width:150px;">
         <p style="padding:5px;margin:0;">Select language:</p>
         <ul>
             <li><a href="javascript:void(0)"
                 onclick="showcontent('java')">Java</a></li>
             <li><a href="javascript:void(0)"
                 onclick="showcontent('cshape')">C#</a></li>
             <li><a href="javascript:void(0)"
                 onclick="showcontent('vb')">VB</a></li>
             <li><a href="javascript:void(0)"
                 onclick="showcontent('python')">Python</a></li>
         </ul>
     </div>
     <div id="content" region="center" title="Language" style="padding:5px;">
     </div>
 </div>
```

```
</body>
</html>
```

其在一个<div>容器中创建了一个边框布局，布局把容器切割为两部分：左边是导航菜单，右边是主要内容。

最后，写一个 onclick 事件处理函数来检索数据，showcontent()函数非常简单：

```
function showcontent(language){
    $('#content').html('Introduction to ' + language + ' language');
}
```

运行效果如图 9-2 所示。（将 jeasyui 文件夹复制至 ch9 文件夹中，注意相应的路径。）

图 9-2　创建边框布局

9.3.2　在面板中创建复杂布局

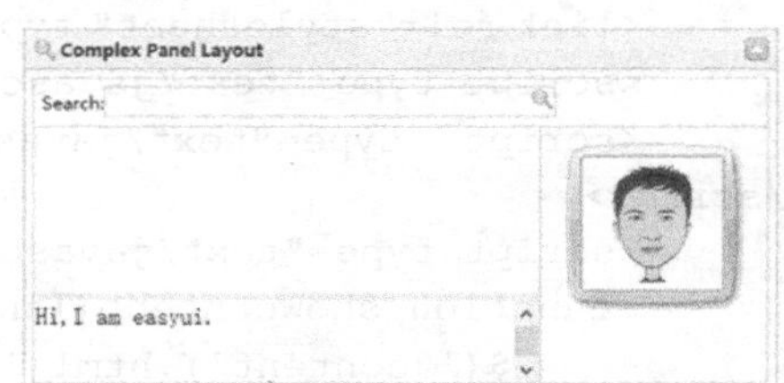

图 9-3　在面板中创建复杂布局

使用面板（Panel）可以创建用于多种用途的自定义布局。在示例中，使用面板和布局插件创建了一个 MSN 消息框，如图 9-3 所示。

在区域面板中使用了多个布局。在消息框的顶部放置一个查询输入框，在右边放置一张人物图片，在中间的区域通过设置 split 属性为 true，把其切割为两部分，允许用户改变区域面板的大小。

示例主要代码如下（源码包参见本书资源 ch9/ jeasyui-layout-panel/ panel_demo.html）：

```
<html>
<head>
 <title>jQuery EasyUI Demo</title>
 <link rel="stylesheet" type="text/css" href="../jeasyui/themes/default/
easyui.css">
 <link rel="stylesheet" type="text/css" href="../jeasyui/themes/icon.css">
 <script type="text/javascript" src="../jeasyui/jquery.min.js"></script>
 <script  type="text/javascript"  src="../jeasyui/jquery.easyui.min.js">
</script>
</head>
<body style="background:#fafafa;">
<h1>Panel</h1>
<div class="easyui-panel" title="Complex Panel Layout" iconCls="icon-search"
collapsible="true" style="padding:5px;width:500px;height:250px;">
 <div class="easyui-layout" fit="true">
     <div region="north" border="false" class="p-search">
        <label>Search:</label><input></input>
     </div>
     <div region="center" border="false">
```

```
                <div class="easyui-layout" fit="true">
                    <div region="east" border="false" class="p-right">
                        <img src="images/msn.gif"/>
                    </div>
                    <div region="center" border="false" style="border:1px solid
#ccc;">
                        <div class="easyui-layout" fit="true">
                            <div region="south" split="true" border="false"
                            style="height:60px;">
                                <textarea style="…">Hi,I am easyui.</textarea>
                            </div>
                            <div region="center" border="false"></div>
                        </div>
                    </div>
                </div>
            </div>
        </div>
    </div>
    </body>
    </html>
```

9.3.3 创建折叠面板

折叠面板（Accordion）包含一系列的面板。所有面板的头部（header）都是可见的，但是一次仅仅显示一个面板的 body 内容。当用户单击面板的头部时，该面板的 body 内容将可见，同时，其他面板的 body 内容将隐藏而不可见。

以下示例创建了 3 个面板，第三个面板包含一个树形菜单。示例主要代码如下（源码包参见本书资源 ch9/ jeasyui-layout-accordion / accordion_demo.html）：

```
    <div class="easyui-accordion" style="width:300px;height:200px;">
        <div title="About Accordion" iconCls="icon-ok" style="…"> … </div>
        <div title="About easyui" iconCls="icon-reload" selected="true" …> …
</div>
        <div title="Tree Menu">
            <ul id="tt1" class="easyui-tree"> … </ul>
        </div>
    </div>
```

创建的折叠面板效果如图 9-4 所示。

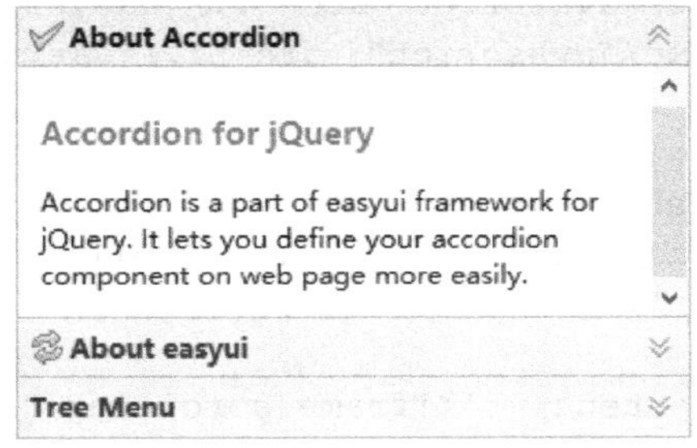

图 9-4　创建折叠面板

9.3.4 创建标签页

标签页（Tabs）有多个可以动态地添加或移除的面板，可以使用 Tabs 来在相同的页面上显示不同的实体。Tabs 一次仅仅显示一个面板，每个面板都有标题、图标和关闭按钮。当 Tabs 被选中时，将显示对应的面板中的内容。

可以使用 HTML 标记创建 Tabs，包含一个 DIV 容器和一些 DIV 面板。示例主要代码如下（源码包参见本书资源 ch9/jeasyui-layout-tabs1/ tabs_demo.html）：

```
<div class="easyui-tabs" style="width:400px;height:100px;">
 <div title="First Tab" style="padding:10px;"> First Tab </div>
 <div title="Second Tab" closable="true" style="padding:10px;"> Second Tab </div>
 <div title="Third Tab" iconCls="icon-reload" closable="true" …> Third Tab </div>
</div>
```

创建的标签页效果如图 9-5 所示。

图 9-5 创建标签页

9.3.5 动态添加标签页

HTML 的<iframe>标签规定了一个内联框架，一个内联框架被用来在当前 HTML 文档中嵌入另一个文档。所有主流浏览器都支持<iframe>标签。

可以使用<iframe>标签动态地添加显示在一个页面上的 Tabs。当单击添加按钮时，一个新的标签页将被添加。如果标签页已经存在，则其将被激活。

示例主要代码如下（源码包参见本书资源 ch9/jeasyui-layout-tabs2/ tabs2_demo.html）：

```
<html>
<head>
 <title>jQuery EasyUI Demo</title>
 <link rel="stylesheet" type="text/css" href="../jeasyui/themes/default/easyui.css">
 <link rel="stylesheet" type="text/css" href="../jeasyui/themes/icon.css">
 <script type="text/javascript" src="../jeasyui/jquery.min.js"></script>
 <script type="text/javascript" src="../jeasyui/jquery.easyui.min.js"></script>
 <script>
     function addTab(title, url){
         if ($('#tt').tabs('exists', title)){
             $('#tt').tabs('select', title);
         } else {
             var content = '<iframe scrolling="auto" frameborder="0"
                         src="'+url+'" style="width:100%;height:100%;">
                         </iframe>';
             $('#tt').tabs('add',{
                 title:title,
```

```
                    content:content,
                    closable:true
                });
            }
        }
    </script>
    </head>
    <body>
     <div style="margin-bottom:10px">
          <a href="#" class="easyui-linkbutton"
             onclick="addTab('google','http://www.google.com')">google</a>
        <a href="#" class="easyui-linkbutton"
             onclick="addTab('jquery','http://jquery.com/')">jquery</a>
        <a href="#" class="easyui-linkbutton"
             onclick="addTab('easyui','http://jeasyui.com/')">easyui</a>
     </div>
     <div id="tt" class="easyui-tabs" style="width:400px;height:250px;">
        <div title="Home"></div>
     </div>
    </body>
    </html>
```

动态添加标签页效果如图 9-6 所示。

图 9-6 动态创建标签页

9.4 jQuery EasyUI 数据网格

9.4.1 转换 HTML 表格为数据网格

以下示例演示如何转换表格（Table）为数据网格（Datagrid）。数据网格的列信息定义在 <thead> 标记中，数据定义在 <tbody> 标记中。这里要确保为所有的数据列设置 field 名称。示例主要代码如下（源码包参见本书资源 ch9/jeasyui-datagrid-datagrid1/ datagrid.html）：

```
    <table id="tt" class="easyui-datagrid" fitColumns="true" style="width:
460px;height:auto;">
          <thead>
              <tr>
                  <th field="name1" width="50">Col 1</th>
                  <th field="name2" width="50">Col 2</th>
                  <th field="name3" width="50">Col 3</th>
                  <th field="name4" width="50">Col 4</th>
```

```
                <th field="name5" width="50">Col 5</th>
            </tr>
        </thead>
        <tbody>
            <tr>
                <td>第 1 列数据 1</td>
                <td>第 2 列数据 1</td>
                <td>第 3 列数据 1</td>
                <td>第 4 列数据 1</td>
                <td>第 5 列数据 1</td>
            </tr>
            <tr>
                <td>Data 1</td>
                <td>Data 2</td>
                <td>Data 3</td>
                <td>Data 4</td>
                <td>Data 5</td>
                <td>Data 6</td>
            </tr>
        </tbody>
    </table>
```

转换 HTML 表格为数据网格的效果如图 9-7 所示。

Col 1	Col 2	Col 3	Col 4	Col 5
第1列数据1	第2列数据1	第3列数据1	第4列数据1	第5列数据1
Data 1	Data 2	Data 3	Data 4	Data 5

图 9-7 转换 HTML 表格为数据网格

9.4.2 取得选中行数据

数据网格组件中有以下两种方法来检索选中行的数据。

（1）getSelected：取得第一个选中行数据，如果没有选中行则返回 null，否则返回记录。

（2）getSelections：取得所有选中行数据，返回元素记录的数组数据。

示例主要代码如下（源码包参见本书资源 ch9/jeasyui-datagrid-datagrid2/ datagrid2_demo.html）：

```
    <html>
    <head>
     <meta http-equiv="Content-Type" content="text/html; charset=UTF-8">
     <title>jQuery EasyUI Demo</title>
     <link rel="stylesheet" type="text/css" href="../jeasyui/themes/default/
easyui.css">
     <link rel="stylesheet" type="text/css" href="../jeasyui/themes/icon.css">
     <script type="text/javascript" src="../jeasyui/jquery.min.js"></script>
     <script  type="text/javascript"  src="../jeasyui/jquery.easyui.min.js">
</script>
     <script>
         function getSelected(){
             var row = $('#tt').datagrid('getSelected');
             if (row){
                 alert('Item ID:'+row.itemid+"\nPrice:"+row.listprice);
```

```
            }
        }
        function getSelections(){
            var ids = [];
            var rows = $('#tt').datagrid('getSelections');
            for(var i=0; i<rows.length; i++){
                ids.push(rows[i].itemid);
            }
            alert(ids.join('\n'));
        }
    </script>
  </head>
  <body>
    <h1>DataGrid</h1>
    <div style="margin-bottom:20px">
        <a href="#" onclick="getSelected()">GetSelected</a>
        <a href="#" onclick="getSelections()">GetSelections</a>
    </div>
    <table id="tt" class="easyui-datagrid" style="width:600px;height:250px"
            url="data/datagrid_data.json"
            title="Load Data" iconCls="icon-save" fitColumns="true">
        <thead>
            <tr>
                <th field="itemid" width="80">Item ID</th>
                <th field="productid" width="80">Product ID</th>
                <th   field="listprice"   width="80"   align="right">List
Price</th>
                <th field="unitcost" width="80" align="right">Unit Cost</th>
                <th field="attr1" width="150">Attribute</th>
                <th field="status" width="60" align="center">Stauts</th>
            </tr>
        </thead>
    </table>
  </body>
  </html>
```

取得选中行数据的效果如图 9-8 所示。

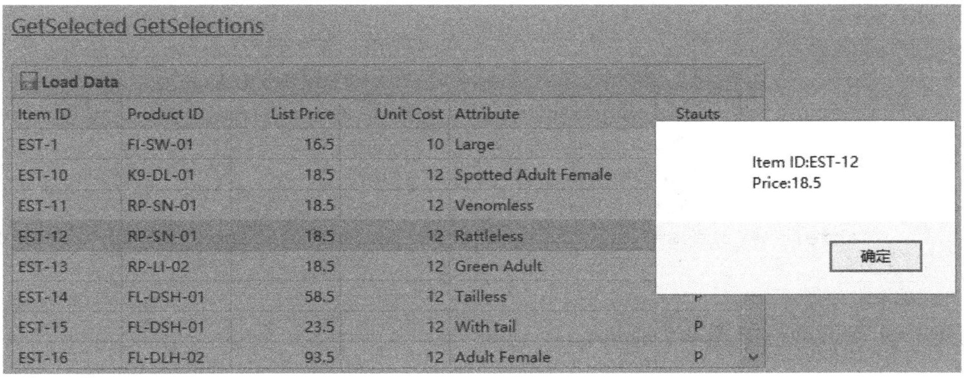

Item ID	Product ID	List Price	Unit Cost	Attribute	Stauts
EST-1	FI-SW-01	16.5	10	Large	
EST-10	K9-DL-01	18.5	12	Spotted Adult Female	
EST-11	RP-SN-01	18.5	12	Venomless	
EST-12	RP-SN-01	18.5	12	Rattleless	
EST-13	RP-LI-02	18.5	12	Green Adult	
EST-14	FL-DSH-01	58.5	12	Tailless	P
EST-15	FL-DSH-01	23.5	12	With tail	P
EST-16	FL-DLH-02	93.5	12	Adult Female	P

图 9-8 取得选中行数据

9.4.3 创建复杂工具栏

数据网格的工具栏（Toolbar）可以包含按钮及其他组件。可以通过一个已存在的 DIV 标签来简单地定义工具栏布局，该 DIV 标签将成为数据网格工具栏的内容。

示例主要代码如下（源码包参见本书资源 ch9/jeasyui-datagrid-datagrid3/ datagrid3_demo.html）：

```
        <table class="easyui-datagrid" style="width:600px;height:250px"
                url="data/datagrid_data.json"
                title="DataGrid - Complex Toolbar" toolbar="#tb"
                singleSelect="true" fitColumns="true">
            <thead>
                <tr>
                    <th field="itemid" width="60">Item ID</th>
                    <th field="productid" width="80">Product ID</th>
                    <th    field="listprice"    align="right"    width="70">List
Price</th>
                    <th field="unitcost" align="right" width="70">Unit Cost</th>
                    <th field="attr1" width="200">Address</th>
                    <th field="status" width="50">Status</th>
                </tr>
            </thead>
        </table>
        <div id="tb" style="padding:5px;height:auto">
         <div style="margin-bottom:5px">
            <a  href="#"class="easyui-linkbutton"  iconCls="icon-add"  plain=
"true"></a>
            <a href="#"  class="easyui-linkbutton"  iconCls="icon-edit"  plain=
"true"></a>
            <a href="#"  class="easyui-linkbutton"  iconCls="icon-save"  plain=
"true"></a>
            <a  href="#"  class="easyui-linkbutton"  iconCls="icon-cut"  plain=
"true"></a>
            <a href="#" class="easyui-linkbutton" iconCls="icon-remove" plain=
"true"></a>
         </div>
         <div>
            Date From: <input class="easyui-datebox" style="width:80px">
            To: <input class="easyui-datebox" style="width:80px">
            Language:
            <input class="easyui-combobox" style="width:100px"
                    url="data/combobox_data.json"
                    valueField="id" textField="text">
            <ahref="#"class="easyui-linkbutton" iconCls="icon-search">Search </a>
         </div>
        </div>
```

创建的复杂工具栏如图 9-9 所示。

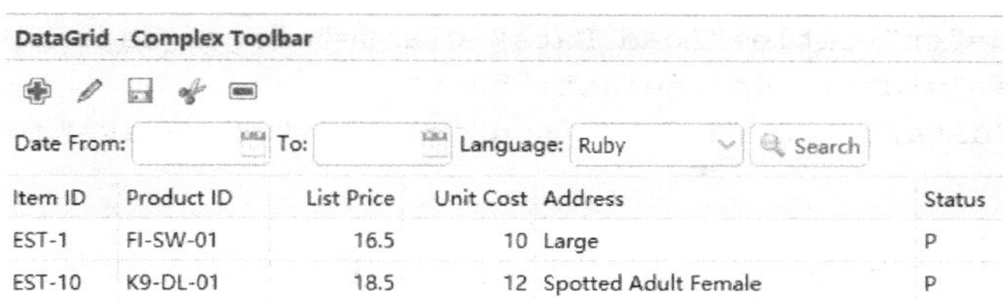

图 9-9 创建复杂工具栏

添加工具栏的示例，还可参考本书资源源码包 ch9/jeasyui-datagrid-datagrid4/ datagrid4_demo.html。

9.4.4 自定义分页

数据网格内置了一个特性很好的分页功能，自定义也比较简单。以下示例将创建一个数据网格，并在分页工具栏中添加一些自定义按钮。

示例主要代码如下（源码包参见本书资源 ch9/jeasyui-datagrid-datagrid5/ datagrid5_demo.html）：

```
<html>
<head>
 …
 <script>
     $(function(){
          var pager = $('#tt').datagrid('getPager');   // 取得数据网格页
          pager.pagination({
               showPageList:false,
               buttons:[{
                    iconCls:'icon-search',
                    handler:function(){
                         alert('search');
                    }
               },{
                    iconCls:'icon-add',
                    handler:function(){
                         alert('add');
                    }
               },{
                    iconCls:'icon-edit',
                    handler:function(){
                         alert('edit');
                    }
               }],
               onBeforeRefresh:function(){
                    alert('before refresh');
                    return true;
               }
          });
     });
 </script>
</head>
<body>
 <h1>DataGrid</h1>
```

```
    <table id="tt" title="Load Data" class="easyui-datagrid"
        style="width:550px;height:250px"
        url="data/datagrid_data.json"iconCls="icon-save" pagination="true">
        <thead>
            <tr>
                <th field="itemid" width="80">Item ID</th>
                <th field="productid" width="80">Product ID</th>
                <th field="listprice"width="80"align="right">ListPrice</th>
                <th field="unitcost" width="80" align="right">Unit Cost</th>
                <th field="attr1" width="100">Attribute</th>
                <th field="status" width="60" align="center">Stauts</th>
            </tr>
        </thead>
    </table>
   </body>
   </html>
```

自定义分页效果如图 9-10 所示。

Load Data

Item ID	Product ID	List Price	Unit Cost	Attribute	Stauts
EST-1	FI-SW-01	16.5	10	Large	P
EST-10	K9-DL-01	18.5	12	Spotted Adult Fe	P
EST-11	RP-SN-01	18.5	12	Venomless	P
EST-12	RP-SN-01	18.5	12	Rattleless	P
EST-13	RP-LI-02	18.5	12	Green Adult	P
EST-14	FL-DSH-01	58.5	12	Tailless	P
EST-15	FL-DSH-01	23.5	12	With tail	P

Page 1 of 3 Displaying 1 to 10 of 28 items

图 9-10 自定义分页

9.5 本章小结

本章主要介绍了 jQuery 和 jQuery EasyUI。

jQuery 是一个高效、精简并且功能丰富的 JavaScript 工具库。它提供的 API 易于使用且兼容众多浏览器，这让诸如 HTML 文档遍历和操作、事件处理、动画和 AJAX 操作更加简单。

jQuery EasyUI 是基于 jQuery 用户界面插件的集合，为一些当前用于交互的 JavaScript 应用提供了必要的功能。jQuery EasyUI 提供了用于创建跨浏览器网页的完整的组件集合，包括功能强大的数据网格、树形表格、面板、下拉列表等。用户可以组合使用这些组件，也可以单独使用其中一个。

EasyUI 支持的两种渲染方式分别为 JavaScript 方式（如$('#p').panel({...})）和 HTML 标记方式（如 class="easyui-panel"）。EasyUI 支持扩展，用户可根据自己的需求扩展控件。

jQuery EasyUI 简单且强大，开发产品时可节省时间和资源。

习题 9

1. 什么是 jQuery？如何使用 jQuery？试编写简单示例加以说明。
2. jQuery 对象和 DOM 对象有什么不同？试编程加以说明。

3．jQuery 选择器有哪些？试编程加以说明。

4.什么是 jQuery AJAX？如何使用 ajax()方法向服务器发出异步请求？试编程加以说明。

5．使用 EasyUI 之前必须先声明 UI 控件，有哪两种声明方法？

6．如何在 JSP 页面中创建 jQuery EasyUI 复杂布局？试编程加以说明。

7．如何在 JSP 页面中使用 jQuery EasyUI 动态添加标签页？试编程加以说明。

8．如何创建并使用 jQuery EasyUI 数据网格？试编程加以说明。

9．请网上查阅 EasyUI 相关资料，完成以下任务。

（1）如何创建 jQuery EasyUI 窗口？试编程加以说明。

（2）如何创建并使用 jQuery EasyUI 树形菜单？试编程加以说明。

（3）如何创建并使用 jQuery EasyUI 表单？试编程加以说明。

第二部分

MyBatis

第 10 章　MyBatis 入门

本章导读

MyBatis 并没有将 Java 对象与数据库表关联起来，而是将 Java 方法与 SQL 语句关联起来。MyBatis 可以使用 XML 或注解进行配置或映射，MyBatis 通过将参数映射到配置的 SQL 中形成最终执行的 SQL 语句，最后将执行 SQL 的结果映射成 Java 对象并返回。本章主要内容有：（1）从 JDBC 到 MyBatis；（2）第一个 MyBatis 示例（基于 Maven）；（3）MyBatis 框架原理；（4）MyBatis 核心组件的生命周期。

10.1　从 JDBC 到 MyBatis

JDBC（Java 数据库连接）是一种用于执行 SQL 语句的 Java API，可以为多种关系数据库提供统一访问，它由一组用 Java 语言编写的类和接口组成。JDBC 提供了一种基准，据此可以构建更高级的工具和接口，使数据库开发人员能够编写数据库应用程序。

下面来简单回顾一下 JDBC 是如何工作的。

（1）加载 JDBC 驱动程序：

```
try{
    Class.forName("org.postgresql.Driver") ;  //加载 PostgreSQL 的驱动类
}catch(ClassNotFoundException e){
    System.out.println("找不到驱动程序类，加载驱动失败！");
    e.printStackTrace() ;
}
```

（2）建立并获取数据库连接：

```
String url = "jdbc:postgresql://127.0.0.1:5432/mybatis_db ";
String usr = "postgres";
String psd = "1";
conn = DriverManager.getConnection(url, usr, psd);
```

（3）创建 Statement 对象：

```
String stringsql = "INSERT INTO t_user (name,gender,email) VALUES (?,?,?)";
PreparedStatement ps = conn.prepareStatement(stringsql);
```

（4）设置 SQL 语句的传入参数：

```
ps.setString(1,name);
ps.setString(2,gender);
ps.setString(3,email);
```

（5）执行 SQL 语句并获得查询结果：

```
int rs = ps.executeUpdate();
```

（6）对查询结果进行转换处理并将处理结果返回：

```
if(rs>0){
out.println("插入成功");
}else
out.println("插入失败");
```

（7）释放连接资源：

```
ps.close();
conn.close();
```

上述7个步骤是JDBC的基本步骤，也是不可或缺的步骤。

如果进行过JDBC的相关开发，从JDBC的实现过程中会发现一些问题。其主要集中在以下几点。

（1）数据库连接过于频繁。每次都需要配置数据源、进行数据连接，结束后还要关闭资源，导致重复代码较多，频繁进行这些操作也会降低一些效率。

（2）SQL语句过于零散，不利于管理。使用JDBC操作数据库时，SQL语句基本上都散落在各个Java类中，这样有3个不足之处：第一，可读性很差，不利于维护以及做性能调优；第二，改动Java代码需要重新编译、打包部署；第三，不利于取出SQL并在数据库客户端执行（取出后还要删掉其中的Java代码，编写好的SQL语句写好后还要通过+号在Java中进行拼凑）。

（3）SQL语句的参数不够灵活。很多情况下，可以通过在SQL语句中设置占位符来达到使用传入参数的目的，这种方式本身有一定的局限性，它是按照一定顺序传入参数的，要与占位符一一匹配。但如果传入的参数是不确定的（如列表查询，根据用户填写的查询条件不同，传入查询的参数也是不同的，有时是一个参数，有时可能是3个参数），那么就要在后台代码中根据请求的传入参数去拼凑相应的SQL语句，这样还是要在Java代码中写SQL语句。

（4）不支持结果映射和结果缓存。其没有映射，每次执行后，必须将结果取出，这个结果如果是常用的一个集合，通常会将其封装，但是由于其没有自动映射的功能，每次都需要手工将结果集映射到实体（Bean）中，比较费时。此外，结果是一次性的，下次再使用或者查询时，需要重新执行SQL，那么如果能够缓存结果，就可以提升SQL查询的性能。

（5）SQL重用度不高。由于SQL语句都有不同的配置，当几个功能的SQL语句差不多的时候，就没有办法实现重用，当需要修改这些SQL时，还需要修改每一处需要修改的SQL，非常不利于维护。

（6）不安全。JDBC中的SQL语句有的地方需要使用拼接字符串的方式实现，这是非常危险的，容易导致SQL注入，如果不做处理，则安全性是非常差的。

这些都是JDBC的弊端，当然，JDBC由于是直接编译实现SQL语句的运行，其执行效率是很高的。在实际工作中，我们很少使用JDBC进行编程，于是提出了对象关系映射（Object Relational Mapping，ORM）。ORM模式是一种用于解决面向对象与关系数据库存在的互不匹配的现象的技术。简单地说，ORM通过使用描述对象和数据库之间映射的元数据，将程序中的对象自动持久化到关系数据库中。

MyBatis和Hibernate都是目前较为流行的ORM框架。

Hibernate是一个面向Java环境的ORM工具，它将对象模型表示的简单的Java对象（Plain Ordinary Java Object，POJO）映射到基于SQL的关系数据结构中。POJO作为普通的JavaBean对象，并不具备持久化操作能力，Hibernate通过定义映射文件将POJO映射到相应的数据库，获得持久化对象PO，通过操作PO，实现对数据库的增、删、改、查操作。

MyBatis 本是Apache的一个开源项目iBatis，2010年，其由Apache Software Foundation迁移到了Google Code，并且改名为MyBatis，2013年11月迁移到Github。 MyBatis官方网址为 https://github.com./mybatis，在其中可以看到MyBatis 的多个子项目。

MyBatis 和 Hibernate 均属于 ORM 框架，用于实现持久化类与数据库之间的相互映射，通过操作持久化类，从而实现对数据库的操作。它们的基本实现过程是相似的，首先，其由配置文件获取数据源；其次，采用工厂模式（即 MyBatis 的 SqlSessionFactory，Hibernate 的 SessionFactory），获取与之交互的接口（即 MyBatis 的 SqlSession，Hibernate 的 Session），进而获得操作持久化类的接口对象（即 MyBatis 的 Mapper，Hibernate 的 Transaction 和 Query）。

Hibernate 的 SQL 大多是自动生成的，无法直接维护 SQL，虽然配备了 Hibernate 查询语言（HQL），但 HQL 在功能上还是不如 SQL 丰富。此外，Hibernate 查询数据表会直接将所有字段查询出来，这样会影响查询效率，而如果 Hibernate 采用自己编写的 SQL 来查询，又破坏了 Hibernate 开发自动生成 SQL 的特点，因此在这一点上，手动编写 SQL 的 MyBatis 更为灵活，可以按需指定查询字段，在查询效率上更优。

10.2　第一个 MyBatis 示例

10.2.1　创建 Maven 项目

1. 创建工程 mybatis_first_demo

在 Eclipse 中创建 Maven Project，如图 10-1 所示，选择 maven-archetype-quickstart，在“Group ID”文本框中输入“com.mialab”，在“Artifact ID”文本框输入“mybatis_first_demo”。（此处所用的 Eclipse 版本详见本书教学资源 tools 文件夹中的 eclipse-jee-oxygen-R-win32.zip。）

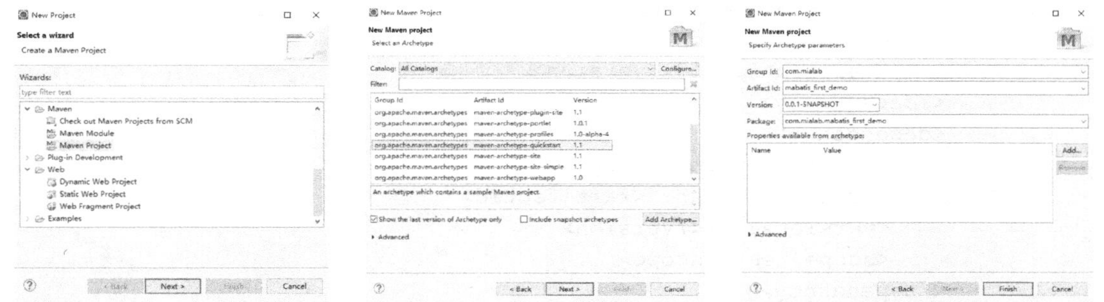

图 10-1　在 Eclipse 中创建 Maven 项目 mybatis_first_demo

2. pom.xml

此 Maven 项目 mybatis_first_demo 中的 pom.xml 内容（详见本书教学资源包）如下：

```
<project xmlns="http://maven.apache.org/POM/4.0.0"
 xmlns:xsi="http://www.w3.org/2001/XMLSchema-instance"
 xsi:schemaLocation="http://maven.apache.org/POM/4.0.0
 http://maven.apache.org/xsd/maven-4.0.0.xsd">
 <modelVersion>4.0.0</modelVersion>
 <groupId>com.mialab</groupId>
 <artifactId>mybatis_first_demo</artifactId>
 <version>0.0.1-SNAPSHOT</version>
 <packaging>jar</packaging>
 <name>mybatis_first_demo</name>
 <url>http://maven.apache.org</url>
```

```
    <properties>
        <java.version>1.8</java.version>
        <project.build.sourceEncoding>UTF-8</project.build.sourceEncoding>
    </properties>
    <dependencies>
        <dependency>
            <groupId>org.mybatis</groupId>
            <artifactId>mybatis</artifactId>
            <version>3.4.4</version>
        </dependency>
        <dependency>
            <groupId>mysql</groupId>
            <artifactId>mysql-connector-java</artifactId>
            <version>5.1.39</version>
        </dependency>
        <dependency>
            <groupId>org.slf4j</groupId>
            <artifactId>slf4j-api</artifactId>
            <version>1.7.12</version>
        </dependency>
        <dependency>
            <groupId>org.slf4j</groupId>
            <artifactId>slf4j-log4j12</artifactId>
            <version>1.7.12</version>
        </dependency>
        <dependency>
            <groupId>log4j</groupId>
            <artifactId>log4j</artifactId>
            <version>1.2.17</version>
        </dependency>
        <dependency>
            <groupId>junit</groupId>
            <artifactId>junit</artifactId>
            <version>4.12</version>
            <scope>test</scope>
        </dependency>
    </dependencies>
    <build>
        <plugins>
            <plugin>
                <artifactId>maven-compiler-plugin</artifactId>
                <version>3.6.2</version>
                <configuration>
                    <source>${java.version}</source>
                    <target>${java.version}</target>
                </configuration>
            </plugin>
        </plugins>
    </build>
</project>
```

POM 用于描述项目如何构建、声明项目依赖等。

10.2.2　准备数据

可以使用 MySQL 客户端工具 Navicat 创建数据库 stu，在数据库 stu 中创建表 student，并插入两条数据，代码如下：

```
DROP TABLE IF EXISTS `student`;
CREATE TABLE `student` (
  `sno` varchar(100) NOT NULL DEFAULT '',
  `name` varchar(100) NOT NULL,
  `sex` varchar(8) DEFAULT NULL,
  `age` int(3) DEFAULT NULL,
  `dept_no` varchar(60) DEFAULT NULL
) ENGINE=InnoDB DEFAULT CHARSET=utf8;
INSERT INTO `student` VALUES ('20171508', '李勇', '男', '20', '2601');
INSERT INTO `student` VALUES ('20171509', '刘娟', '女', '19', '2602');
```

10.2.3　MyBatis 配置

在 mybatis_first_demo 工程中 src 文件夹的 main 目录下，创建 resources 子文件夹，再在 src/main/resources 中创建 mybatis-config.xml 文件，如图 10-2 所示。

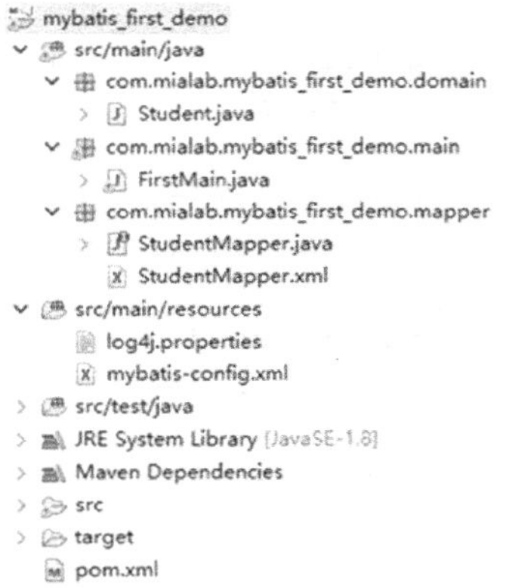

图 10-2　mybatis_first_demo 工程目录结构

mybatis-config.xml 文件内容如下：

```
<?xml version="1.0" encoding="UTF-8" ?>
<!DOCTYPE configuration
 PUBLIC "-//mybatis.org//DTD Config 3.0//EN"
 "http://mybatis.org/dtd/mybatis-3-config.dtd">
<configuration>
 <settings>
     <setting name="logImpl" value="LOG4J" />
 </settings>
 <environments default=" development ">
     <environment id=" development ">
         <transactionManager type="JDBC" />
         <dataSource type="POOLED">
             <property name="driver" value="com.mysql.jdbc.Driver" />
             <property name="url" value="jdbc:mysql://127.0.0.1:3306/stu" />
             <property name="username" value="root" />
             <property name="password" value="1" />
         </dataSource>
     </environment>
 </environments>
 <mappers>
     <mapper
resource="com/mialab/mybatis_first_demo/mapper/StudentMapper.xml" />
 </mappers>
</configuration>
```

MyBatis 配置文件的根元素是<configuration>，子元素<settings>中的 logImpl 属性配置指定使用 Log4j 输出日志，子元素<environments>用来配置连接的数据库，子元素<mappers>中配置了一个包含完整类路径的 StudentMapper.xml，这是 MyBatis 的 SQL 语句映射文件。

10.2.4 创建实体类

根据 stu 数据库中的表 student 来创建实体类 Student。具体操作如下：如图 10-2 所示，右击 src/main/java，从弹出的快捷菜单中选择“New→Package”选项，输入包名为 com.mialab.mybatis_first_demo.domain，在 domain 包下创建类 Student.java，代码如下：

```
package com.mialab.mybatis_first_demo.domain;
public class Student {
 private String sno;
 private String name;
 private String sex;
 private int age;
 private String dept_no;
 public String getSno() {
     return sno;
 }
 public void setSno(String sno) {
     this.sno = sno;
 }
 public String getName() {
     return name;
 }
 public void setName(String name) {
     this.name = name;
 }
 public String getSex() {
     return sex;
 }
 public void setSex(String sex) {
     this.sex = sex;
 }
 public int getAge() {
     return age;
 }
 public void setAge(int age) {
     this.age = age;
 }
 public String getDept_no() {
     return dept_no;
 }
 public void setDept_no(String dept_no) {
     this.dept_no = dept_no;
 }
 @Override
 public String toString() {
     return "Student [sno=" + sno + ", name=" + name + ", sex=" + sex + ",
         age=" + age + ", dept_no=" + dept_no + "]";
 }
```

```
    }
```

10.2.5 创建映射接口和 SQL 映射文件

1. 创建映射接口

如图 10-2 所示，在 src/main/java 中创建 Package“com.mialab.mybatis_first_demo.mapper”，再在 mapper 包中创建接口 StudentMapper.java，具体代码如下：

```
package com.mialab.mybatis_first_demo.mapper;
import com.mialab.mybatis_first_demo.domain.Student;
public interface StudentMapper {
 public Student getStudent(String sno);
}
```

注意：接口的方法要和 XML 映射文件中的 id 保持一致。

2. 创建 SQL 映射文件

在包 com.mialab.mybatis_first_demo.mapper 下创建 StudentMapper.xml 文件。

```
<?xml version="1.0" encoding="UTF-8" ?>
<!DOCTYPE mapper
PUBLIC "-//mybatis.org//DTD Mapper 3.0//EN"
"http://mybatis.org/dtd/mybatis-3-mapper.dtd">
<mapper namespace="com.mialab.mybatis_first_demo.mapper.StudentMapper">
 <select id="getStudent" resultType="com.mialab.mybatis_first_demo.domain.
Student">
     select * from student where sno = #{sno}
 </select>
</mapper>
```

此 XML 映射文件的根元素是<mapper>，其属性 **namespace 值为 StudentMapper 接口的完整类路径**，如图 10-2 所示。

10.2.6 配置 Log4j

如图 10-2 所示，在 src/main/resources 中创建 log4j.properties 文件，其内容如下：

```
#全局的日志输出
log4j.rootLogger= TRACE, stdout
#MyBatis 日志配置
com.mialab.mybatis_first_demo.mapper.StudentMapper=TRACE
#控制台输出配置
log4j.appender.stdout=org.apache.log4j.ConsoleAppender
log4j.appender.stdout.layout=org.apache.log4j.PatternLayout
log4j.appender.stdout.layout.ConversionPattern=%5p [%t] - %m%n
```

使用 log4j.properties 配置文件的目的是方便查看控制台输出，以易于调试。

10.2.7 测试

如图 10-2 所示，在 src/main/java 中创建 Package “com.mialab.mybatis_first_demo.main”。在此包中创建测试类 FirstMain.java，其主要代码如下：

```
    public class FirstMain {
     private static SqlSessionFactory sqlSessionFactory;
     public static void main(String[] args) {
         SqlSession session = null;
         try {
             // 读取 mybatis-config.xml 文件
             InputStream
inputStream=Resources.getResourceAsStream("mybatis-config.xml");
             // 初始化 MyBatis，创建 SqlSessionFactory 类的实例
             sqlSessionFactory = new SqlSessionFactoryBuilder().build(input
Stream);
             // 创建 Session 实例
             session = sqlSessionFactory.openSession();
             StudentMapper mapper = session.getMapper(StudentMapper.class);
             Student student = mapper.getStudent("20171509");
             System.out.println(student);
         } catch (Exception e) {
             e.printStackTrace();
         } finally {
             if(session!=null)
                 session.close();
         }
     }
    }
```

如图 10-2 所示，右击 com.mialab.mybatis_first_demo.main 包下的 FirstMain.java，在弹出的快捷菜单中选择“Run As → Java Application”选项，在控制台上可以得到以下结果：

```
Student [sno=20171509, name=刘娟, sex=女, age=19, dept_no=2602]
```

如果在运行 FirstMain 类的 main 方法时遇到问题，可右击 mybatis_first_demo 工程，选择“Project→Clean...”选项，弹出如图 10-3 所示的对话框，选中“mybatis_first_demo”复选框，再单击“Clean”按钮即可。

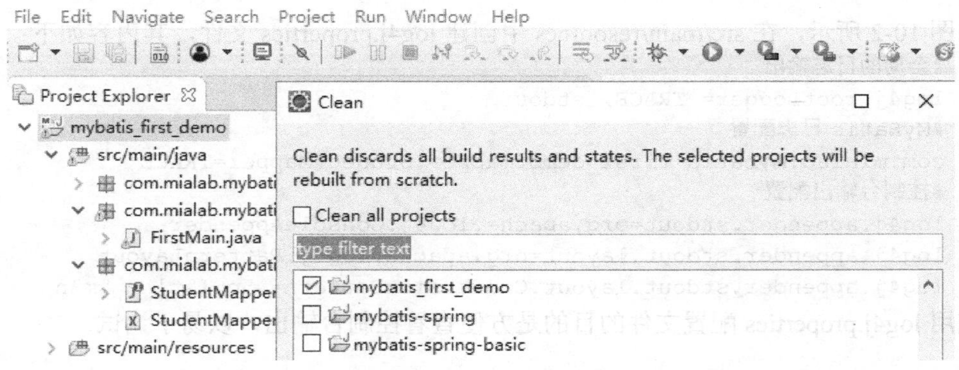

图 10-3 清理 mybatis_first_demo 工程

10.3 MyBatis 框架原理

与 Hibernate 框架相比，MyBatis 学习成本相对较低。在 MyBatis 中，SQL 语句是单独存

放在 XML 文件中的，这样使得 SQL 语句的修改和优化比较方便，使用 MyBatis 框架也变得较为灵活，因而，MyBatis 框架可适用于需求变化较多的项目。使用 MyBatis 框架可以让程序员集中精力于 SQL 语句的开发上。当前互联网电商项目多使用 MyBatis 作为持久层框架，这样不仅增强了灵活性，还可以提高数据库访问的速度。

10.3.1 MyBatis 整体架构

MyBatis 整体架构可分为以下 3 层：接口层、核心处理层和基础支撑层。

（1）接口层：其核心是 SqlSession 接口，该接口中定义了 MyBatis 暴露给应用程序调用的 API，也就是上层应用与 MyBatis 交互的桥梁。接口层在接收到调用请求时，会调用核心处理层的相应模块来完成具体的数据库操作。

（2）核心处理层：在核心处理层中实现了 MyBatis 的核心处理流程，其中包括 MyBatis 的初始化以及完成一次数据库操作涉及的全部流程。核心处理层实现的主要功能有配置解析、SQL 解析、参数映射、SQL 执行、结果集映射等。

（3）基础支撑层：包含整个 MyBatis 的基础模块，为核心处理层的功能提供了良好的支撑。其主要模块有反射模块、类型转换模块、日志模块、事务管理模块、缓存模块、解析器模块、资源加载模块、数据源模块等。

10.3.2 MyBatis 运行原理

每一个 MyBatis 应用都以一个 SqlSessionFactory 实例为中心，SqlSessionFactory 的生命周期应存在于整个 MyBatis 应用，与此 MyBatis 应用共存亡。SqlSessionFactory 的作用是创建 SqlSession 接口对象。SqlSessionFactory 实例可以通过 SqlSessionFactoryBuilder 类的 build 方法来获得。具体来说，在 SqlSessionFactoryBuilder 类的 build 方法中，通过解析 XML 配置文件或者只使用代码（本质上都是先构建 Configuration 对象），来构建 SqlSessionFactory 实例，这里使用了建造者模式（Builder Pattern）。在前面的示例 mybatis_first_demo 工程中，是通过 XML 文件（配置文件和映射文件）来构建 SqlSessionFactory 对象的，而以下的代码是不借助 XML 文件而只使用代码来创建 SqlSessionFactory 的实例。

```
    … // 这里是关于 DataSource 对象的构建，此处代码略
    TransactionFactory transactionFactory = new JdbcTransactionFactory();
    Environment environment = new Environment("development", transaction
Factory,dataSource);
    Configuration configuration = new Configuration(environment);
    configuration.addMapper(StudentMapper.class);
    SqlSessionFactory sqlSessionFactory = new SqlSessionFactoryBuilder().
build(configuration);
```

MyBatis 使用 SqlSession 对象来封装对数据库的一次会话访问，通过 SqlSession 对象实现事务的控制和数据查询。SqlSession 包含了执行 SQL 所需要的所有方法，可以通过 SqlSession 实例直接运行映射的 SQL 语句，完成对数据的增、删、改、查和事务提交等（其中，增、删、改操作要执行 commit 操作），事务提交后，关闭 SqlSession。

MyBatis 的整个执行流程如图 10-4 所示。

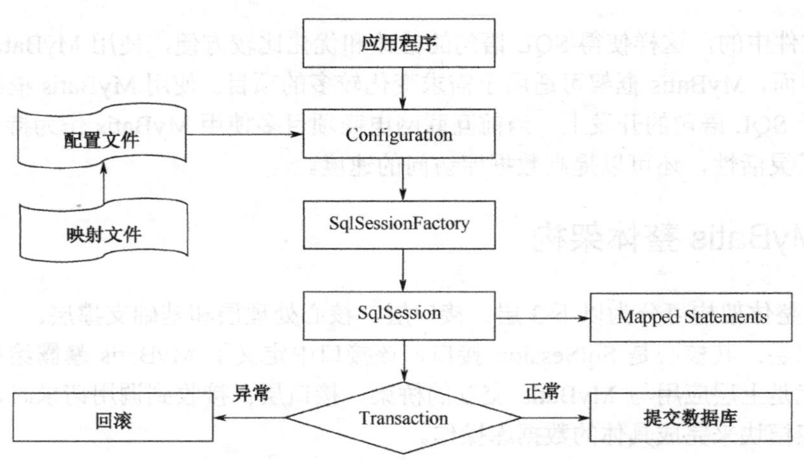

图 10-4 MyBatis 的执行流程

以下是添加 student 表中一条记录的示例代码，用以说明 MyBatis 的执行流程。这里为了简单起见，没有在 insertStudentTest()方法中处理异常（捕捉异常），而是在定义 insertStudentTest()方法时抛出异常，表明异常处理是由调用 insertStudentTest()方法的方法来处理的。

```
public void insertUserTest() throws Exception {
    //① 读取配置文件
    InputStream inputStream = Resources.getResourceAsStream("mybatis-config.
xml");
    //② 根据配置文件创建 SqlSessionFactory
    SqlSessionFactory sqlSessionFactory = new SqlSessionFactoryBuilder().
build(inputStream);
    //③ 通过 SqlSessionFactory 获得 SqlSession 接口对象
    //获得会话，事务处理开始
    SqlSession sqlSession = sqlSessionFactory.openSession();
    //④ 通过 SqlSession 接口对象获得 StudentMapper 实例
    StudentMapper mapper = session.getMapper(StudentMapper.class);
    //这里设置的 Student 对象作为执行的 SQL 语句的参数
    //StudentMapper 接口中声明的方法：public int addStudent(Student student);
    // addStudent(Student student)方法映射相应的 SQL 语句
    Student student = new Student();
    student.setSno("20171622");
    student.setName("李白");
    student.setAge(88);
    student.setSex("男");
    student.setDept_no("2609");
    //⑤ StudentMapper 实例执行映射的 SQL 语句，并返回映射结果
    //这里添加表记录成功，返回影响的记录数
    mapper.addStudent(student);
    //切记：增、删、改操作时，要执行 commit 操作
    sqlSession.commit();
    //⑥ 关闭 SqlSession
    sqlSession.close();
}
```

需要强调的是，SqlSession 的 getMapper 方法是联系应用程序和 MyBatis 的纽带，应用程序访问 getMapper 时，MyBatis 会根据传入的接口类型和对应的 XML 配置文件生成一个代理

对象，这个代理对象就称为 Mapper 对象。应用程序获得 Mapper 对象后，就通过它来访问数据库。

10.4 MyBatis 核心组件的生命周期

10.4.1 SqlSessionFactoryBuilder

一旦通过 SqlSessionFactoryBuilder 创建了 SqlSessionFactory，SqlSessionFactoryBuilder 就不需要存在了。因此，SqlSessionFactoryBuilder 实例的最佳生命周期是只存在于创建 SqlSessionFactory 方法中（即本地方法变量）。

10.4.2 SqlSessionFactory

SqlSessionFactory 实例一旦被创建，应该在开发者的应用程序执行期间都存在。倘若 SqlSessionFactory 实例在应用程序运行期间被重复创建多次，这样的代码会被嗅出“腐化软件的气味”。关于 SqlSessionFactory 实例的创建，可以考虑使用单例模式、静态单例模式或者依赖注入。

10.4.3 SqlSession

SqlSession 是一个会话，相当于 JDBC 的一个 Connection 对象，它的生命周期应该是在请求数据库处理事务的过程中。每个线程都应该有它自己的 SqlSession 实例。SqlSession 的实例不能共享使用，它也是线程不安全的。因此，其最佳的存在范围是请求或方法范围。绝对不能将 SqlSession 实例的引用放在一个类的静态字段甚至是实例字段中，也不能将 SqlSession 实例的引用放在任何类型的管理范围中，如 Servlet 框架的 HttpSession 对象中。如果现在正使用某种 Web 框架，则要考虑将 SqlSession 放在一个和 HTTP 请求对象相似的范围内。换句话说，收到 HTTP 请求后，开发者可以打开一个 SqlSession，但返回响应后就要关闭它。关闭 Session 很重要，应该确保使用 finally 块来关闭。

10.4.4 Mapper Instances

映射器是用来绑定映射语句的接口。映射器接口对象可以从 SqlSession 中获得。从技术上来说，当被请求时，任意映射器对象的生存最大范围与 SqlSession 对象是相同的。不管怎样，映射器对象的生存最佳范围是方法范围。也就是说，它们应该在使用它们的方法中被请求，然后就抛弃掉。例如：

```
SqlSession session = sqlSessionFactory.openSession();
try {
 BlogMapper mapper = session.getMapper(BlogMapper.class);
 // do work
} finally {
 session.close();
}
```

10.5 本章小结

MyBatis 是支持普通 SQL 查询、存储过程和高级映射的优秀持久层框架。MyBatis 消除了几乎所有的 JDBC 代码和参数的手工设置，以及对结果集的检索。MyBatis 可以使用简单的 XML 或注解进行配置和原始映射，利用接口（映射成 SQL 语句）将 Java 的 POJO 映射成数据库中的记录。

由于 MyBatis 具有高度灵活、可优化、已维护等特点，目前已成为大型移动互联网项目的首选框架。相对于 Hibernate 等“全自动”ORM 机制而言，MyBatis 以 SQL 开发的工作量和数据库移植性上的让步，为系统设计提供了更大的自由空间。

习题 10

1. 与 JDBC 相比，MyBatis 的优点有哪些？与 Hibernate 相比呢？
2. 如何使用 MyBatis 框架查询数据库中的记录？试编程加以说明。
3. MyBatis 整体架构是怎样的？请加以说明。
4. MyBatis 运行原理是怎样的？请加以说明。
5. MyBatis 核心组件的生命周期是怎样的？

第 11 章　配置和映射

本章导读

MyBatis 的真正强大之外在于它的映射语句，这也是它的魅力所在。由于它的异常强大，映射器的 XML 文件就显得相对简单。如果将其与具有相同功能的 JDBC 代码进行对比，会立即发现其省掉了将近 95%的代码。MyBatis 就是针对 SQL 构建的，并且比普通的方法效果更好。本章主要内容有：（1）实现表数据的增、删、改、查；（2）MyBatis 主配置文件；（3）XML 映射文件；（4）高级结果映射。

11.1　示例：实现表数据的增、删、改、查

1．创建工程 mybatis_DML_demo

在 Eclipse 中创建 Maven Project，如图 11-1 所示，选择 maven-archetype-quickstart，在“Group ID”选项中输入“com.mialab”，在“Artifact ID”选项输入“mybatis_DML_demo”。

最终完成的 mybatis_DML_demo 工程目录和 student 初始表数据如图 11-1 和图 11-2 所示。

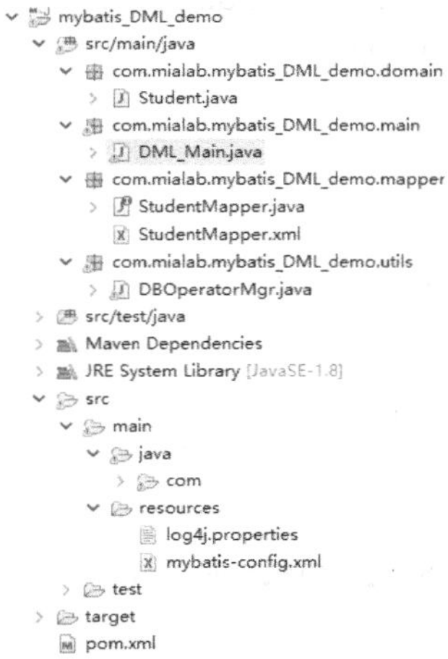

```
mybatis_DML_demo
  src/main/java
    com.mialab.mybatis_DML_demo.domain
      Student.java
    com.mialab.mybatis_DML_demo.main
      DML_Main.java
    com.mialab.mybatis_DML_demo.mapper
      StudentMapper.java
      StudentMapper.xml
    com.mialab.mybatis_DML_demo.utils
      DBOperatorMgr.java
  src/test/java
  Maven Dependencies
  JRE System Library [JavaSE-1.8]
  src
    main
      java
        com
      resources
        log4j.properties
        mybatis-config.xml
    test
  target
  pom.xml
```

图 11-1　mybatis_ DML _demo 工程目录结构

sno	name	sex	age	dept_no
20171508	李勇	男	20	2601
20171509	刘娟	女	19	2602
20171622	李白	男	88	2609

图 11-2　student 表中的数据

mybatis-config.xml、log4j.properties 创建及配置、POJO 类的创建及实现代码都类似以前示例，这里不再赘述。详见本书教学资源包中的 mybatis_ DML _demo 工程。

2．创建数据库操作的工具类：DBOperatorMgr.java

在 src/main/java 中创建 Package“com.mialab.mybatis_DML_demo.utils”，如图 11-1 所示。在 utils 包中创建数据库操作工具类 DBOperatorMgr.java，具体代码如下。

```
public class DBOperatorMgr {
 static Logger logger = Logger.getLogger(DBOperatorMgr.class.getName());
 private static DBOperatorMgr dbMgr;
 private SqlSessionFactory sqlSessionFactory;
 private DBOperatorMgr() {
```

```
        String resource = "mybatis-config.xml";
        InputStream inputStream;
        try {
            inputStream = Resources.getResourceAsStream(resource);
            sqlSessionFactory = new SqlSessionFactoryBuilder().build(input
Stream);
        } catch (Exception e) {
            logger.error(e.toString());
        }
    }

    public static DBOperatorMgr getInstance() {
        if (dbMgr == null) {
            dbMgr = new DBOperatorMgr();
        }
        return dbMgr;
    }

    public SqlSessionFactory getSqlSessionFactory() {
        return sqlSessionFactory;
    }
}
```

3．创建映射接口

在 src/main/java 中创建 Package“com.mialab.mybatis_DML_demo.mapper”，如图 11-1 所示。在 mapper 包中创建接口 StudentMapper.java，具体代码如下。

```
public interface StudentMapper {
  public Student getStudent(String sno);
  public int addStudent(Student student);
  public List<Student> getSudentAll();
  public int updateStudent(Student student);
  public int deleteStudent(String sno);
}
```

注意：接口的方法要和 XML 映射文件的 id 保持一致。

4．创建 XML 映射文件

在包 com.mialab.mybatis_DML_demo.mapper 中创建 StudentMapper.xml 文件。

```
<?xml version="1.0" encoding="UTF-8" ?>
<!DOCTYPE mapper
PUBLIC "-//mybatis.org//DTD Mapper 3.0//EN"
"http://mybatis.org/dtd/mybatis-3-mapper.dtd">
<mapper namespace="com.mialab.mybatis_DML_demo.mapper.StudentMapper">
  <select id="getStudent" resultType=
          "com.mialab.mybatis_DML_demo.domain.Student">
      select * from student where sno = #{sno}
  </select>
  <insert id="addStudent" parameterType="student">
      insert into
      student(sno,name,sex,age,dept_no)
      values(#{sno},#{name},#{sex},#{age},#{dept_no})
  </insert>
```

```
<resultMap id="studentResultMap" type="student">
    <id property="sno" column="sno" />
    <result property="name" column="name" />
    <result property="sex" column="sex" />
    <result property="age" column="age" />
    <result property="dept_no" column="dept_no" />
</resultMap>
<select id="getSudentAll" resultMap="studentResultMap">
    select * from student
</select>

<update id="updateStudent" parameterType="student">
    update student set name
    = #{name}, sex = #{sex}, age = #{age}, dept_no = #{dept_no}
    where sno =
    #{sno}
</update>

<delete id="deleteStudent" parameterType="String">
    delete from student
    where sno = #{sno}
</delete>
</mapper>
```

此 XML 映射文件的根元素是<mapper>，其属性 namespace 值为 StudentMapper 接口的完整类路径，如图 11-1 所示。这里要注意的是，mybatis-config.xml 文件中有设置“student”别名的内容，具体如下：

```
<!-- 设置别名 -->
<typeAliases>
    <typeAlias alias="student"
        type="com.mialab.mybatis_DML_demo.domain.Student" />
</typeAliases>
```

5. 测试

如图 11-1 所示，在 src/main/java 中创建 Package“com.mialab.mybatis_DML_demo.main”。在此包中创建测试类 DML_Main.java，主要代码如下：

```
public class DML_Main {
 public static void main(String[] args) {
    //testInsert();
    testSelectAll();
    //testSelect("20171509");
    //testUpdate();
    //testDelete("20171622");
 }

 private static void testDelete(String sno) {
    Logger log = Logger.getLogger(DML_Main.class);
    SqlSession session = null;
    try {
        session                                                    =
```

```
DBOperatorMgr.getInstance().getSqlSessionFactory().openSession();
                StudentMapper mapper = session.getMapper(StudentMapper.class);

                mapper.deleteStudent(sno);
                session.commit();
            } catch(Exception ex) {
                session.rollback();
                ex.printStackTrace();
            } finally {
                if (session != null) {
                    session.close();
                }
            }
        }

        private static void testSelectAll() {
            Logger log = Logger.getLogger(DML_Main.class);
            SqlSession session = null;
            try {
                session  =  DBOperatorMgr.getInstance().getSqlSessionFactory().
openSession();
                StudentMapper mapper = session.getMapper(StudentMapper.class);
                List<Student> stu_list = mapper.getSudentAll();
                for(Student stu:stu_list) {
                    //System.out.println(stu);
                    log.info(stu);
                }
            } finally {
                if (session != null) {
                    session.close();
                }
            }
        }

        private static void testInsert() {
            Logger log = Logger.getLogger(DML_Main.class);
            SqlSession session = null;
            try {
                session  =  DBOperatorMgr.getInstance().getSqlSessionFactory().
openSession();
                StudentMapper mapper = session.getMapper(StudentMapper.class);
                Student student = new Student();
                student.setSno("20171622");
                student.setName("李白");
                student.setAge(88);
                student.setSex("男");
                student.setDept_no("2609");
                log.info(student);
                mapper.addStudent(student);
                session.commit();
            } catch(Exception ex) {
                session.rollback();
                ex.printStackTrace();
```

```
        } finally {
            if (session != null) {
                session.close();
            }
        }
    }

    private static void testSelect(String sno) {
        Logger log = Logger.getLogger(DML_Main.class);
        SqlSession session = null;
        try {
            session = DBOperatorMgr.getInstance().getSqlSessionFactory().
openSession();
            StudentMapper mapper = session.getMapper(StudentMapper.class);
            Student student = mapper.getStudent(sno);
            //System.out.println(student);
            log.info(student);
        } finally {
            if (session != null) {
                session.close();
            }
        }
    }

    private static void testUpdate() {
        Logger log = Logger.getLogger(DML_Main.class);
        SqlSession session = null;
        try {
            session = DBOperatorMgr.getInstance().getSqlSessionFactory().
openSession();
            StudentMapper mapper = session.getMapper(StudentMapper.class);
            Student student = new Student();
            student.setSno("20171622");
            student.setName("苏东坡");
            student.setAge(68);
            student.setSex("女");
            student.setDept_no("2612");
            log.info(student);
            mapper.updateStudent(student);
            session.commit();
        } catch(Exception ex) {
            session.rollback();
            ex.printStackTrace();
        } finally {
            if (session != null) {
                session.close();
            }
        }
    }
}
```

先测试 testSelectAll()方法。如图 11-1 所示，右击 com.mialab.mybatis_DML_demo.main 包中的 DML_Main.java，在弹出的快捷菜单中选择“Run As → Java Application”选项，在控

制台中可以得到以下的结果：

```
INFO [main] - Student [sno=20171508, name=李勇, sex=男, age=20, dept_no=2601]
INFO [main] - Student [sno=20171509, name=刘娟, sex=女, age=19, dept_no=2602]
INFO [main] - Student [sno=20171622, name=李白, sex=男, age=88, dept_no=2609]
```

可分别对增、删、改、查的方法 testInsert()、testUpdate()等进行测试。

如果在运行 DML_Main 类的 main 方法时遇到问题，可右击 mybatis_DML_demo 工程，选择 “Project→Clean...” 选项，在弹出的对话框中，选中“mybatis_first_demo”工程，单击“Clean”按钮，重新运行即可。

这里要注意的是，如果按照通常构建 Maven 项目的方法，右击 mybatis_ DML _demo 工程的 pom.xml 文件，选择“Run As”→“Maven Clean”选项。Clean 成功后，再选择“Run As”→“Maven Install”选项来构建 target“mybatis_DML_demo-0.0.1-SNAPSHOT.jar”。当再次运行 DML_Main 类的主方法时，由于 StudentMapper.xml 未打包到 target，因此会抛出异常。控制台会出现以下错误信息。

```
ERROR [main] - org.apache.ibatis.exceptions.PersistenceException:
### Error building SqlSession.
### The error may exist in
com/mialab/mybatis_DML_demo/mapper/StudentMapper.xml
### Cause: org.apache.ibatis.builder.BuilderException:
Error parsing SQL Mapper Configuration.
Cause: java.io.IOException: Could not find resource
com/mialab/mybatis_DML_demo/mapper/StudentMapper.xml
```

解决方法：右击 mybatis_DML_demo 工程，选择“Project”→“Clean...”选项，重新运行即可。（经测试，把 XML 映射文件放置到 src/main/resources 文件夹下，再运行“Maven Install”命令，XML 映射文件可打包到 target 中。这里的路径要正确。）

11.2 MyBatis 主配置文件

MyBatis 的 XML 配置文件包含了影响 MyBatis 行为很深的设置和属性信息，也可称之为主配置文件。此 XML 文档的层级结构如下：

- ✧ configuration →配置
- ✧ properties →属性
- ✧ settings →设置
- ✧ typeAliases →类型命名
- ✧ typeHandlers →类型处理器
- ✧ objectFactory →对象工厂
- ✧ plugins →插件
- ✧ environments →配置环境
- ✧ environment →环境变量
- ✧ transactionManager →事务管理器
- ✧ dataSource →数据源
- ✧ databaseIdProvider →数据库厂商标识
- ✧ mappers →映射器

必须注意的是，MyBatis 配置项的顺序不能颠倒。如果其顺序颠倒了，则会在 MyBatis 启动阶段发生异常，导致程序无法运行。例如，可能会产生如下的报错信息：

```
    org.xml.sax.SAXParseException: The content of element type "configuration"
must match
    "(properties?,settings?,typeAliases?,…,
environments?,databaseIdProvider?,mappers?)"
```

1. properties（属性）

MyBatis 提供了 3 种方式来使用 properties，分别如下。

（1）property 子元素。

（2）properties 文件。

（3）程序代码传递。

properties 属性可以通过 property 子元素来传递，也可以配置在典型的 Java 属性配置文件中。例如：

```
<properties resource="org/mybatis/example/jdbc.properties">
  <property name="username" value="boy"/>
  <property name="password" value="1"/>
</properties>
```

其中的属性可以在整个配置文件中使用，使用可替换的属性（参数）来实现动态配置。例如：

```
<dataSource type="POOLED">
  <property name="driver" value="${driver}"/>
  <property name="url" value="${url}"/>
  <property name="username" value="${username}"/>
  <property name="password" value="${password}"/>
</dataSource>
```

其中，username 和 password 将会由 property 子元素中设置的值来替换。driver 和 url 属性将会由包含进来的 jdbc.properties文件中的值来替换。

属性也可以被传递到 SqlSessionBuilder.build() 方法中。例如：

```
SqlSessionFactory factory = sqlSessionFactoryBuilder.build(reader, props);
// ... or ...
SqlSessionFactory factory = sqlSessionFactoryBuilder.build(reader, environment,
props);
```

如果属性多于一个，则MyBatis将按照如下的顺序进行属性加载。

① 在 properties 元素体内指定的属性被读取。

② 从类路径下资源或者 properties 元素的 url 属性中加载的属性被读取，它会覆盖已经存在的完全一样的属性。

③ 作为方法参数传递的属性最后被读取，它也会覆盖任一已经存在的完全一样的属性，这些属性可能是从 properties 元素体内和资源或者 url 属性中加载的。

因此，最高优先级的属性是那些作为方法参数的属性，其次是资源/url 属性，最后是 properties 元素中指定的属性。

2. settings （设置）

settings 会改变 MyBatis 在运行时的行为方式，是 MyBatis 中极为重要的调整设置。表 11-1 描述了 MyBatis 的设置信息，以及其含义和默认值。

表 11-1 settings 设置说明

设置参数	描　述	有效值	默认值
cacheEnabled	该配置影响所有映射器中配置缓存的全局开关	true | false	true
lazyLoadingEnabled	延迟加载的全局开关。当开启时，所有关联对象都会延迟加载。特定关联关系中可通过设置 fetchType 属性来覆盖该项的开关状态	true | false	false
aggressiveLazyLoading	当启用时，对任意延迟属性的调用会使带有延迟加载属性的对象完整加载；反之，每种属性将会按需加载	true | false	false（版本 3.4.1 以后）
multipleResultSetsEnabled	是否允许单一语句返回多结果集（需要兼容驱动）	true | false	true
useColumnLabel	使用列标签代替列名。不同的驱动在这方面会有不同的表现，具体可参考相关驱动文档或通过测试这两种不同的模式来观察所用驱动的结果	true | false	true
useGeneratedKeys	允许 JDBC 支持自动生成主键，需要驱动兼容。如果设置为 true，则这个设置强制使用自动生成主键，尽管一些驱动不能兼容但仍可正常工作（如 Derby）	true | false	false
autoMappingBehavior	指定 MyBatis 应如何自动映射到字段或属性。NONE 表示取消自动映射；PARTIAL 表示只会自动映射，没有定义嵌套结果集和映射结果集；FULL 会自动映射任意复杂的结果集（无论是否嵌套）	NONE、PARTIAL、FULL	PARTIAL
defaultExecutorType	配置默认的执行器。SIMPLE 是普通的执行器；REUSE 会重用预处理语句；BATCH 执行器将重用语句并执行批量更新	SIMPLE、REUSE、BATCH	SIMPLE
defaultStatementTimeout	设置超时时间，它决定驱动等待数据库响应的秒数	任何正整数	Not Set (null)
safeRowBoundsEnabled	允许在嵌套语句中使用分页	true | false	false
mapUnderscoreToCamelCase	是否开启自动驼峰命名规则映射，即从经典数据库列名 A_COLUMN 到经典 Java 属性名 aColumn 的类似映射	true | false	false
localCacheScope	MyBatis 利用本地缓存机制防止循环引用和加速重复嵌套查询。默认值为 SESSION，这种情况下会缓存一个会话中执行的所有查询。若设置值为 STATEMENT，则本地会话仅用在语句执行上，对相同 SqlSession 的不同调用将不会共享数据	SESSION | STATEMENT	SESSION
jdbcTypeForNull	当没有为参数提供特定的 JDBC 类型时，为空值指定 JDBC 类型。某些驱动需要指定列的 JDBC 类型，多数情况直接用一般类型即可，如 NULL、VARCHAR 或 OTHER	NULL、VARCHAR、OTHER	OTHER
callSettersOnNulls	指定当结果集中值为 null 的时候是否调用映射对象的 setter（map 对象时为 put）方法，这对于有 Map.keySet() 依赖或 null 值初始化的时候是有用的。注意，原始类型（int、boolean 等）是不能设置成 null 的	true | false	false
logPrefix	指定 MyBatis 增加到日志名称的前缀	任何字符串	Not Set
logImpl	指定 MyBatis 所用日志的具体实现，未指定时将自动查找	SLF4J | LOG4J 等	Not Set
proxyFactory	指定 Mybatis 创建具有延迟加载能力的对象所用到的代理工具	CGLIB | JAVASSIST	CGLIB

3. typeAliases （别名）

类型别名是为 Java 类型命名的一个短的名称。它只和 XML 配置有关，存在的意义仅在于用来减少类完全限定名的冗余。例如：

```
<typeAliases>
  <typeAlias alias="Blog" type="domain.blog.Blog"/>
</typeAliases>
```

使用这个配置，“Blog”可以用在任何使用“domain.blog.Blog”的地方。

也可以指定一个包名，MyBatis 会在包名下面搜索需要的 Java Bean，例如：

```
<typeAliases>
    <package name="com.mialab.mybatisdemo.domain"/>
</typeAliases>
```

这样，每一个包中的 Java Bean 在没有注解的情况下，会使用 Bean 的首字母小写的非限定类名来作为它的别名。

如果在包中的实体类中发现了@Alias 注解，则将使用注解的值作为它的别名。例如，com.mialab.mybatisdemo.domain 包下存在 Student 实体类，并且有以下代码。

```
@Alias("student")
public class Student {
    ...
}
```

那么，可用“student”来作为“com.mialab.mybatisdemo.domain.Student”的别名。

4．typeHandlers 类型处理器

无论是 MyBatis 在预处理语句中设置一个参数，还是从结果集中取出一个值，类型处理器都被用来将获取的值以合适的方式转换成 Java 类型。

5．objectFactory 对象工厂

MyBatis 每次创建结果对象的新实例时，它都会使用一个对象工厂实例来完成。默认的对象工厂需要做的仅仅是实例化目标类，要么通过默认构造方法来实例化，要么在参数映射存在的时候通过参数构造方法来实例化。如果想覆盖对象工厂的默认行为，则可以通过创建自己的对象工厂来实现。

6．plugins（插件）

MyBatis 允许用户在已映射语句执行过程中的某一点进行拦截调用。默认情况下，MyBatis 允许使用插件来拦截的接口和方法包括以下几个。

```
    Executor (update, query, flushStatements, commit, rollback, getTransaction,
close, isClosed)
    ParameterHandler (getParameterObject, setParameters)
    ResultSetHandler (handleResultSets, handleCursorResultSets, handleOutput
Parameters)
    StatementHandler (prepare, parameterize, batch, update, query)
```

这 4 个接口及其包含的方法的细节可以通过查看每个方法的签名来了解，或者直接查看 MyBatis 发行包中的源代码。假设用户想做的不仅仅是监控方法的调用，那么应该很好地了解正在重写的方法的行为。因为在试图修改或重写已有方法行为的时候，很可能会破坏 MyBatis 的核心模块，这些都是底层的类和方法，所以使用插件的时候要特别当心。

7．environments （配置环境）

MyBatis 可以配置多种环境，这种机制使得 MyBatis 可以将 SQL 映射应用于多种数据库中。例如，开发、测试和生产环境需要有不同的配置。尽管可以配置多个环境，但是每个 SqlSessionFactory 实例只能选择一个环境，即每个数据库对应一个 SqlSessionFactory 实例。所以如果想连接两个数据库，就需要创建两个 SqlSessionFactory 实例，每个数据库对应一个。

8．mappers （映射器）

既然 MyBatis 的行为已经由上述元素配置，现在需要开发者自己编写 SQL 映射语句。mappers

会告诉 MyBatis 到哪里去找映射文件，进而找到这些 SQL 语句。实际开发中，可以使用相对于类路径的资源引用或完全限定资源定位符（包括 file:///URLs）、类名或者包名等。例如：

```
<!-- Using classpath relative resources -->
<mappers>
    <mapper resource="org/mybatis/builder/AuthorMapper.xml"/>
    <mapper resource="org/mybatis/builder/BlogMapper.xml"/>
</mappers>
<!-- Using url fully qualified paths -->
<mappers>
    <mapper url="file:///var/mappers/AuthorMapper.xml"/>
    <mapper url="file:///var/mappers/BlogMapper.xml"/>
</mappers>
<!-- Using mapper interface classes -->
<mappers>
    <mapper class="org.mybatis.builder.AuthorMapper"/>
    <mapper class="org.mybatis.builder.BlogMapper"/>
</mappers>
<!-- Register all interfaces in a package as mappers -->
<mappers>
    <package name="org.mybatis.builder"/>
</mappers>
```

这些语句简单地告诉了 MyBatis 去哪里查找映射文件。

11.3 XML 映射文件

一个完整的 SQL 映射文件，如图 11-3 所示，可以分为以下几部分。

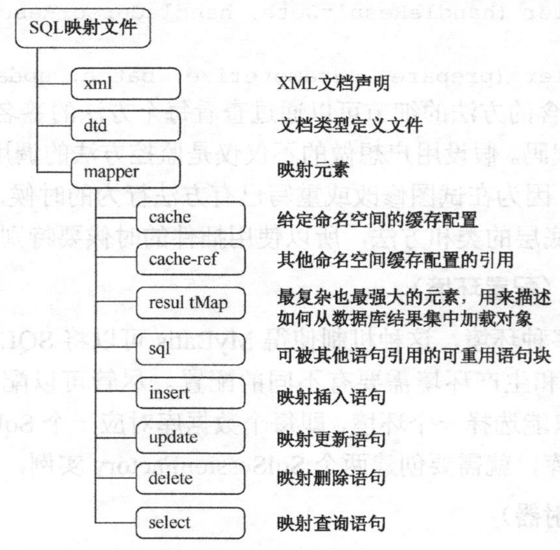

图 11-3 SQL 映射文件

xml 和 dtd 部分是必须填写且不需要配置的部分，每次使用时，只需要将这部分复制到文件顶部即可。

mapper 元素是整个映射文件的容器，所有的 SQL 映射都包含在这个元素中，mapper 本身有一个参数 namespace，即命名空间，这个命名空间就是文件所对应的接口文件的 Java 类，只有当这个命名空间被配置时，才可以直接通过访问 Java 接口的方法实现 SQL 调用。

映射器是 MyBatis 最复杂且最重要的组件。它由一个接口加上 XML 文件（或者注解）组成。在映射器中可以配置参数、SQL 语句、存储过程、缓存等内容，并且通过简易的映射规则映射到指定的 POJO 或者其他对象上，映射器能有效消除 JDBC 底层的代码。

MyBatis 的映射器也可以使用注解完成，但可读性较差，企业中应用不广，官方亦不推荐使用。MyBatis 注解方式的基本用法将会在本书的第 13 章加以介绍。

下面简单介绍常用映射元素的使用方法。

1. select

查询语句是 MyBatis 中最常用的元素之一，多数应用也都是查询比修改要频繁。对每个插入、更新或删除操作，通常对应多个查询操作。这是 MyBatis 的基本原则之一，也是将焦点和精力放到查询和结果映射的原因。对简单类别的查询元素是非常简单的。例如：

```
<select id="selectPerson" parameterType="int" resultType="hashmap">
 SELECT * FROM PERSON WHERE ID = #{id}
</select>
```

这个语句被称为 selectPerson，使用一个 int（或 Integer）类型的参数，并返回一个 HashMap 类型的对象，其中的键是列名，值是列对应的值。

注意参数标识 #{id}，其告诉 MyBatis 创建一个 PreparedStatement（预处理语句）参数。使用 JDBC，这样的一个参数在 SQL 中会由一个“?”来标识，并被传递到一个新的预处理语句中，类似于以下的 JDBC 代码（不是 MyBatis 的代码）：

```
String selectPerson = "SELECT * FROM PERSON WHERE ID=?";
PreparedStatement ps = conn.prepareStatement(selectPerson);
ps.setInt(1,id);
```

当然，这需要很多单独的 JDBC 的代码来提取结果并将它们映射到对象实例中，这就是 MyBatis 节省时间的原因。我们需要深入了解参数和结果映射。select 元素有很多属性允许用户配置，以决定每条语句的作用细节。

关于 select 元素主要属性的描述参见表 11-2。

表 11-2 select 元素的主要属性描述

属　性	描　述
id	在命名空间中唯一的标识符，可以被用来引用这条语句
parameterType	将会传入这条语句的参数类的完全限定名或别名。这个属性是可选的，因为 MyBatis 可以通过 TypeHandler 推断出具体传入语句的参数，默认值为 unset
resultType	从这条语句中返回的期望类型的类的完全限定名或别名。**注意，如果是集合情形，则应该是集合可以包含的类型，而不能是集合本身**。可以使用 resultType 或 resultMap，但不能同时使用
resultMap	外部 resultMap 的命名引用。结果集的映射是 MyBatis 最强大的特性，若能对其有一个很好地理解，则许多复杂映射的情形都能迎刃而解。可以使用 resultMap 或 resultType，但不能同时使用
flushCache	如果设置为 true，则任何时候只要语句被调用，都会导致本地缓存和二级缓存被清空，默认值为 false
useCache	如果设置为 true，则将会导致本条语句的结果被二级缓存，select 元素默认值为 true
timeout	这个设置是在抛出异常之前，驱动程序等待数据库返回请求结果的秒数。默认值为 unset（依赖驱动）

续表

属　　性	描　　述
fetchSize	这是尝试影响驱动程序每次批量返回的结果行数和这个设置值相等。默认值为 unset（依赖驱动）
statementType	值为 STATEMENT、PREPARED 或 CALLABLE 之一。这会让 MyBatis 分别使用 JDBC 中的 Statement、PreparedStatement 或 CallableStatement，默认值为 PREPARED
resultSetType	值为 FORWARD_ONLY、SCROLL_SENSITIVE 或 SCROLL_INSENSITIVE 之一，默认值为 unset（依赖驱动），是结果集的类型
databaseId	如果配置了 databaseIdProvider，则 MyBatis 会加载所有的不带 databaseId 或匹配当前 databaseId 的语句；如果带或者不带 database Id 的语句都有，则不带的会被忽略

2．insert、update、delete

数据操纵语句 insert、update 和 delete 在它们的实现中非常相似。

```
<insert
  id="insertAuthor"
  parameterType="domain.blog.Author"
  flushCache="true"
  statementType="PREPARED"
  keyProperty=""
  keyColumn=""
  useGeneratedKeys=""
  timeout="20">
<update
  id="updateAuthor"
  parameterType="domain.blog.Author"
  flushCache="true"
  statementType="PREPARED"
  timeout="20">
<delete
  id="deleteAuthor"
  parameterType="domain.blog.Author"
  flushCache="true"
  statementType="PREPARED"
  timeout="20">
```

insert、update 和 delete 元素的主要属性描述如表 11-3 所示。

表 11-3　insert、update 和 delete 元素的主要属性描述

属　　性	描　　述
id	命名空间中的唯一标识符，可被用来代表这条语句
parameterType	将要传入语句的参数的完全限定类名或别名。这个属性是可选的，因为 MyBatis 可以通过 TypeHandler 推断出具体传入语句的参数，默认值为 unset
parameterMap	这是引用外部 parameterMap 的已经被废弃的方法。使用内联参数映射和 parameterType 属性
flushCache	将其设置为 true，任何时候只要语句被调用，都会导致本地缓存和二级缓存都被清空，默认值为 true（对应插入、更新和删除语句）
statementType	STATEMENT、PREPARED 或 CALLABLE 之一。这会让 MyBatis 分别使用 Statement、PreparedStatement 或 CallableStatement，默认值为 PREPARED
useGeneratedKeys	（仅对 insert 和 update 有用）这会令 MyBatis 使用 JDBC 的 getGeneratedKeys 方法来获取由数据库内部生成的主键（如像 MySQL 和 SQL Server 这样的关系数据库管理系统的自动递增字段），默认值为 false

续表

属　　性	描　　述
keyProperty	（仅对 insert 和 update 有用）唯一标记一个属性，MyBatis 会通过 getGeneratedKeys 的返回值或者通过 insert 语句的 selectKey 子元素设置其键值，默认为 unset。如果希望得到多个生成的列，则也可以是逗号分隔的属性名称列表
keyColumn	（仅对 insert 和 update 有用）通过生成的键值设置表中的列名，这个设置仅在某些数据库（如 PostgreSQL）中是必需的，当主键列不是表中的第一列时需要设置。如果希望得到多个生成的列，则也可以是逗号分隔的属性名称列表

下面是 insert、update 和 delete 语句的示例。

```
<insert id="addStudent" parameterType="student">
 insert into
 student(sno,name,sex,age,dept_no)
 values(#{sno},#{name},#{sex},#{age},#{dept_no})
</insert>
<update id="updateStudent" parameterType="student">
 update student set name
 = #{name}, sex = #{sex}, age = #{age}, dept_no = #{dept_no}
 where sno =
 #{sno}
</update>
<delete id="deleteStudent" parameterType="String">
 delete from student
 where sno = #{sno}
</delete>
```

3．sql

这个元素可以被用来定义可重用的 SQL 代码段，可以包含在其他语句中。例如：

```
    <sql  id="userColumns">  ${alias}.id,${alias}.username,${alias}.password
</sql>
```

这个 SQL 片段可以被包含在其他语句中。例如：

```
<select id="selectUsers" resultType="map">
select
<include refid="userColumns"><property name="alias" value="t1"/></include>,
from some_table t1
</select>
```

4．parameters（参数）

在 MyBatis 中，参数是非常强大的元素。类似于之前的语句，简单参数示例如下。

```
<select id="selectUsers" parameterType="int" resultType="User">
select id, username, password from users where id = #{id}
</select>
```

这个示例说明了一个非常简单的命名参数映射。参数类型被设置为 int，这里的参数名是 id，也可以是其他名称。参数为简单数据类型的，都与此类似。以下示例中参数类型是一个对象，MyBatis 的处理方式不同于简单数据类型。

```
    <insert id="insertUser"parameterType="User" >
    insert into users (id, username, password) values (#{id}, #{username},
#{password})
    </insert>
```

如果 User 类型的参数对象传递到了语句中，则 id、username 和 password 属性将会被查找，它们的值就会被传递到预处理语句的参数中。

5. resultMap

resultMap 是映射中最复杂也是最强大的元素，用来描述如何从数据库结果集中来加载对象。resultMap 定义的主要是一个结果集的映射关系，也就是 SQL 到 Java Bean 的映射关系定义。

下面是简单映射语句的示例，但没有明确的 resultMap。例如：

```
<select id="selectUsers"  resultType="map">
 select id, username, hashedPassword from some_table where id = #{id}
</select>
```

resultType="map"表示返回的数据是一个 Map 集合（使用列名作为 key、列值作为 value）。虽然数据被封装成 Map 集合返回，但是 Map 集合并不能很好地描述一个领域模型。可以使用 JavaBeans 或 POJOs 来作为领域模型描述数据。MyBatis 对两者都支持。来看下面这个 JavaBean：

```
package com.someapp.model;
public class User {
 private int id;
 private String username;
 private String hashedPassword;
 public int getId() {
     return id;
 }
 public void setId(int id) {
     this.id = id;
 }
 public String getUsername() {
     return username;
 }
 public void setUsername(String username) {
     this.username = username;
 }
 public String getHashedPassword() {
     return hashedPassword;
 }
 public void setHashedPassword(String hashedPassword) {
     this.hashedPassword = hashedPassword;
 }
}
```

基于 JavaBean 的规范，上面这个类有 3 个属性：id、username 和 hashedPassword。这些在 select 语句中会精确匹配到列名。

这样的一个 JavaBean 可以被映射到结果集，就像映射到 HashMap 一样简单。

```
<select id="selectUsers" parameterType="int" resultType="com.someapp.model.
User">
    select id, username, hashedPassword from some_table where id = #{id}
</select>
```

以下使用了类型别名，使用它们时可以不输入类的全路径。例如：

```
<!-- 在 XML 配置文件中-->
<typeAlias type="com.someapp.model.User" alias="User"/>
<!-- 在 SQL 映射的 XML 文件中-->
```

```
<select id="selectUsers"  parameterType="int"  resultType="User">
  select id, username, hashedPassword from some_table where id = #{id}
</select>
```

在这些情况下，MyBatis 会在幕后自动创建一个 resultMap，基于属性名来映射列到 JavaBean 的属性上。如果列名没有精确匹配，则可以在列名上使用 select 字句的别名（一个标准的 SQL 特性）来匹配标签。例如：

```
<select id="selectUsers" parameterType="int" resultType="User">
  select
    user_id as "id",
    user_name as "userName",
    hashed_password as "hashedPassword"
  from some_table
  where id = #{id}
</select>
```

resultMap 最常见的使用方式如下所示，这也是解决列名不匹配的另外一种方式。

```
<resultMap id="userResultMap" type="User">
  <id property="id" column="user_id" />
  <result property="username" column="user_name"/>
  <result property="password" column="hashed_password"/>
</resultMap>
```

引用此语句时使用 resultMap 属性即可（注意，这里去掉了 resultType 属性）。例如：

```
<select id="selectUsers" parameterType="int" resultMap="userResultMap">
  select user_id, user_name, hashed_password from some_table where id = #{id}
</select>
```

6. resultMap 使用示例

（1）准备数据。

在先前创建的数据库 stu 中创建表 student_2，并插入若干条数据，代码如下：

```
DROP TABLE IF EXISTS'student_2';
CREATE TABLE 'student_2' (
  `stu_sno` varchar(100) NOT NULL DEFAULT '',
  'stu_name' varchar(100) NOT NULL,
  'stu_sex' varchar(8) DEFAULT NULL,
  'stu_age' int(3) DEFAULT NULL,
  'stu_dept_no' varchar(60) DEFAULT NULL
) ENGINE=InnoDB DEFAULT CHARSET=utf8;
INSERT INTO 'student_2' VALUES ('20171505', 'Marry', '女', '18', '2602');
INSERT INTO 'student_2' VALUES ('20171506', '王宝宝', '男', '22', '2602');
INSERT INTO 'student_2' VALUES ('20171508', '李勇', '男', '20', '2605');
INSERT INTO 'student_2' VALUES ('20171509', '刘娟', '女', '19', '2605');
```

（2）创建工程 mybatis_ResultMap_demo。

在 Eclipse 中创建 Maven Project，选择 maven-archetype-quickstart，在“Group ID”文本框中输入“com.mialab”，在“Artifact ID”文本框中输入“mybatis_ ResultMap _demo”。

最终完成的 mybatis_ ResultMap _demo 工程目录及文件如图 11-4 所示，student_2 初始表数据如图 11-5 所示。（代码详见本书教学资源包中的 mybatis_ ResultMap _demo 工程。）

```
mybatis_ResultMap_demo
  src/main/java
    com.mialab.mybatis_ResultMap_demo.domain
      Student.java
    com.mialab.mybatis_ResultMap_demo.main
      ResultMap_Main.java
    com.mialab.mybatis_ResultMap_demo.mapper
      StudentMapper.java
      StudentMapper.xml
    com.mialab.mybatis_ResultMap_demo.utils
      DBOperatorMgr.java
  src/test/java
  JRE System Library [JavaSE-1.8]
  Maven Dependencies
  src
    main
      resources
        log4j.properties
        mybatis-config.xml
    test
  target
  pom.xml
```

图 11-4 mybatis_ ResultMap _demo 工程目录及文件

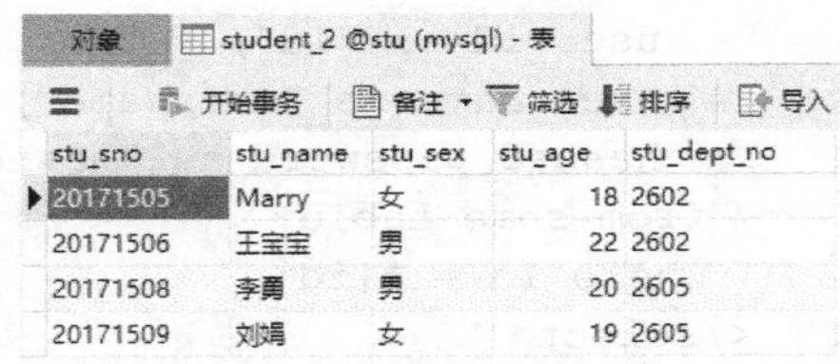

stu_sno	stu_name	stu_sex	stu_age	stu_dept_no
20171505	Marry	女	18	2602
20171506	王宝宝	男	22	2602
20171508	李勇	男	20	2605
20171509	刘娟	女	19	2605

图 11-5 student_2 表中的数据

（3）创建实体对象映射数据库表。

Student 对象用于映射 student_2 表，Student.java 主要代码如下（此处 get 方法、set 方法和 toString 方法略，详见本书教学资源包中的 mybatis_ ResultMap _demo 工程）：

```
public class Student {
 private String sno;
 private String name;
 private String sex;
 private int age;
 private String dept_no;
 …
}
```

（4）创建映射接口和映射文件。

接口 StudentMapper.java 的主要代码如下：

```
public interface StudentMapper {
     public List<Student> getSudentAll();
}
```

映射文件 StudentMapper.xml 的主要代码如下：

```
<mapper
namespace="com.mialab.mybatis_ResultMap_demo.mapper.StudentMapper">
 <resultMap id="studentResultMap" type="student">
     <id property="sno" column="stu_sno" />
     <result property="name" column="stu_name" />
     <result property="sex" column="stu_sex" />
     <result property="age" column="stu_age" />
     <result property="dept_no" column="stu_dept_no" />
 </resultMap>
 <select id="getSudentAll" resultMap="studentResultMap">
     select * from student_2
     </select>
</mapper>
```

这里 column 属性表示数据库表的列名，property 表示数据库列映射到返回类型的属性。<resultMap id="studentResultMap" type="student">中的 id 是 resultMap 的唯一标识符，type 则表示 resultMap 的实际返回类型。"student"是类型别名，表示的是"com.mialab.mybatis_

ResultMap_demo.domain.Student"，在 mybatis-config.xml 中有声明。

<id property="sno" column="stu_sno" />中的 id 表示这个对象的主键（或者唯一标识），property 表示 POJO 的属性名称，column 表示数据库表的列名。这样，POJO 就和数据库 SQL 的结果一一对应起来了。

（5）测试。

main 方法的主要测试代码如下：

```
StudentMapper mapper = session.getMapper(StudentMapper.class);
List<Student> stu_list = mapper.getSudentAll();
for(Student stu:stu_list) {
    log.info(stu);
}
```

运行 ResultMap_Main 的 main 方法，控制台显示内容如下：

```
    DEBUG [main] - ==>  Preparing: select * from student_2
    DEBUG [main] - ==> Parameters:
    DEBUG [main] - <==      Total: 4
    INFO [main] - Student [sno=20171505, name=Marry, sex=女, age=18, dept_ no=
2602]
    INFO [main] - Student [sno=20171506, name=王宝宝, sex=男, age=22, dept_ no=
2602]
    INFO [main] - Student [sno=20171508, name=李勇, sex=男, age=20, dept_no=2605]
    INFO [main] - Student [sno=20171509, name=刘娟, sex=女, age=19, dept_no=2605]
```

11.4　高级结果映射

在关系型数据库中，经常要处理一对一、一对多和多对多的关系。使用 MyBatis 的高级结果映射（Advanced Result Maps），亦可以说使用 MyBatis 的关联映射，能够比较轻松地处理这类问题，大大简化了持久层数据的访问。

11.4.1　示例说明

这里的示例是 3 个 Maven 项目：OneToOne_demo 工程、OneToMany_demo 工程、ManyToMany_demo 工程，分别使用 MyBatis 的一对一映射、一对多映射和多对多映射进行示例说明。

1. 准备数据

可以使用 MySQL 客户端工具 Navicat 创建数据库 hrm，在数据库 hrm 中分别创建表 employee_basic（员工基本情况表）、employee_school（员工学历表）、dept（部门信息表）、job（工作岗位情况表）和 employee_job（员工工作岗位表）。

数据库表 employee_basic、employee_school、dept、job、employee_job 的逻辑结构分别如表 11-4~表 11-8 所示。

表 11-4　employee_basic（员工基本情况表）

列　名	数据类型	可否为空	说　明
emp_no	varchar(8)	NOT NULL	员工号（主键）
emp_name	varchar(12)	NOT NULL	员工姓名

续表

列　　名	数据类型	可否为空	说　　明
dept_id	varchar(3)	NOT NULL	岗位任务描述
emp_gender	varchar(6)	NULL	所在的部门号（外键），指向部门信息表的 dept_id
emp_email	varchar(28)	NULL	电子邮箱
emp_nation	varchar(16)	NULL	民族
emp_marriage	varchar(6)	NULL	婚姻状况
emp_health	varchar(10)	NULL	健康状况
emp_zzmm	varchar(8)	NULL	政治面貌
emp_blood	varchar(6)	NULL	血型
emp_state	varchar(10)	NULL	员工状态

表 11-5　employee_school（员工学历表）

列　　名	数据类型	可否为空	说　　明
emp_id	varchar(8)	NOT NULL	员工号（外键），指向员工基本情况表的 emp_no
emp_xueli	varchar(8)	NULL	学历
emp_major	varchar(16)	NULL	专业
by_date	date	NULL	毕业时间
by_school	varchar(30)	NULL	毕业学校

表 11-6　dept（部门信息表）

列　　名	数据类型	可否为空	说　　明
dept_id	varchar(3)	NOT NULL	部门代号（主键）
dept_name	varchar(16)	NULL	部门名称
dept_manager	varchar(18)	NULL	部门经理代号（员工号）

表 11-7　job（工作岗位情况表）

列　　名	数据类型	可否为空	说　　明
job_id	varchar(4)	NOT NULL	工作岗位代号（主键）
job_name	varchar(16)	NOT NULL	工作岗位名称
job_task	varchar(60)	NULL	岗位任务描述

表 11-8　employee_job（员工工作岗位表）

列　　名	数据类型	可否为空	说　　明
id	int(3)	NOT NULL	序号（主键）
emp_no	varchar(8)	NULL	员工编号（外键）
job_id	varchar(4)	NULL	工作岗位（角色）编号（外键）

相应的 SQL 脚本如下（详见本书教学资源包 ch11 文件夹中的 hrm.sql 文件）：

```
DROP TABLE IF EXISTS 'dept';
CREATE TABLE 'dept' (
  'dept_id' varchar(3) NOT NULL,
```

```
      'dept_name' varchar(16) DEFAULT NULL,
      'dept_manager' varchar(8) DEFAULT NULL,
      PRIMARY KEY ('dept_id')
    ) ENGINE=InnoDB DEFAULT CHARSET=utf8;
    INSERT INTO 'dept' VALUES ('101', '人力资源部', 'HW9803');
    INSERT INTO 'dept' VALUES ('102', '财务部', 'HW9804');
    INSERT INTO 'dept' VALUES ('103', '技术部', 'HW9806');

    DROP TABLE IF EXISTS 'employee_basic';
    CREATE TABLE 'employee_basic' (
      'emp_no' varchar(8) NOT NULL COMMENT '员工号',
      'emp_name' varchar(12) NOT NULL,
      'dept_id' varchar(3) NOT NULL COMMENT '所在的部门号（外键）',
      'emp_gender' varchar(6) DEFAULT NULL,
      'emp_email' varchar(28) DEFAULT NULL,
      'emp_nation' varchar(16) DEFAULT NULL COMMENT '民族',
      'emp_marriage' varchar(6) DEFAULT NULL,
      'emp_health' varchar(10) DEFAULT NULL,
      'emp_zzmm' varchar(8) DEFAULT NULL COMMENT '政治面貌',
      'emp_blood` varchar(6) DEFAULT NULL COMMENT '血型',
      'emp_state' varchar(10) DEFAULT NULL COMMENT '员工状态',
      PRIMARY KEY ('emp_no'),
      KEY 'dept_id' ('dept_id'),
      CONSTRAINT 'employee_basic_ibfk_1' FOREIGN KEY ('dept_id') REFERENCES
'dept' ('dept_id')
    ) ENGINE=InnoDB DEFAULT CHARSET=utf8;

    INSERT INTO 'employee_basic' VALUES ('HW9801', '张丹枫', '101', '男',
'zhang@163.com', '汉族', '已婚', '良好', '党员', 'O型', '在职');
    INSERT INTO 'employee_basic' VALUES ('HW9802', '史密斯', '101', '男',
'smith@163.com', '满族', '已婚', '良好', '群众', 'A型', '在职');
    INSERT INTO 'employee_basic'VALUES ('HW9803', '余小男', '101', '女',
'yu@163.com', '汉族', '未婚', '良好', '群众', 'B型', '在职');
    INSERT INTO 'employee_basic' VALUES ('HW9804', '李莉莉', '102', '女',
'lili@163.com', '回族', '未婚', '良好', '团员', 'B型', '在职');
    INSERT INTO 'employee_basic' VALUES ('HW9805', 'Marry', '102', '女',
'marry@163.com', '汉族', '未婚', '良好', '群众', 'O型', '在职');
    INSERT INTO 'employee_basic' VALUES ('HW9806', '郭靖', '103', '男',
'guo@163.com', '汉族', '已婚', '良好', '党员', 'O型', '在职');
    INSERT INTO 'employee_basic' VALUES ('HW9807', '王强', '103', '男',
'wangqiang@163.com', '汉族', '已婚', '良好', '党员', 'O型', '在职');
    INSERT INTO 'employee_basic' VALUES ('HW9808', '李娜', '103', '女',
'nana@163.com', '汉族', '未婚', '良好', '群众', 'O型', '在职');
    INSERT INTO 'employee_basic' VALUES ('HW9809', '许天仪', '103', '男',
'xu@163.com', '汉', '已婚', '一般', '党员', 'O型', '退休');

    DROP TABLE IF EXISTS 'employee_job';
    CREATE TABLE 'employee_job' (
      'id'int(3) NOT NULL AUTO_INCREMENT,
```

```
    'emp_no' varchar(8) DEFAULT NULL,
    'job_id' varchar(4) DEFAULT NULL,
    PRIMARY KEY ('id'),
    KEY 'emp_no' ('emp_no'),
    KEY 'job_id' ('job_id'),
    CONSTRAINT 'employee_job_ibfk_1' FOREIGN KEY ('emp_no') REFERENCES
'employee_basic' ('emp_no'),
    CONSTRAINT 'employee_job_ibfk_2' FOREIGN KEY ('job_id') REFERENCES 'job'
('job_id')
  ) ENGINE=InnoDB AUTO_INCREMENT=12 DEFAULT CHARSET=utf8;

  INSERT INTO 'employee_job' VALUES ('1', 'HW9801', '2601');
  INSERT INTO 'employee_job' VALUES ('2', 'HW9802', '2601');
  INSERT INTO 'employee_job' VALUES ('3', 'HW9803', '2602');
  INSERT INTO 'employee_job' VALUES ('4', 'HW9804', '2601');
  INSERT INTO 'employee_job' VALUES ('5', 'HW9805', '2602');
  INSERT INTO 'employee_job' VALUES ('6', 'HW9806', '2605');
  INSERT INTO 'employee_job' VALUES ('7', 'HW9806', '2606');
  INSERT INTO 'employee_job' VALUES ('8', 'HW9807', '2603');
  INSERT INTO 'employee_job' VALUES ('9', 'HW9808', '2603');
  INSERT INTO 'employee_job' VALUES ('10', 'HW9808', '2604');
  INSERT INTO 'employee_job' VALUES ('11', 'HW9809', '2601');

  DROP TABLE IF EXISTS 'employee_school';
  CREATE TABLE 'employee_school' (
    'emp_id' varchar(8) NOT NULL COMMENT '员工号（外键），指向员工基本情况表的
emp_no',
    'emp_xueli' varchar(8) DEFAULT NULL,
    'emp_major' varchar(16) DEFAULT NULL,
    'by_date' date DEFAULT NULL,
    'by_school'varchar(30) DEFAULT NULL,
    KEY 'emp_id' (`emp_id'),
    CONSTRAINT 'employee_school_ibfk_1' FOREIGN KEY ('emp_id') REFERENCES
'employee_basic' ('emp_no')
  ) ENGINE=InnoDB DEFAULT CHARSET=utf8;

  INSERT INTO 'employee_school' VALUES ('HW9801', '本科', '管理学',
'1987-03-18', '西安交通大学');
  INSERT INTO 'employee_school' VALUES ('HW9802', '硕士', '人力资源管理',
'2015-06-09', '安徽财经大学');
  INSERT INTO 'employee_school' VALUES ('HW9803', '硕士', '人力资源管理',
'2010-06-25', '清华大学');
  INSERT INTO `employee_school' VALUES ('HW9804', '硕士', '会计学',
'2008-03-12', '上海财经大学');
  INSERT INTO 'employee_school' VALUES ('HW9805', '本科', '会计学',
'1990-07-08', '中国人民大学');
  INSERT INTO 'employee_school' VALUES ('HW9806', '博士', '计算机科学与技术',
'2012-06-22', '上海交通大学');
  INSERT INTO 'employee_school' VALUES ('HW9807', '博士', '软件工程',
```

```
'2009-03-22', '东南大学');
      INSERT INTO 'employee_school' VALUES ('HW9808', '硕士', '应用数学',
'2006-06-28', '复旦大学');
      INSERT INTO 'employee_school' VALUES ('HW9809', '本科', '计算机科学与技术',
'1976-06-26', '中国人民大学');

      DROP TABLE IF EXISTS 'job';
      CREATE TABLE 'job' (
        'job_id' varchar(4) NOT NULL,
        'job_name' varchar(16) NOT NULL,
        'job_task' varchar(60) DEFAULT NULL,
        PRIMARY KEY ('job_id')
      ) ENGINE=InnoDB DEFAULT CHARSET=utf8;
      INSERT INTO 'job' VALUES ('2601', '职员', '一般职员，非技术岗');
      INSERT INTO 'job' VALUES ('2602', '主管', '指导检查下属职员');
      INSERT INTO 'job' VALUES ('2603', 'Java开发工程师', '负责Java开发');
      INSERT INTO 'job' VALUES ('2604', 'Android开发工程师', '负责Android开发');
      INSERT INTO 'job' VALUES ('2605', '架构师', '负责系统架构');
      INSERT INTO 'job' VALUES ('2606', '技术总监', '公司技术总负责');
```

Navicat Premium 中的 hrm 数据库如图 11-6 所示。

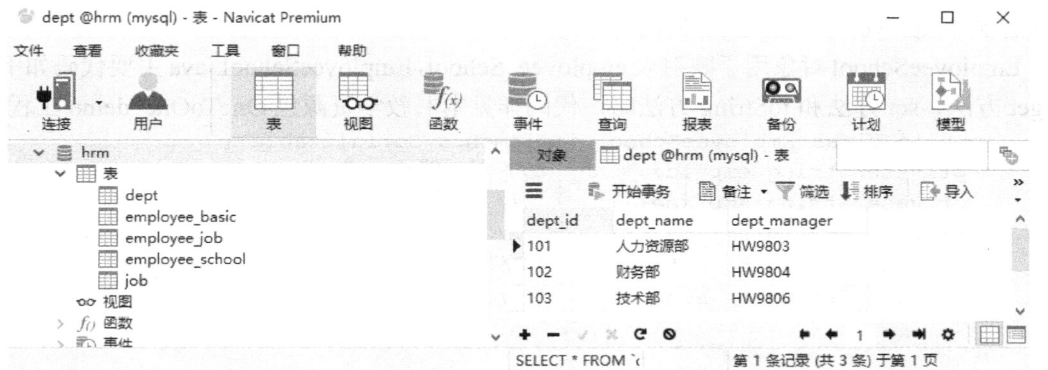

图 11-6　Navicat Premium 中的 hrm 数据库

2．创建 Maven 工程及其他

这里在 Eclipse 中分别创建了 3 个 Maven 工程：OneToOne_demo、OneToMany_demo 和 ManyToMany_demo（详见本书教学资源包 ch11 文件夹）。

工程中关于 pom.xml 文件的配置，mybatis-config.xml、log4j.properties 文件创建及配置，工具类 DBOperatorMgr.java 的创建及实现代码都类似以前实例，这里不再赘述。

以下将对这 3 个示例如何使用 MyBatis 的高级结果映射，来实现一对一关系、一对多关系和多对多关系做比较详细的讲解。

11.4.2　一对一映射

1．OneToOne_demo 工程目录及文件

最终完成的 OneToOne_demo 工程目录及文件如图 11-7 所示。

```
OneToOne_demo
  src/main/java
    com.mialab.OneToOne_demo.domain
      Employee.java
      EmployeeSchool.java
    com.mialab.OneToOne_demo.main
      OneToOne_Main.java
    com.mialab.OneToOne_demo.mapper
      EmployeeMapper.java
      EmployeeMapper.xml
      EmployeeSchoolMapper.xml
    com.mialab.OneToOne_demo.utils
      DBOperatorMgr.java
  src/main/resources
    log4j.properties
    mybatis-config.xml
  src/test/java
  JRE System Library [JavaSE-1.8]
  Maven Dependencies
  src
  target
  pom.xml
```

图 11-7　OneToOne_demo 工程目录及文件

2. 创建实体类

考虑到一个员工有且只有一份学历表，员工实体和学历表实体是一对一的关系，于是创建实体类 Employee 和 EmployeeSchool。

EmployeeSchool 对象用于映射表 employee_School，EmployeeSchool.java 主要代码如下（此处 get 方法、set 方法和 toString 方法略，代码详见本书教学资源包 OneToOne_demo 工程）：

```
public class EmployeeSchool implements Serializable {
 private String emp_id;          // 员工编号
 private String emp_xueli;       // 学历
 private String emp_major;       // 专业
 private String by_date;         // 毕业时间
 private String by_school;       // 毕业学校
   ...
}
```

Employee.java 主要代码如下（此处 get 方法、set 方法和 toString 方法略）：

```
public class Employee implements Serializable {
 private String emp_no;
 private String emp_name;
 private String dept_id;            // 部门编号
 private String emp_gender;         // 员工性别
 private String emp_email;
 private EmployeeSchool employeeSchool;    // 员工学历对象
 …
}
```

3. 创建映射接口

在 EmployeeMapper.java 中定义接口的方法：

```
public interface EmployeeMapper {
 Employee selectEmployeeById(String id);
 }
```

4．XML 映射文件

映射文件 EmployeeSchoolMapper.xml 的主要代码如下：

```
    <mapper
namespace="com.mialab.OneToOne_demo.mapper.EmployeeSchoolMapper">
     <!-- 根据 emp_id 查询员工学历，返回 EmployeeSchool 对象 -->
     <select id="selectEmployeeSchoolById" parameterType="String"
         resultType="com.mialab.OneToOne_demo.domain.EmployeeSchool">
         SELECT * from employee_school where emp_id = #{emp_id}
     </select>
    </mapper>
```

映射文件 EmployeeMapper.xml 的主要代码如下：

```
    <mapper namespace="com.mialab.OneToOne_demo.mapper.EmployeeMapper">
     <!-- 根据 emp_no 查询 Employee，返回 resultMap -->
     <select id="selectEmployeeById" parameterType="String"
         resultMap="employeeMap">
         SELECT * from employee_basic where emp_no = #{emp_no}
     </select>

     <!-- 映射 Employee 对象的 resultMap -->
     <resultMap type="com.mialab.OneToOne_demo.domain.Employee"
         id="employeeMap">
         <id property="emp_no" column="emp_no" />
         <result property="emp_name" column="emp_name" />
         <result property="dept_id" column="dept_id" />
         <result property="emp_gender" column="emp_gender" />
         <result property="emp_email" column="emp_email" />

         <!-- 一对一关联映射：association -->
         <association property="employeeSchool" column="emp_no"
             select=
    "com.mialab.OneToOne_demo.mapper.EmployeeSchoolMapper.selectEmployeeSch
oolById"
             javaType="com.mialab.OneToOne_demo.domain.EmployeeSchool" />
     </resultMap>
     </mapper>
```

5．测试

OneToOne_Main 类的测试代码如下：

```
    public class OneToOne_Main {
     public static void main(String[] args) {
         testSelect("HW9803");
     }

     private static void testSelect(String emp_no) {
         Logger log = Logger.getLogger(OneToOne_Main.class);
         SqlSession session = null;
         try {
             session = DBOperatorMgr.getInstance().
                         getSqlSessionFactory().openSession();
             EmployeeMapper mapper = session.getMapper(EmployeeMapper.class);
             Employee employee = mapper.selectEmployeeById(emp_no);
```

```
                //System.out.println(employee);
                log.info(employee);
                log.info(employee.getEmployeeSchool());
            } finally {
                if (session != null) {
                    session.close();
                }
            }
        }
    }
```

运行OneToOne_Main类的main方法，控制台输出内容如下：

```
    INFO  [main]  - Employee  [emp_no=HW9803,  emp_name=余小男,  dept_id=101,
emp_gender=女, emp_email=yu@163.com]
    INFO [main] - EmployeeSchool [emp_id=HW9803, emp_xueli=硕士, emp_major=人
力资源管理, by_date=2010-06-25, by_school=清华大学]
```

11.4.3 一对多映射

1. OneToMany _demo 工程目录及文件

最终完成的 OneToMany _demo 工程目录及文件如图 11-8 所示。

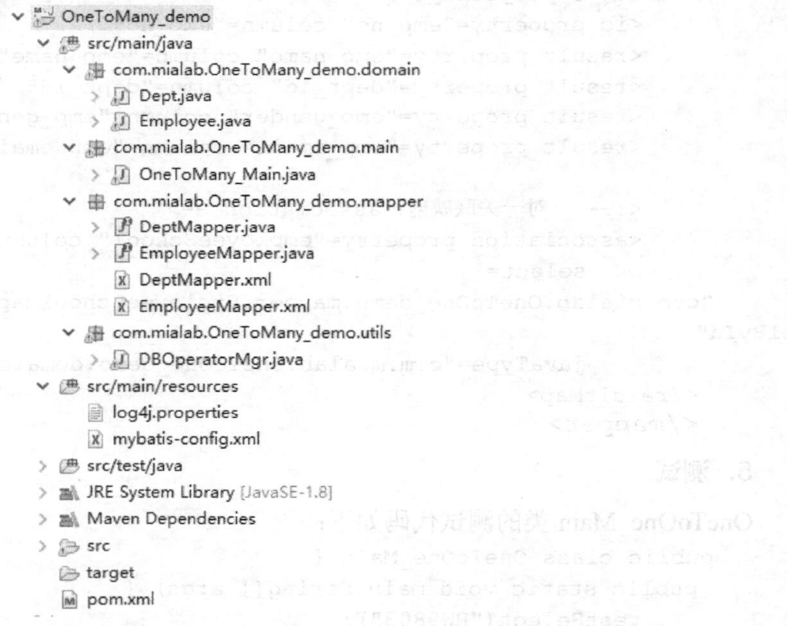

图 11-8 OneToMany_demo 工程目录及文件

2. 创建实体类

考虑到一个部门有多个员工，而每个员工归属于一个确定的部门，部门和员工是一对多的关系，于是创建实体类 Dept.java 和 Employee.java。

Dept.java 主要代码如下（此处 get 方法、set 方法和 toString 方法略，代码详见本书教学资源包 OneToMany_demo 工程）：

```
    public class Dept implements Serializable {
     private String dept_id;                  // 部门编号
```

```
    private String dept_name;        // 部门名称
    private String dept_manager;        // 部门经理
    private List<Employee> employees;     // 员工对象集合
    …
}
```

Employee.java 主要代码如下（此处 get 方法、set 方法和 toString 方法略）：

```
public class Employee implements Serializable {
    private String emp_no;
    private String emp_name;
    private String dept_id;
    private String emp_gender;
    private String emp_email;
    private Dept dept;
    …
}
```

3．创建映射接口

在 DeptMapper.java 中定义接口的方法：

```
public interface DeptMapper {
    Dept  selectDeptById(String dept_id);
}
```

在 EmployeeMapper.java 中定义接口的方法：

```
public interface EmployeeMapper {
    Employee  selectEmployeeById(String emp_no);
}
```

4．XML 映射文件

映射文件 DeptMapper.xml 的主要代码如下：

```
<mapper namespace="com.mialab.OneToMany_demo.mapper.DeptMapper">
    <!-- 根据 dept_id 查询部门信息，返回 resultMap -->
    <select id="selectDeptById"  parameterType="String"  resultMap="deptResult
Map">
        SELECT * FROM dept WHERE dept_id = #{dept_id}
    </select>

    <!-- 映射 Dept 对象的 resultMap -->
    <resultMap type="com.mialab.OneToMany_demo.domain.Dept" id="deptResultMap">
        <id property="dept_id" column="dept_id" />
        <result property="dept_name" column="dept_name" />
        <result property="dept_manager" column="dept_manager" />
        <!-- 一对多关联映射：collection fetchType="lazy"，表示延迟加载 -->
        <collection property="employees" javaType="ArrayList"
            column="dept_id"
             ofType="com.mialab.OneToMany_demo.domain.Employee"
            select=
    "com.mialab.OneToMany_demo.mapper.EmployeeMapper.selectEmployeeByDeptId"
            fetchType="lazy">
            <id property="emp_no" column="emp_no" />
            <result property="emp_name" column="emp_name" />
            <result property="dept_id" column="dept_id" />
            <result property="emp_gender" column="emp_gender" />
            <result property="emp_email" column="emp_email" />
```

```
        </collection>
    </resultMap>
</mapper>
```

映射文件 EmployeeMapper.xml 的主要代码如下：

```
<mapper namespace="com.mialab.OneToMany_demo.mapper.EmployeeMapper">
  <!-- 根据 emp_no 查询员工信息，多表连接，返回 resultMap -->
  <select id="selectEmployeeById" parameterType="String"
      resultMap="employeeResultMap">
      SELECT * FROM employee_basic e,dept d
      WHERE e.dept_id = d.dept_id
      AND e.emp_no = #{emp_no}
  </select>

  <!-- 根据部门 dept_id 查询员工信息，返回 resultMap -->
  <select id="selectEmployeeByDeptId" parameterType="String"
      resultMap="employeeResultMap">
      SELECT * FROM employee_basic WHERE dept_id = #{dept_id}
  </select>

  <!-- 映射 Student 对象的 resultMap -->
  <resultMap type="com.mialab.OneToMany_demo.domain.Employee"
            id="employeeResultMap">
      <id property="emp_no" column="emp_no" />
      <result property="emp_name" column="emp_name" />
      <result property="dept_id" column="dept_id" />
      <result property="emp_gender" column="emp_gender" />
      <result property="emp_email" column="emp_email" />

      <!-- 多对一关联映射：association -->
      <association property="dept"
              javaType="com.mialab.OneToMany_demo.domain.Dept">
          <id property="dept_id" column="dept_id" />
          <result property="dept_name" column="dept_name" />
          <result property="dept_manager" column="dept_manager" />
      </association>
  </resultMap>
</mapper>
```

5. 测试

OneToMany_Main 类的测试代码如下：

```
public class OneToMany_Main {
  public static void main(String[] args) {
      //测试一对多
      testSelectDeptById("101");
      //测试多对一
      //testSelectEmployeeById("HW9803");
  }

  private static void testSelectDeptById(String dept_id) {
      Logger log = Logger.getLogger(OneToMany_Main.class);
      SqlSession session = null;
      try {
```

```
            session =
DBOperatorMgr.getInstance().getSqlSessionFactory().openSession();
            DeptMapper mapper = session.getMapper(DeptMapper.class);
            Dept dept = mapper.selectDeptById(dept_id);
            //System.out.println(dept);
            log.info(dept);
            // log.info(dept.getEmployees());
            List<Employee> employees = dept.getEmployees();
            for(Employee employee : employees) {
                System.out.println(employee);
                //log.info(employee);
            }
        } finally {
            if (session != null) {
                session.close();
            }
        }
    }

    private static void testSelectEmployeeById(String emp_no) {
        Logger log = Logger.getLogger(OneToMany_Main.class);
        SqlSession session = null;
        try {
            session =
DBOperatorMgr.getInstance().getSqlSessionFactory().openSession();
            EmployeeMapper mapper = session.getMapper(EmployeeMapper.class);
            Employee employee = mapper.selectEmployeeById(emp_no);
            //System.out.println(employee);
            log.info(employee);
            log.info(employee.getDept());
        } finally {
            if (session != null) {
                session.close();
            }
        }
    }
```

测试一对多时控制台输出内容如下：

```
    Dept [dept_id=101, dept_name=人力资源部, dept_manager=HW9803]
    Employee [emp_no=HW9801, emp_name=张丹枫, dept_id=101, emp_gender=男, emp_email=zhang@163.com]
    Employee [emp_no=HW9802, emp_name=史密斯, dept_id=101, emp_gender=男, emp_email=smith@163.com]
    Employee [emp_no=HW9803, emp_name=余小男, dept_id=101, emp_gender=女, emp_email=yu@163.com]
```

测试多对一时控制台输出内容如下：

```
    DEBUG [main] - ==>  Preparing: SELECT * FROM employee_basic e,dept d WHERE e.dept_id = d.dept_id AND e.emp_no = ?
    DEBUG [main] - ==> Parameters: HW9803(String)
    DEBUG [main] - <==      Total: 1
```

```
    INFO [main] - Employee [emp_no=HW9803, emp_name=余小男, dept_id=101, emp_gender=女, emp_email=yu@163.com]
    INFO [main] - Dept [dept_id=101, dept_name=人力资源部, dept_manager=HW9803]
```

11.4.4 多对多关联

1. ManyToMany _demo 工程目录及文件

最终完成的 ManyToMany _demo 工程目录及文件如图 11-9 所示。

```
ManyToMany_demo
  src/main/java
    com.mialab.ManyToMany_demo.domain
      Employee.java
      Job.java
    com.mialab.ManyToMany_demo.main
      ManyToMany_Main.java
    com.mialab.ManyToMany_demo.mapper
      EmployeeMapper.java
      JobMapper.java
      EmployeeMapper.xml
      JobMapper.xml
    com.mialab.ManyToMany_demo.utils
      DBOperatorMgr.java
  src/main/resources
    log4j.properties
    mybatis-config.xml
  src/test/java
  JRE System Library [JavaSE-1.8]
  Maven Dependencies
  src
  target
  pom.xml
```

图 11-9 ManyToMany_demo 工程目录及文件

2. 创建实体类

这里假设一个员工可有多于一个的工作岗位，一个工作岗位可由一个或更多的员工来承担，员工和工作岗位是多对多的关系，于是创建实体类 Employee.java 和 Job.java。

Job.java 主要代码如下（此处 get 方法、set 方法和 toString 方法略，代码详见本书教学资源包 ManyToMany_demo 工程）：

```
public class Job implements Serializable {
  private String job_id;
  private String job_name;
  private String job_task;
  private List<Employee> employees;
  …
}
```

Employee.java 主要代码如下（此处 get 方法、set 方法和 toString 方法略）：

```
public class Employee implements Serializable {
  private String emp_no;
  private String emp_name;
  private String dept_id;
  private String emp_gender;
  private String emp_email;
```

```
    private List<Job> jobs;
    …
    }
```

必须注意的是，这里多对多的关联是通过被拆分为两个一对多的关联来实现的。

3．创建映射接口

在 JobMapper.java 中定义接口的方法：

```
    public interface JobMapper {
     Job  selectJobById(String job_id);
    }
```

在 EmployeeMapper.java 中定义接口的方法：

```
    public interface EmployeeMapper {
     Employee  selectEmployeeById(String emp_no);
    }
```

4．XML 映射文件

映射文件 JobMapper.xml 的主要代码如下：

```
    <mapper namespace="com.mialab.ManyToMany_demo.mapper.JobMapper">
     <resultMaptype="com.mialab.ManyToMany_demo.domain.Job"id="jobResultMap
">
        <id property="job_id" column="job_id" />
        <result property="job_name" column="job_name" />
        <result property="job_task" column="job_task" />

        <!-- 多对多映射的关键：collection -->
        <collection property="employees" javaType="ArrayList"
            column="job_id"
            ofType="com.mialab.ManyToMany_demo.domain.Employee"
            select=
    "com.mialab.ManyToMany_demo.mapper.EmployeeMapper.selectEmployeeByJobId"
            fetchType="lazy">
            <id property="emp_no" column="emp_no" />
            <result property="emp_name" column="emp_name" />
            <result property="dept_id" column="dept_id" />
            <result property="emp_gender" column="emp_gender" />
            <result property="emp_email" column="emp_email" />
        </collection>
     </resultMap>

     <selectid="selectJobById"parameterType="String"resultMap="jobResultMap
">
        SELECT * FROM job WHERE job_id = #{job_id}
     </select>

     <select id="selectJobByEmpId" parameterType="String"
        resultType="com.mialab.ManyToMany_demo.domain.Job">
        SELECT * FROM job WHERE job_id IN (
        SELECT job_id FROM
        employee_job WHERE emp_no = #{emp_no}
        )
     </select>
    </mapper>
```

映射文件 JobMapper.xml 的主要代码如下：

```
<mapper namespace="com.mialab.ManyToMany_demo.mapper.EmployeeMapper">
 <resultMap type="com.mialab.ManyToMany_demo.domain.Employee"
     id="employeeResultMap">
     <id property="emp_no" column="emp_no" />
     <result property="emp_name" column="emp_name" />
     <result property="dept_id" column="dept_id" />
     <result property="emp_gender" column="emp_gender" />
     <result property="emp_email" column="emp_email" />

     <!-- 多对多映射的关键：collection -->
     <collection property="jobs" javaType="ArrayList"
         column="emp_no"
         ofType="com.mialab.ManyToMany_demo.domain.Job"
         select=
"com.mialab.ManyToMany_demo.mapper.JobMapper.selectJobByEmpId"
         fetchType="lazy">
         <id property="job_id" column="job_id" />
         <result property="job_name" column="job_name" />
         <result property="job_task" column="job_task" />
     </collection>
 </resultMap>

 <select id="selectEmployeeById" parameterType="String"
     resultMap="employeeResultMap">
     SELECT * FROM employee_basic WHERE emp_no = #{emp_no}
 </select>

 <select id="selectEmployeeByJobId" parameterType="String"
     resultType="com.mialab.ManyToMany_demo.domain.Employee">
     SELECT * FROM employee_basic WHERE emp_no IN (
     SELECT emp_no FROM
     employee_job WHERE job_id = #{job_id}
     )
 </select>
 </mapper>
```

5. 测试

ManyToMany_Main 类的测试代码如下：

```
public class ManyToMany_Main {
 public static void main(String[] args) {
     //testSelectEmployeeById("HW9806");
     testSelectJobById("2601");
 }

 private static void testSelectJobById(String job_id) {
     Logger log = Logger.getLogger(ManyToMany_Main.class);
     SqlSession session = null;
     try {
         session =

DBOperatorMgr.getInstance().getSqlSessionFactory().openSession();
```

```
            JobMapper mapper = session.getMapper(JobMapper.class);
            Job job = mapper.selectJobById(job_id);
            // System.out.println(job);
            log.info(job);
            // log.info(job.getEmployees());
            List<Employee> employees = job.getEmployees();
            for (Employee employee : employees) {
                System.out.println(employee);
                // log.info(employee);
            }
        } finally {
            if (session != null) {
                session.close();
            }
        }
    }

    private static void testSelectEmployeeById(String emp_no) {
        Logger log = Logger.getLogger(ManyToMany_Main.class);
        SqlSession session = null;
        try {
            session =

DBOperatorMgr.getInstance().getSqlSessionFactory().openSession();
            EmployeeMapper mapper = session.getMapper(EmployeeMapper.class);
            Employee employee = mapper.selectEmployeeById(emp_no);
            // System.out.println(employee);
            log.info(employee);
            // log.info(employee.getJobs());
            List<Job> jobs = employee.getJobs();
            for (Job job : jobs) {
                System.out.println(job);
                // log.info(job);
            }
        } finally {
            if (session != null) {
                session.close();
            }
        }
    }
}
```

先测试 testSelectJobById()方法，表示一个工作岗位对应的若干员工，控制台输出内容如下：

```
Job [job_id=2601, job_name=职员, job_task=一般职员，非技术岗]
Employee [emp_no=HW9801, emp_name=张丹枫, dept_id=101, emp_gender=男, …]
Employee [emp_no=HW9802, emp_name=史密斯, dept_id=101, emp_gender=男, …]
Employee [emp_no=HW9804, emp_name=李莉莉, dept_id=102, emp_gender=女, …]
Employee [emp_no=HW9809, emp_name=许天仪, dept_id=103, emp_gender=男, …]
```

再注释掉以下语句：

```
// testSelectJobById("2601");
```

最后，再测试 testSelectEmployeeById()方法，表示的是某些员工可能有多于一个的工作岗位（工作角色），控制台输出内容如下：

```
Employee [emp_no=HW9806, emp_name=郭靖, dept_id=103, emp_gender=男, …]
```

```
Job [job_id=2605, job_name=架构师, job_task=负责系统架构]
Job [job_id=2606, job_name=技术总监, job_task=公司技术总负责]
```

11.5 本章小结

MyBatis的配置文件对整个MyBatis体系产生了深远的影响。在MyBatis的主配置文件中，配置项的顺序不能颠倒。如果颠倒了它们的顺序，MyBatis在解析XML文件的时候就会发生异常，导致程序无法运行。本章对MyBatis的配置元素也进行了较为详细的介绍。

映射器是MyBatis最复杂且最重要的组件。它由一个接口加上XML文件（或者注解）组成。在映射器中可以配置参数、各类SQL语句、存储过程、缓存、级联等复杂的内容，并且通过简单的映射规则映射到指定的POJO或者其他对象上，映射器能有效地消除JDBC底层的代码。

MyBatis的映射器也可以使用注解来完成，但是它在企业中应用不广，主要是因为注解不仅可读性较差，而且丢失了XML上下文相互引用的功能。

习题 11

1. 使用MyBatis编程实现表数据的增、删、改、查操作。
2. MyBatis的主配置文件有哪些配置元素（配置项）？如何使用？
3. MyBatis是如何处理一对一关系的？试编程说明。
4. MyBatis是如何处理一对多关系的？试编程说明。
5. MyBatis是如何处理多对多关系的？试编程说明。
6. 什么是映射器？SQL映射文件有哪些常用元素？其作用分别是什么？

第12章　动态 SQL

本章导读

动态 SQL 是 MyBatis 的强大特性之一。使用过 JDBC 或其他类似框架的人都会知道，根据不同条件拼接 SQL 语句时不仅不能忘记必要的空格，还要注意省略列名列表最后的逗号。这是一件很麻烦又让人感到痛苦的事情，而 MyBatis 的动态 SQL 特性可以让用户彻底摆脱这种痛苦。MyBatis 提供了对 SQL 语句的动态组装能力，大量的判断都可以在 MyBatis 的映射 XML 文件中配置，大大减少了编写代码的工作量。

本章主要内容有：MyBatis 中常用的动态 SQL 元素用法说明及使用示例，这里常用的动态 SQL 元素包括 if、choose（when、otherwise）、where、set、trim、foreach、bind 等。

12.1　示例：使用动态 SQL

1. 准备数据

仍使用前述数据库 hrm，其中有 5 张表：employee_basic、employee_school、dept、job、employee_job。本章示例主要使用了表 employee_basic 中的数据。

2. 创建工程 DynamicSQL_demo

在 Eclipse 中创建 Maven Project，选择 maven→archetype→quickstart，在“Group ID”文本框中输入“com.mialab”，在“Artifact ID”文本框中输入“DynamicSQL_demo”。

employee_basic 初始表数据和最终完成的 DynamicSQL_demo 工程目录结果如图 12-1 和图 12-2 所示。（代码详见本书教学资源包中的 DynamicSQL_demo 工程。）

emp_no	emp_name	dept_id	emp_gender	emp_email	emp_nation	emp_marriage	emp_health	emp_zzmm	emp_blood	emp_state
HW9801	张丹枫	101	男	zhang@163.com	汉族	已婚	良好	党员	O型	在职
HW9802	史密斯	101	男	smith@163.com	满族	已婚	良好	群众	A型	在职
HW9803	余小男	101	女	yu@163.com	汉族	未婚	良好	群众	B型	在职
HW9804	李莉莉	102	女	lili@163.com	回族	未婚	良好	团员	B型	在职
HW9805	Marry	102	女	marry@163.com	汉族	未婚	良好	群众	O型	在职
HW9806	郭靖	103	男	guo@163.com	汉族	已婚	良好	党员	O型	在职
HW9807	王强	103	男	wangqiang@163.com	汉族	已婚	良好	党员	O型	在职
HW9808	李娜	103	女	nana@163.com	汉族	未婚	良好	群众	O型	在职
HW9809	许天仪	103	男	xu@163.com	汉	已婚	一般	党员	O型	退休

图 12-1　employee_basic 表中的数据

3. 创建实体对象映射数据库表

Employee 对象用于映射 employee_basic 表，Employee.java 主要代码如下（此处 get 方法、set 方法和 toString 方法略，详见本书教学资源包中的 DynamicSQL_demo 工程）：

```
public class Employee implements Serializable {
    private String emp_no;
    private String emp_name;
    private String dept_id;
```

```
        private String emp_gender;
        private String emp_email;
        private String emp_nation;
        private String emp_marriage;
        private String emp_health;
        private String emp_zzmm;
        private String emp_blood;
    private String emp_state;
    …
}
```

```
DynamicSQL_demo
  src/main/java
    com.mialab.DynamicSQL_demo.domain
      Employee.java
    com.mialab.DynamicSQL_demo.main
      DynamicSQL_Main.java
    com.mialab.DynamicSQL_demo.mapper
      EmployeeMapper.java
      EmployeeMapper.xml
    com.mialab.DynamicSQL_demo.utils
      DBOperatorMgr.java
  src/main/resources
    log4j.properties
    mybatis-config.xml
  src/test/java
  JRE System Library [JavaSE-1.8]
  Maven Dependencies
  src
  target
  pom.xml
```

图 12-2 DynamicSQL_demo 工程目录结构

4. 创建映射接口和映射文件

接口 EmployeeMapper.java 的主要代码如下：

```
public interface EmployeeMapper {
    Employee selectEmployeeById(String emp_no);

    // 使用 if，单个条件
    List<Employee> selectEmployeeByDeptId(HashMap<String, Object> params);
    // 使用 if，多个条件
    List<Employee> selectEmployeeByMultiple(HashMap<String, Object> params);

    // 使用 choose (when、otherwise)
    List<Employee> findEmployee_1(HashMap<String, Object> params);
    // 使用 where
    List<Employee> findEmployee_2(HashMap<String, Object> params);
    // 使用 trim
    List<Employee> findEmployee_5(HashMap<String, Object> params);

    // 动态更新语句中使用 set
    void updateEmployee(Employee employee);
    // 使用 foreach
    List<Employee> findEmployee_3(List<String> emp_nos);
    // 使用 bind
    List<Employee> findEmployee_4(Employee employee);
}
```

映射文件 EmployeeMapper.xml 中的具体内容将在后文讲述。

5. 测试

DynamicSQL_Main 类的 main 方法的主要代码如下：

```
public static void main(String[] args) {
    // 使用 if，有单个查询条件，如可根据 dept_id（部门编号）来查询员工信息
    testSelectEmployeeByDeptId("101");
    // 使用 if，有多个查询条件，如可根据民族、血型、已婚未婚等条件来查询员工信息
    // testSelectEmployeeByMultiple();

    // 使用 choose（when、otherwise） 查找员工
    // testFindEmployee_1();
    // 使用 where
    // testFindEmployee_2();

    // 动态更新语句中使用 set
    // testUpdateEmployee("HW9809");
    // 使用 foreach
    // testFindEmployee_3();
    // 使用 bind
    // testFindEmployee_4();

    // 使用 trim
    // testFindEmployee_5();
}
```

main 方法分别调用了 DynamicSQL_Main 类中定义的 testSelectEmployeeByDeptId("101")、testSelectEmployeeByMultiple()、testFindEmployee_1()、testFindEmployee_2()等方法，具体运行测试结果将在后文分析讲述。（注意，运行测试时，main 方法每次只调用一个方法。）

12.2 if

if 标签的使用见 DynamicSQL_demo 工程的 EmployeeMapper.xml 文件。

```
<!-- 映射 Employee 对象的 resultMap -->
<resultMap type="com.mialab.DynamicSQL_demo.domain.Employee"
    id="employeeMap">
    <id property="emp_no" column="emp_no" />
    <result property="emp_name" column="emp_name" />
    <result property="dept_id" column="dept_id" />
    <result property="emp_gender" column="emp_gender" />
    <result property="emp_email" column="emp_email" />
    <result property="emp_nation" column="emp_nation" />
    <result property="emp_marriage" column="emp_marriage" />
    <result property="emp_health" column="emp_health" />
    <result property="emp_zzmm" column="emp_zzmm" />
    <result property="emp_blood" column="emp_blood" />
    <result property="emp_state" column="emp_state" />
</resultMap>

<!-- if -->
```

```
    <select id="selectEmployeeByDeptId" resultMap="employeeMap">
        SELECT * FROM employee_basic WHERE emp_health = '良好'
        <!-- 可选条件，如果传进来的参数有 dept_id 属性，则加上 dept_id 查询条件 -->
        <if test="dept_id != null ">
            and dept_id = #{dept_id}
        </if>
    </select>

    <!--  if 的使用，也可以有多个查询条件 -->
    <select id="selectEmployeeByMultiple" resultMap="employeeMap">
        SELECT * FROM employee_basic WHERE emp_health = '良好'
        <!-- 如可根据民族、血型以及已婚未婚等条件来查询员工信息 -->
        <if test="emp_nation != null and emp_blood != null
                            and emp_marriage != null">
            and emp_nation = #{emp_nation} and emp_blood = #{emp_blood} and
            emp_marriage = #{emp_marriage}
        </if>
    </select>
```

DynamicSQL_Main 类中测试 if 使用（单个可选条件）的代码如下：

```
    public static void main(String[] args) {
        // 使用 if，有单个查询条件，如可根据 dept_id（部门编号）来查询员工信息
        testSelectEmployeeByDeptId("101");
    }

    // 测试 if，单个条件
    private static void testSelectEmployeeByDeptId(String dept_id) {
        Logger log = Logger.getLogger(DynamicSQL_Main.class);
        SqlSession session = null;
        try {
            session = DBOperatorMgr.getInstance().getSqlSessionFactory().open
Session();

            EmployeeMapper mapper = session.getMapper(EmployeeMapper.class);

            HashMap<String, Object> params = new HashMap<String, Object>();
            // 设置 dept_id 属性
            params.put("dept_id", dept_id);

            List<Employee> employees = mapper.selectEmployeeByDeptId(params);
            // log.info(employees);
            for (Employee employee : employees) {
                System.out.println(employee);
                // log.info(employee);
            }
        } finally {
            if (session != null) {
                session.close();
            }
        }
    }
```

测试运行 main 方法，控制台输出内容如下：

```
    DEBUG [main] - ==>  Preparing: SELECT * FROM employee_basic WHERE emp_health
= '良好' and dept_id = ?
```

```
DEBUG [main] - ==> Parameters: 101(String)
DEBUG [main] - <==      Total: 3
Employee [emp_no=HW9801, emp_name=张丹枫, dept_id=101, emp_gender=男, …]
Employee [emp_no=HW9802, emp_name=史密斯, dept_id=101, emp_gender=男, …]
Employee [emp_no=HW9803, emp_name=余小男, dept_id=101, emp_gender=女, …]
```

需要注意的是，在 EmployeeMapper.java 中定义的接口方法 selectEmployeeByDeptId 和 selectEmployeeByDeptId 中的参数都是 HashMap 对象，在测试方法中把查询条件封装到 HashMap 对象上，再传递给 MyBatis。这里的测试方法是 testSelectEmployeeByDeptId()和 testSelectEmployeeByMultiple()。

```
public interface EmployeeMapper {
 // 使用 if，单个条件
 List<Employee> selectEmployeeByDeptId(HashMap<String, Object> params);
 // 使用 if，多个条件
 List<Employee> selectEmployeeByMultiple(HashMap<String, Object> params);
}
```

DynamicSQL_Main 类中测试 if 使用（多个可选条件）的代码如下：

```
public static void main(String[] args) {
// 使用 if，多个条件
List<Employee> selectEmployeeByMultiple(HashMap<String, Object> params);
}

// 测试 if，多个条件
private static void testSelectEmployeeByMultiple() {
    Logger log = Logger.getLogger(DynamicSQL_Main.class);
    SqlSession session = null;
    try {
        session = DBOperatorMgr.getInstance().getSqlSessionFactory().open
Session();
        EmployeeMapper mapper = session.getMapper(EmployeeMapper.class);

        HashMap<String, Object> params = new HashMap<String, Object>();
        // 设置 emp_nation 属性、emp_blood 属性、emp_marriage 属性
        params.put("emp_nation", "汉族");
        params.put("emp_blood", "O 型");
        params.put("emp_marriage", "未婚");

        List<Employee> employees = mapper.selectEmployeeByMultiple(params);
        // log.info(employees);
        for (Employee employee : employees) {
            System.out.println(employee);
            // log.info(employee);
        }
    } finally {
        if (session != null) {
            session.close();
        }
    }
}
```

测试运行 main 方法，控制台输出内容如下：

```
DEBUG [main] - ==>  Preparing: SELECT * FROM employee_basic WHERE emp_health
= '良好' and emp_nation = ? and emp_blood = ? and emp_marriage = ?
```

```
DEBUG [main] - ==> Parameters: 汉族(String), O型(String), 未婚(String)
DEBUG [main] - <==      Total: 2
Employee [emp_no=HW9805, emp_name=Marry, dept_id=102, emp_gender=女, …]
Employee [emp_no=HW9808, emp_name=李娜, dept_id=103, emp_gender=女, …]
```

如果注释以下语句：

```
//params.put("emp_blood", "O型");
```

或者注释：

```
//params.put("emp_nation", "汉族");
```

分别会发生什么情况呢？控制台输出又会怎样呢？

12.3 choose、when、otherwise

if 标签提供了基本判断，但无法实现类似于多分支选择的逻辑。要想实现这样的逻辑，就需要用到 choose、when、otherwise 标签。choose、when、otherwise 标签的使用见 DynamicSQL_demo 工程的 EmployeeMapper.xml 文件。

```
<!-- 使用 choose (when、otherwise) 查找员工 -->
<select id="findEmployee_1" resultMap="employeeMap">
 SELECT * FROM employee_basic WHERE emp_health = '良好'
 <!-- 如果传入了 dept_id，则根据 dept_id 查询-->
 <!-- 如果没有传入 dept_id ，则根据 emp_marriage 和 emp_gender 查询-->
 <!-- 否则查询 emp_zzmm 是党员的员工信息 -->
 <choose>
     <when test="dept_id != null">
         and dept_id = #{dept_id}
     </when>
     <when test="emp_marriage != null and emp_gender != null">
         and emp_marriage = #{emp_marriage} and emp_gender =
             #{emp_gender}
     </when>
     <otherwise>
         and emp_zzmm = '党员'
     </otherwise>
 </choose>
</select>
```

choose 元素中包含了 when 和 otherwise 两个标签，一个 choose 至少有一个 when 标签，有 0 个或者 1 个 otherwise 标签。

在 EmployeeMapper.java 中定义的接口方法如下：

```
public interface EmployeeMapper {
 // 使用 choose (when、otherwise)
 List<Employee> findEmployee_1(HashMap<String, Object> params);
}
```

DynamicSQL_Main 类中测试使用 choose、when、otherwise 标签的代码如下：

```
public static void main(String[] args) {
    // 使用 choose (when、otherwise) 查找员工
    testFindEmployee_1();
}

// 使用 choose (when、otherwise)
```

```
    private static void testFindEmployee_1() {
        Logger log = Logger.getLogger(DynamicSQL_Main.class);
        SqlSession session = null;
        try {
            session  = DBOperatorMgr.getInstance().getSqlSessionFactory().
openSession();
            EmployeeMapper mapper = session.getMapper(EmployeeMapper.class);
            HashMap<String, Object> params = new HashMap<String, Object>();
            // 设置 dept_id 属性，或 emp_marriage 属性和 emp_gender 属性
            // params.put("dept_id", "101");
            params.put("emp_marriage", "未婚");
            params.put("emp_gender", "女");

            List<Employee> employees = mapper.findEmployee_1(params);
            // log.info(employees);
            for (Employee employee : employees) {
                System.out.println(employee);
                // log.info(employee);
            }
        } finally {
            if (session != null) {
                session.close();
            }
        }
    }
```

测试运行 main 方法，控制台输出内容如下：

```
    DEBUG [main] - ==>  Preparing: SELECT * FROM employee_basic WHERE emp_health
= '良好' and emp_marriage = ? and emp_gender = ?
    DEBUG [main] - ==> Parameters: 未婚(String), 女(String)
    DEBUG [main] - <==      Total: 4
    Employee [emp_no=HW9803, emp_name=余小男, dept_id=101, emp_gender=女, …]
    Employee [emp_no=HW9804, emp_name=李莉莉, dept_id=102, emp_gender=女, …]
    Employee [emp_no=HW9805, emp_name=Marry, dept_id=102, emp_gender=女, …]
    Employee [emp_no=HW9808, emp_name=李娜, dept_id=103, emp_gender=女, …]
```

测试代码中的执行路径是 choose 标签下的第二分支。如果以第一分支或者默认分支作为执行路径，又该如何？控制台输出又会怎样呢？

12.4 where、set、trim

1. where 标签的用法

where 标签的作用：如果该标签包含的元素中有返回值，则插入一个 where；如果 where 后面的字符串是以 AND 和 OR 开头的，则将它们剔除。

where 标签的使用见 DynamicSQL_demo 工程的 EmployeeMapper.xml 文件。

```
    <!-- 使用 where -->
    <select id="findEmployee_2" resultMap="employeeMap">
        SELECT * FROM employee_basic
        <where>
            <if test="emp_health != null ">
                emp_health = #{emp_health}
```

```
            </if>
            <if test="dept_id != null ">
                and dept_id = #{dept_id}
            </if>
            <if test="emp_marriage != null and emp_gender != null">
                and emp_marriage = #{emp_marriage} and emp_gender =
                #{emp_gender}
            </if>
        </where>
    </select>
```

在 EmployeeMapper.java 中定义的接口方法如下：

```
    public interface EmployeeMapper {
        // 使用 where
        List<Employee> findEmployee_2(HashMap<String, Object> params);
    }
```

DynamicSQL_Main 类中测试使用 where 标签的代码如下：

```
    public static void main(String[] args) {
        // 使用 where
         testFindEmployee_2();
    }

    // 使用 where
    private static void testFindEmployee_2() {
        Logger log = Logger.getLogger(DynamicSQL_Main.class);
        SqlSession session = null;
        try {
            session  =  DBOperatorMgr.getInstance().getSqlSessionFactory().
openSession();
            EmployeeMapper mapper = session.getMapper(EmployeeMapper.class);
            HashMap<String, Object> params = new HashMap<String, Object>();
            // 设置 emp_health 属性、dept_id 属性
            // 或 emp_marriage 属性和 emp_gender 属性
            params.put("emp_health", "良好");
            params.put("dept_id", "101");
            params.put("emp_marriage", "未婚");
            params.put("emp_gender", "女");

            List<Employee> employees = mapper.findEmployee_2(params);
            // log.info(employees);
            for (Employee employee : employees) {
                System.out.println(employee);
                // log.info(employee);
            }
        } finally {
            if (session != null) {
                session.close();
            }
        }
    }
```

测试运行 main 方法，控制台输出内容如下：

```
    DEBUG [main] - ==>  Preparing: SELECT * FROM employee_basic WHERE emp_health
= ? and dept_id = ? and emp_marriage = ? and emp_gender = ?
```

```
    DEBUG [main] - ==> Parameters: 良好(String), 101(String), 未婚(String), 女(String)
    DEBUG [main] - <==      Total: 1
    Employee [emp_no=HW9803, emp_name=余小男, dept_id=101, emp_gender=女, emp_email=yu@163.com, emp_nation=汉族, emp_marriage=未婚, emp_health=良好, emp_zzmm=群众, emp_blood=B型, emp_state=在职]
```

如果注释掉以下语句，情况又会如何？控制台输出又会怎样呢？

```
    // params.put("emp_gender", "女");
```

2. set 标签的用法

set 标签的作用：如果该标签包含的元素中有返回值，则插入一个 set；如果 set 后面的字符串是以逗号结尾的，则将这个逗号剔除。

set 标签的使用见 DynamicSQL_demo 工程的 EmployeeMapper.xml 文件。

```
    <!-- 动态更新语句中使用 set -->
    <update id="updateEmployee" parameterType=
        "com.mialab.DynamicSQL_demo.domain.Employee">
        update employee_basic
        <set>
            <if test="emp_name != null">emp_name=#{emp_name},</if>
            <if test="dept_id != null">dept_id=#{dept_id},</if>
            <if test="emp_gender != null">emp_gender=#{emp_gender},</if>
            <if test="emp_email != null">emp_email=#{emp_email},</if>
            <if test="emp_nation != null">emp_nation=#{emp_nation},</if>
            <if test="emp_marriage != null">emp_marriage=#{emp_marriage},</if>
            <if test="emp_health != null">emp_health=#{emp_health},</if>
            <if test="emp_zzmm != null">emp_zzmm=#{emp_zzmm},</if>
            <if test="emp_blood != null">emp_blood=#{emp_blood},</if>
            <if test="emp_state != null">emp_state=#{emp_state}</if>
        </set>
        where emp_no=#{emp_no}
    </update>
```

在 EmployeeMapper.java 中定义的接口方法如下：

```
   public interface EmployeeMapper {
        // 动态更新语句中使用 set
        void updateEmployee(Employee employee);
   }
```

DynamicSQL_Main 类中测试使用 set 标签的代码如下：

```
    public static void main(String[] args) {
        // 动态更新语句中使用 set
        testUpdateEmployee("HW9809");
    }

    // 动态更新语句中使用 set
    private static void testUpdateEmployee(String emp_no) {
        Logger log = Logger.getLogger(DynamicSQL_Main.class);
        SqlSession session = null;
        try {
            session = DBOperatorMgr.getInstance().getSqlSessionFactory().openSession();
            EmployeeMapper mapper = session.getMapper(EmployeeMapper.class);
```

```
        Employee employee = mapper.selectEmployeeById(emp_no);
        System.out.println(employee);
        // 设置需要修改的属性
        employee.setEmp_name("马宏魁");
        employee.setEmp_nation("回");
        employee.setEmp_blood("B 型");
        employee.setEmp_email("ma@163.com");
        employee.setEmp_zzmm("群众");
        employee.setEmp_state("退休");
        mapper.updateEmployee(employee);
        session.commit();
        System.out.println(employee);
        // log.info(employee);
    } catch (Exception ex) {
        session.rollback();
        ex.printStackTrace();
    } finally {
        if (session != null) {
            session.close();
        }
    }
}
```

测试运行 main 方法，employee_basic 表中的记录发生改变，并得到控制台输出的如下内容：

```
DEBUG [main] - ==> Parameters: 马宏魁(String), 103(String), 男(String), ma@163.com(String), 回(String), 已婚(String), 一般(String), 群众(String), B 型(String), 退休(String), HW9809(String)
DEBUG [main] - <==    Updates: 1
```

3. trim 标签的用法

where 标签对应的 trim 实现如下。

```
<trim prefix="WHERE" prefixOverrides="AND |OR ">
…
</trim>
```

注意：这里的 AND 和 OR 后面的空格不能省略。

trim 标签有以下属性。

① prefix：当 trim 元素内包含内容时，会给内容添加 prefix 指定的前缀。

② prefixOverrides：当 trim 元素内包含内容时，会把内容中匹配的前缀字符串去掉。

trim 标签的使用见 DynamicSQL_demo 工程的 EmployeeMapper.xml 文件。

```
<!-- 使用 trim -->
<select id="findEmployee_5" resultMap="employeeMap">
    SELECT * FROM employee_basic
    <trim prefix="where" prefixOverrides="and ">
        <if test="dept_id != null ">
            and dept_id = #{dept_id}
        </if>
        <if test="emp_marriage != null and emp_gender != null">
            and emp_marriage = #{emp_marriage} and emp_gender =
            #{emp_gender}
        </if>
    </trim>
</select>
```

在 EmployeeMapper.java 中定义的接口方法如下：

```
public interface EmployeeMapper {
    // 使用 trim
    List<Employee> findEmployee_5(HashMap<String, Object> params);
}
```

DynamicSQL_Main 类中测试使用 trim 标签的代码如下：

```
public static void main(String[] args) {
    // 使用 trim
    testFindEmployee_5();
}

// 使用 trim
private static void testFindEmployee_5() {
    Logger log = Logger.getLogger(DynamicSQL_Main.class);
    SqlSession session = null;
    try {
        session = DBOperatorMgr.getInstance().getSqlSessionFactory().
openSession();
        EmployeeMapper mapper = session.getMapper(EmployeeMapper.class);
        HashMap<String, Object> params = new HashMap<String, Object>();
        // 设置 dept_id 属性、emp_marriage 属性、emp_gender 属性
        params.put("dept_id", "101");
        params.put("emp_marriage", "未婚");
        params.put("emp_gender", "女");
        List<Employee> employees = mapper.findEmployee_5(params);
        // log.info(employees);
        for (Employee emp : employees) {
            System.out.println(emp);
        }
    } catch (Exception ex) {
        session.rollback();
        ex.printStackTrace();
    } finally {
        if (session != null) {
            session.close();
        }
    }
}
```

测试运行 main 方法，控制台输出内容如下：

```
DEBUG [main] - ==>  Preparing: SELECT * FROM employee_basic where dept_id
= ? and emp_marriage = ? and emp_gender = ?
DEBUG [main] - ==> Parameters: 101(String), 未婚(String), 女(String)
DEBUG [main] - <==      Total: 1
Employee [emp_no=HW9803, emp_name=余小男, dept_id=101, emp_gender=女,
emp_email=yu@163.com, emp_nation=汉族, emp_marriage=未婚, emp_health=良好,
emp_zzmm=群众, emp_blood=B 型, emp_state=在职]
```

12.5 foreach

foreach 标签的使用见 DynamicSQL_demo 工程的 EmployeeMapper.xml 文件。

```
    <!-- 使用 foreach -->
    <select id="findEmployee_3" resultMap="employeeMap">
        SELECT * FROM employee_basic WHERE emp_no in
        <foreach item="item" index="index" collection="list"
                open="("separator="," close=")">
            #{item}
        </foreach>
    </select>
```

在 EmployeeMapper.java 中定义的接口方法如下：

```
    public interface EmployeeMapper {
        // 使用 foreach
        List<Employee> findEmployee_3(List<String> emp_nos);
    }
```

DynamicSQL_Main 类中测试使用 foreach 标签的代码如下：

```
    public static void main(String[] args) {
        // 使用 foreach
        testFindEmployee_3();
    }
    // 使用 foreach
    private static void testFindEmployee_3() {
        Logger log = Logger.getLogger(DynamicSQL_Main.class);
        SqlSession session = null;
        try {
            session  =  DBOperatorMgr.getInstance().getSqlSessionFactory().
openSession();
            EmployeeMapper mapper = session.getMapper(EmployeeMapper.class);
            List<String> emp_nos = new ArrayList<String>();
            emp_nos.add("HW9803");
            emp_nos.add("HW9804");
            emp_nos.add("HW9808");
            List<Employee> employees = mapper.findEmployee_3(emp_nos);
            // log.info(employees);
            for (Employee employee : employees) {
                System.out.println(employee);
                // log.info(employee);
            }
        } catch (Exception ex) {
            session.rollback();
            ex.printStackTrace();
        } finally {
            if (session != null) {
                session.close();
            }
        }
    }
```

测试运行 main 方法，控制台输出内容如下：

```
    DEBUG [main] - ==>  Preparing: SELECT * FROM employee_basic WHERE emp_no
in ( ? , ? , ? )
    DEBUG  [main]  -  ==>  Parameters:  HW9803(String),  HW9804(String),
HW9808(String)
    DEBUG [main] - <==      Total: 3
    Employee [emp_no=HW9803, emp_name=余小男, dept_id=101, emp_gender=女, …]
```

```
Employee [emp_no=HW9804, emp_name=李莉莉, dept_id=102, emp_gender=女, …]
Employee [emp_no=HW9808, emp_name=李娜, dept_id=103, emp_gender=女, …]
```

12.6　bind

bind 标签可以使用 OGNL 表达式创建一个变量并将其绑定到上下文中。

bind 标签的使用见 DynamicSQL_demo 工程的 EmployeeMapper.xml 文件。

```
    <!-- 使用 bind -->
    <select id="findEmployee_4" resultMap="employeeMap">
        <bind name="employeeNameLike" value="'%' + emp_name + '%'" />
        SELECT * FROM employee_basic WHERE emp_name LIKE #{employeeNameLike}
  </select>
```

在 EmployeeMapper.java 中定义的接口方法如下：

```
  public interface EmployeeMapper {
      // 使用 bind
      List<Employee> findEmployee_4(Employee employee);
  }
```

DynamicSQL_Main 类中测试使用 bind 标签的代码如下：

```
    public static void main(String[] args) {
        // 使用 bind
        testFindEmployee_4();
    }

    // 使用 bind
    private static void testFindEmployee_4() {
        Logger log = Logger.getLogger(DynamicSQL_Main.class);
        SqlSession session = null;
        try {
            session  =  DBOperatorMgr.getInstance().getSqlSessionFactory().
openSession();
            EmployeeMapper mapper = session.getMapper(EmployeeMapper.class);
            Employee employee = new Employee();
            employee.setEmp_name("李");
            List<Employee> employees = mapper.findEmployee_4(employee);
            // log.info(employees);
            for (Employee emp : employees) {
                System.out.println(emp);
                // log.info(employee);
            }
        } catch (Exception ex) {
            session.rollback();
            ex.printStackTrace();
        } finally {
            if (session != null) {
                session.close();
            }
        }
    }
```

测试运行 main 方法，控制台输出内容如下：

```
DEBUG [main] - ==>  Preparing: SELECT * FROM employee_basic WHERE emp_name
```

```
LIKE ?
    DEBUG [main] - ==> Parameters: %李%(String)
    DEBUG [main] - <==      Total: 2
    Employee [emp_no=HW9804, emp_name=李莉莉, dept_id=102, emp_gender=女, emp_email=lili@163.com, emp_nation=回族, emp_marriage=未婚, emp_health=良好, emp_zzmm=团员, emp_blood=B型, emp_state=在职]
    Employee [emp_no=HW9808, emp_name=李娜, dept_id=103, emp_gender=女, emp_email=nana@163.com, emp_nation=汉族, emp_marriage=未婚, emp_health=良好, emp_zzmm=群众, emp_blood=O型, emp_state=在职]
```

12.7 本章小结

MyBatis 提供了对 SQL 语句的动态组装能力，使用 XML 的几个简单的元素，便能完成动态 SQL 的功能，大量的判断都可以在 MyBatis 的映射 XML 文件中配置，以达到许多需要大量代码才能实现的功能，大大减少了代码量。这体现了 MyBatis 具有的灵活性、高度可配置性和良好的可维护性。

习题 12

1．MyBatis 中常用的动态 SQL 元素有哪些？

2．MyBatis 动态 SQL 元素中的 if、choose、when、otherwise 如何使用？试编程加以说明。

3．MyBatis 动态 SQL 元素中的 where、set、trim 如何使用？试编程加以说明。

4．MyBatis 动态 SQL 元素中的 foreach、bind 如何使用？试编程加以说明。

第 13 章 MyBatis 其他

本章导读

MyBatis 注解方式就是将 SQL 语句直接写在接口上，其优点在于对于需求比较简单的系统，效率较高；缺点是当 SQL 有变化时需要重新编译代码。使用缓存可以使应用更快地获取数据，避免频繁的数据库交互，在查询越多、缓存命中率越高的情况下，使用缓存的作用就越明显。本章主要内容有：（1）MyBatis 注解方式；（2）MyBatis 缓存配置，包括一级缓存和二级缓存。

13.1 MyBatis 注解方式

以下介绍如何使用注解方式。

13.1.1 使用注解方式实现表数据的增、删、改、查

【示例】使用注解方式实现表数据的增、删、改、查。

1．创建工程 annotation _DML_demo

在 Eclipse 中创建 Maven Project，选择 maven→archetype→quickstart，在“Group ID”文本框中输入“com.mialab”，在“Artifact ID”文本框中输入“annotation_DML_demo”。最终完成的 annotation_DML_demo 工程目录结构如图 13-1 所示，student 初始表中的数据如图 13-2 所示。

```
annotation_DML_demo
  src/main/java
    com.mialab.annotation_DML_demo.domain
      Student.java
    com.mialab.annotation_DML_demo.main
      AnnotationDML_Main.java
    com.mialab.annotation_DML_demo.mapper
      StudentMapper.java
    com.mialab.annotation_DML_demo.utils
      DBOperatorMgr.java
  src/main/resources
    log4j.properties
    mybatis-config.xml
```

图 13-1　annotation_DML_demo 工程目录结构

sno	name	sex	age	dept_no
20171508	李勇	男	20	2601
20171509	刘娟	女	19	2602
20171699	周小莉	女	20	2629
20171510	余小男	女	22	2602

图 13-2　student 初始表中的数据

pom.xml、mybatis-config.xml、数据库操作工具类 DBOperatorMgr.java、log4j.properties 创建及配置、POJO 类的创建及实现代码都类似以前示例，这里不再赘述，详见本书教学资源包 ch13 目录下的 annotation_DML_demo 工程。

2．创建映射接口 StudentMapper

在 src/main/java 中创建 Package“com.mialab.annotation_DML_demo.mapper”，如图 13-1 所示。在 mapper 包中创建接口 StudentMapper.java，具体代码如下。

```
    public interface StudentMapper {
     @Select("SELECT * FROM STUDENT WHERE sno = #{sno}")
     @Results({
         @Result(id=true,column="sno",property="sno"),
         @Result(column="name",property="name"),
         @Result(column="sex",property="sex"),
         @Result(column="age",property="age"),
         @Result(column="dept_no",property="dept_no")
     })
     Student getStudent(String sno);

     @Insert("INSERT INTO STUDENT(sno,name,sex,age,dept_no) VALUES(#{sno},
#{name},#{sex},#{age},#{dept_no})")
     int addStudent(Student student);

     @Select("SELECT * FROM STUDENT")
     List<Student> getSudentAll();

     @Update("UPDATE STUDENT SET name = #{name},sex = #{sex},age = #{age},
dept_no = #{dept_no} WHERE sno = #{sno}")
     int updateStudent(Student student);

     @Delete("DELETE FROM STUDENT WHERE sno = #{sno}")
     int deleteStudent(String sno);
    }
```

@Select 映射查询的 SQL 语句，@Insert 映射插入的 SQL 语句，@Update 映射更新的 SQL 语句，@Delete 映射删除的 SQL 语句；@Result 表示列和属性之间的单独结果映射，它的属性 id 是一个布尔值，表示是否被用于主键映射；@Results 则表示多个结果映射列表。

3．测试

在 src/main/java 中创建 Package“com.mialab.annotation_DML_demo.main”，如图 13-1 所示，在此包中创建测试类 AnnotationDML_Main.java，其主要代码如下：

```
    public class AnnotationDML_Main {
     public static void main(String[] args) {
         // testInsert();
          testSelectAll();
         //testSelect("20171509");
         // testUpdate();
         // testDelete("20171628");
     }

     private static void testDelete(String sno) {
         Logger log = Logger.getLogger(AnnotationDML_Main.class);
         SqlSession session = null;
         try {
             session =

  DBOperatorMgr.getInstance().getSqlSessionFactory().openSession();
             StudentMapper mapper = session.getMapper(StudentMapper.class);
             mapper.deleteStudent(sno);
             session.commit();
```

```
        } catch(Exception ex) {
            session.rollback();
            ex.printStackTrace();
        } finally {
            if (session != null) {
                session.close();
            }
        }
    }

    private static void testUpdate() {
        Logger log = Logger.getLogger(AnnotationDML_Main.class);
        SqlSession session = null;
        try {
            session =

  DBOperatorMgr.getInstance().getSqlSessionFactory().openSession();
            StudentMapper mapper = session.getMapper(StudentMapper.class);
            Student student = new Student();
            student.setSno("20171628");
            student.setName("苏东坡");
            student.setAge(68);
            student.setSex("女");
            student.setDept_no("2612");
            log.info(student);
            mapper.updateStudent(student);
            session.commit();
        } catch(Exception ex) {
            session.rollback();
            ex.printStackTrace();
        } finally {
            if (session != null) {
                session.close();
            }
        }
    }

    private static void testInsert() {
        Logger log = Logger.getLogger(AnnotationDML_Main.class);
        SqlSession session = null;
        try {
            session =

  DBOperatorMgr.getInstance().getSqlSessionFactory().openSession();
            StudentMapper mapper = session.getMapper(StudentMapper.class);
            Student student = new Student();
            student.setSno("20171628");
            student.setName("杜甫");
            student.setAge(69);
            student.setSex("男");
            student.setDept_no("2612");
            log.info(student);
            mapper.addStudent(student);
```

```
            session.commit();
        } catch(Exception ex) {
            session.rollback();
            ex.printStackTrace();
        } finally {
            if (session != null) {
                session.close();
            }
        }
    }

    private static void testSelect(String sno) {
        Logger log = Logger.getLogger(AnnotationDML_Main.class);
        SqlSession session = null;
        try {
            session =
DBOperatorMgr.getInstance().getSqlSessionFactory().openSession();
            StudentMapper mapper = session.getMapper(StudentMapper.class);
            Student student = mapper.getStudent(sno);
            //System.out.println(student);
            log.info(student);
        } finally {
            if (session != null) {
                session.close();
            }
        }
    }

    private static void testSelectAll() {
        Logger log = Logger.getLogger(AnnotationDML_Main.class);
        SqlSession session = null;
        try {
            session =

DBOperatorMgr.getInstance().getSqlSessionFactory().openSession();
            StudentMapper mapper = session.getMapper(StudentMapper.class);
            List<Student> stu_list = mapper.getSudentAll();
            for(Student stu:stu_list) {
                //System.out.println(stu);
                log.info(stu);
            }
        } finally {
            if (session != null) {
                session.close();
            }
        }
    }
}
```

先测试 testSelectAll()方法。如图 13-1 所示，右击 com.mialab. annotation_DML_demo.main 包下的 AnnotationDML_Main.java，在弹出的快捷菜单中选择“Run As → Java Application”选项，在控制台上可以得到以下的结果：

```
    INFO [main] - Student [sno=20171508, name=李勇, sex=男, age=20, dept_no=2601]
    INFO [main] - Student [sno=20171509, name=刘娟, sex=女, age=19, dept_no=2602]
    INFO [main] - Student [sno=20171699, name= 周 小 莉 , sex= 女 , age=20,
dept_no=2629]
    INFO [main] - Student [sno=20171510, name= 余 小 男 , sex= 女 , age=22,
dept_no=2602]
```

可分别对增、删、改、查的方法 testInsert()、testDelete()、testUpdate()等进行测试。

13.1.2　使用注解的动态 SQL

【示例】使用注解的动态 SQL。

1．创建工程 annotation_DSQL_demo

在 Eclipse 中创建 Maven Project，选择 maven→archetype→quickstart，在“Group ID”文本框中输入“com.mialab”，在“Artifact ID”文本框中输入“annotation_DSQL_demo”。最终完成的 annotation_DSQL_demo 工程目录结构如图 13-3 所示，student 初始表中的数据如图 13-4 所示。

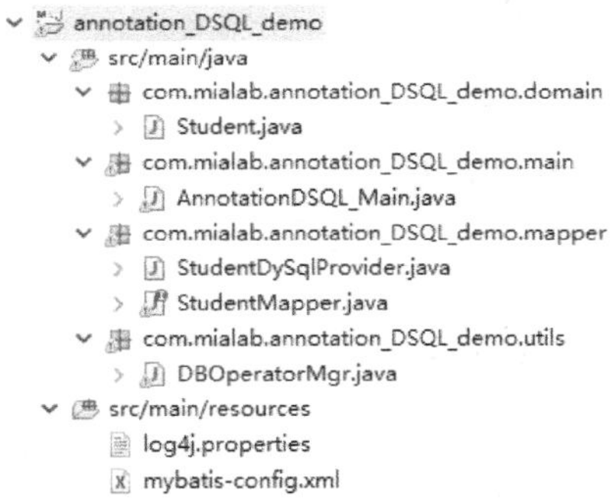

```
annotation_DSQL_demo
  src/main/java
    com.mialab.annotation_DSQL_demo.domain
      Student.java
    com.mialab.annotation_DSQL_demo.main
      AnnotationDSQL_Main.java
    com.mialab.annotation_DSQL_demo.mapper
      StudentDySqlProvider.java
      StudentMapper.java
    com.mialab.annotation_DSQL_demo.utils
      DBOperatorMgr.java
  src/main/resources
    log4j.properties
    mybatis-config.xml
```

图 13-3　annotation_DSQL_demo 工程目录结构

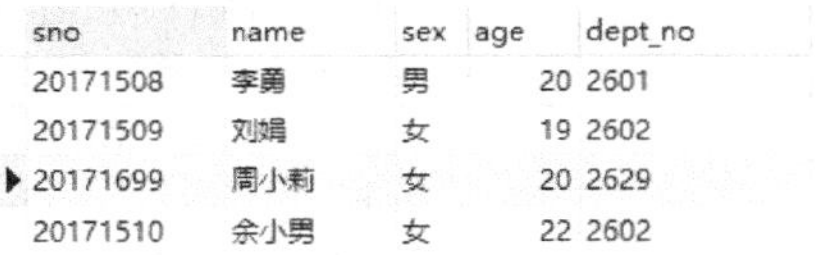

sno	name	sex	age	dept_no
20171508	李勇	男	20	2601
20171509	刘娟	女	19	2602
20171699	周小莉	女	20	2629
20171510	余小男	女	22	2602

图 13-4　student 初始表中的数据

pom.xml、mybatis-config.xml、数据库操作工具类 DBOperatorMgr.java、log4j.properties 创建及配置、POJO 类的创建及实现代码都类似以前示例，这里不再赘述。详见本书教学资源包 ch13 目录下的 annotation_DSQL_demo 工程。

2．创建映射接口

在 src/main/java 中创建 Package“com.mialab. annotation_DSQL_demo.mapper”，如图 13-3 所示。在 mapper 包中创建接口 StudentMapper.java 和类 StudentDySqlProvider.java。

接口 StudentMapper.java 主要代码如下：

```
    public interface StudentMapper {
     @SelectProvider(type = StudentDySqlProvider.class, method =
        "selectStudentByMultiple")
     List<Student> selectStudentByMultiple(Map<String, Object> params);

     @InsertProvider(type = StudentDySqlProvider.class, method = "addStudent")
     int addStudent(Student student);

     // 根据学号 sno 查询
     @SelectProvider(type = StudentDySqlProvider.class, method = "getStudent
ById")
```

```
    Student getStudentById(Map<String, Object> param);

    @SelectProvider(type = StudentDySqlProvider.class, method = "getSudent
All")
    List<Student> getSudentAll();

    @UpdateProvider(type = StudentDySqlProvider.class, method = "update
Student")
    int updateStudent(Student student);

    @DeleteProvider(type = StudentDySqlProvider.class, method = "delete
Student")
    int deleteStudent(Map<String, Object> param);
   }
```

@SelectProvider 表示 Select 语句的动态 SQL 映射，允许指定一个类名和一个方法在执行时返回运行的查询语句。其有两个属性：type 和 method。type 属性是类的完全限定名，method 是该类中的方法名。与@SelectProvider 类似，@InsertProvider 表示 Insert 语句的动态 SQL 映射，@UpdateProvider 表示 Update 语句的动态 SQL 映射，@DeleteProvider 表示 Delete 语句的动态 SQL 映射。它们也同样具有两个属性：type 和 method。

StudentDySqlProvider.java 主要代码如下：

```
   public class StudentDySqlProvider {
    public String updateStudent(Student student) {
        return new SQL() {
            {
                UPDATE("student");
                if (student.getName() != null) {
                    SET("name = #{name}");
                }
                if (student.getSex() != null) {
                    SET("sex = #{sex}");
                }
                if (student.getAge() != null) {
                    SET("age = #{age}");
                }
                if (student.getDept_no() != null) {
                    SET("dept_no = #{dept_no}");
                }
                WHERE(" sno = #{sno} ");
            }
        }.toString();
    }

    public String deleteStudent(Map<String, Object> param) {
        return new SQL() {
            {
                DELETE_FROM("student");
                if (param.get("sno") != null) {
                    WHERE(" sno = #{sno} ");
                }
                if (param.get("name") != null) {
                    WHERE("name = #{name}");
```

```
            }
            if (param.get("dept_no") != null) {
                WHERE("dept_no = #{dept_no}");
            }
        }
    }.toString();
}

public String selectStudentByMultiple(Map<String, Object> param) {
    return new SQL() {
        {
            SELECT("*");
            FROM("student");
            if (param.get("sno") != null) {
                WHERE("sno = #{sno} ");
            }
            if (param.get("name") != null) {
                WHERE(" name = #{name} ");
            }
            if (param.get("sex") != null) {
                WHERE("sex = #{sex}");
            }
            if (param.get("age") != null) {
                WHERE("age = #{age}");
            }
            if (param.get("dept_no") != null) {
                WHERE("dept_no = #{dept_no}");
            }
        }
    }.toString();
}

public String getStudentById(Map<String, Object> param) {
    return new SQL() {
        {
            SELECT("*");
            FROM("student");
            if (param.get("sno") != null) {
                WHERE("sno = #{sno} ");
            }
        }
    }.toString();
}

public String getSudentAll() {
    return new SQL() {
        {
            SELECT("*");
            FROM("student");
        }
    }.toString();
}
```

```
    public String addStudent(Student student) {
        return new SQL() {
            {
                INSERT_INTO("student");
                if (student.getSno() != null) {
                    VALUES("sno", "#{sno}");
                }
                if (student.getName() != null) {
                    VALUES("name", "#{name}");
                }
                if (student.getSex() != null) {
                    VALUES("sex", "#{sex}");
                }
                if (student.getAge() != null) {
                    VALUES("age", "#{age}");
                }
                if (student.getDept_no() != null) {
                    VALUES("dept_no", "#{dept_no}");
                }
            }
        }.toString();
    }
}
```

3．测试

测试类 AnnotationDSQL_Main.java 的主要代码如下：

```
public class AnnotationDSQL_Main {
    public static void main(String[] args) {
        testSelectStudentByMultiple();
        // testAddStudent();
        // testGetStudentById();
        // testGetSudentAll();
        // testUpdateStudent();
        // testDeleteStudent();
    }

    private static void testDeleteStudent() {
        Logger log = Logger.getLogger(AnnotationDSQL_Main.class);
        SqlSession session = null;
        try {
            session =
DBOperatorMgr.getInstance().getSqlSessionFactory().openSession();
            StudentMapper mapper = session.getMapper(StudentMapper.class);

            HashMap<String, Object> params = new HashMap<String, Object>();
            params.put("sno", "20171622");
            mapper.deleteStudent(params);
            session.commit();
        } catch(Exception ex) {
            session.rollback();
            ex.printStackTrace();
        } finally {
```

```
            if (session != null) {
                session.close();
            }
        }
    }

    private static void testUpdateStudent() {
        Logger log = Logger.getLogger(AnnotationDSQL_Main.class);
        SqlSession session = null;
        try {
            session =

DBOperatorMgr.getInstance().getSqlSessionFactory().openSession();
            StudentMapper mapper = session.getMapper(StudentMapper.class);
            Student student = new Student();
            student.setSno("20171622");
            student.setName("苏晓萍");
            student.setAge(68);
            student.setSex("女");
            student.setDept_no("2602");
            log.info(student);
            mapper.updateStudent(student);
            session.commit();
        } catch(Exception ex) {
            session.rollback();
            ex.printStackTrace();
        } finally {
            if (session != null) {
                session.close();
            }
        }
    }

    private static void testGetSudentAll() {
        Logger log = Logger.getLogger(AnnotationDSQL_Main.class);
        SqlSession session = null;
        try {
            session =

DBOperatorMgr.getInstance().getSqlSessionFactory().openSession();
            StudentMapper mapper = session.getMapper(StudentMapper.class);
            List<Student> students = mapper.getSudentAll();
            // log.info(students);
            for (Student stu : students) {
                System.out.println(stu);
                // log.info(stu);
            }
        } catch (Exception ex) {
            session.rollback();
            ex.printStackTrace();
        } finally {
            if (session != null) {
                session.close();
```

```
            }
        }
    }

    private static void testGetStudentById() {
        Logger log = Logger.getLogger(AnnotationDSQL_Main.class);
        SqlSession session = null;
        try {
            session =

DBOperatorMgr.getInstance().getSqlSessionFactory().openSession();
            StudentMapper mapper = session.getMapper(StudentMapper.class);
            HashMap<String, Object> params = new HashMap<String, Object>();
            params.put("sno", "20171509");
            Student student = mapper.getStudentById(params);
            log.info(student);
            // System.out.println(student);
        } catch (Exception ex) {
            session.rollback();
            ex.printStackTrace();
        } finally {
            if (session != null) {
                session.close();
            }
        }
    }

    private static void testAddStudent() {
        Logger log = Logger.getLogger(AnnotationDSQL_Main.class);
        SqlSession session = null;
        try {
            session =

DBOperatorMgr.getInstance().getSqlSessionFactory().openSession();
            StudentMapper mapper = session.getMapper(StudentMapper.class);
            Student student = new Student();
            student.setSno("20171510");
            student.setName("余小男");
            student.setAge(22);
            student.setSex("女");
            student.setDept_no("2602");
            log.info(student);
            mapper.addStudent(student);
            session.commit();
        } catch (Exception ex) {
            session.rollback();
            ex.printStackTrace();
        } finally {
            if (session != null) {
                session.close();
            }
        }
    }
```

```
    private static void testSelectStudentByMultiple() {
        Logger log = Logger.getLogger(AnnotationDSQL_Main.class);
        SqlSession session = null;
        try {
            session =
DBOperatorMgr.getInstance().getSqlSessionFactory().openSession();
            StudentMapper mapper = session.getMapper(StudentMapper.class);
            HashMap<String, Object> params = new HashMap<String, Object>();
            params.put("sex", "女");
            params.put("dept_no", "2602");
            List<Student> students = mapper.selectStudentByMultiple(params);
            // log.info(students);
            for (Student stu : students) {
                System.out.println(stu);
                // log.info(stu);
            }
        } catch (Exception ex) {
            session.rollback();
            ex.printStackTrace();
        } finally {
            if (session != null) {
                session.close();
            }
        }
    }
}
```

先测试 testSelectStudentByMultiple()方法。运行测试类 AnnotationDSQL_Main 的 main 方法，在控制台上可以得到以下结果：

```
Student [sno=20171509, name=刘娟, sex=女, age=19, dept_no=2602]
Student [sno=20171510, name=余小男, sex=女, age=22, dept_no=2602]
```

可分别对测试类 AnnotationDSQL_Main 中的其他方法 testGetStudentById()、testAddStudent()、testGetSudentAll()、testDeleteStudent()等进行测试，在此不再赘述。

13.2 MyBatis 缓存配置

MyBatis 作为持久化框架，提供了非常强大的查询缓存特性，可以非常方便地配置和定制使用。

13.2.1 一级缓存（SqlSession 层面）

每当我们使用 MyBatis 开启一次和数据库的会话，MyBatis 就会创建一个 SqlSession 对象表示一次数据库会话。在对数据库的一次会话中，我们有可能会反复地执行完全相同的查询语句，如果不采取一些措施，每一次查询都会查询一次数据库，而若用户在极短的时间内做了完全相同的查询，那么它们的结果极有可能完全相同，由于查询一次数据库的代价很大，这有可能造成很大的资源浪费。

为了解决这一问题，减少资源的浪费，MyBatis 会在表示会话的 SqlSession 对象中建立一个简单的缓存，将每次查询到的结果缓存起来，当下次查询的时候，如果判断先前有完全一样的查询，则会直接从缓存中将结果取出，返回给用户，不需要再进行一次数据库查询。MyBatis 会在一次会话的表示某个 SqlSession 对象中，创建一个本地缓存，对于每一次查询，都会尝试根据查询的条件去本地缓存中查找是否存在结果，如果在缓存中，就直接从缓存中取出，然后返回给用户；否则，从数据库中读取数据，将查询结果存入缓存并返回给用户。

对于会话级别的数据缓存，人们称之为一级数据缓存，简称一级缓存。

由于 MyBatis 使用 SqlSession 对象表示一次数据库的会话，那么，会话级别的一级缓存也应该是在 SqlSession 中控制的。SqlSession 只是一个 MyBatis 对外的接口，SqlSession 将它的工作交给了执行器这个角色来完成，负责完成对数据库的各种操作。

当创建了一个 SqlSession 对象时，MyBatis 会为这个 SqlSession 对象创建一个新的执行器，而缓存信息就会被维护在这个执行器中，MyBatis 将缓存和对缓存相关的操作封装到了 Cache 接口中。

Executor 接口的实现类 BaseExecutor 中拥有一个 Cache 接口的实现类——PerpetualCache，则对于 BaseExecutor 对象而言，它将使用 PerpetualCache 对象维护缓存。

一级缓存的作用域是 SqlSession 范围内的，当一个 SqlSession 结束后，SqlSession 中的一级缓存也就不存在了。MyBatis 默认开启一级缓存，不需要进行任何配置。

【示例】MyBatis 的一级缓存。

1．创建工程 CacheDemo_1

在 Eclipse 中创建 Maven Project，选择 maven→archetype→quickstart，在“Group ID”文本框中输入“com.mialab”，在“Artifact ID”文本框中输入“CacheDemo_1”。最终完成的 CacheDemo_1 工程目录结构如图 13-5 所示，student 初始表中的数据如图 13-6 所示。

```
CacheDemo_1
  src/main/java
    com.mialab.CacheDemo_1.domain
      Student.java
    com.mialab.CacheDemo_1.main
      CacheDemo_Main.java
    com.mialab.CacheDemo_1.mapper
      StudentMapper.java
      StudentMapper.xml
    com.mialab.CacheDemo_1.utils
      DBOperatorMgr.java
  src/main/resources
    log4j.properties
    mybatis-config.xml
```

图 13-5　CacheDemo_1 工程目录结构

sno	name	sex	age	dept_no
20171508	李勇	男	20	2601
20171509	刘娟	女	19	2602
20171699	周小莉	女	20	2629
20171510	余小男	女	22	2602

图 13-6　student 初始表中的数据

pom.xml、mybatis-config.xml、数据库操作工具类 DBOperatorMgr.java、log4j.properties 创建及配置、POJO 类的创建及实现代码都类似以前示例，这里不再赘述。详见本书教学资源包 ch13 目录中的 CacheDemo_1 工程。

2．测试

测试类 CacheDemo_Main.java 的主要代码如下：

```
public class CacheDemo_Main {
  public static void main(String[] args) {
      testSelect_1();
```

```
        // testSelect_2();
    }
    private static void testSelect_1() {
        Logger log = Logger.getLogger(CacheDemo_Main.class);
        SqlSession session = null;
        try {
            session =
DBOperatorMgr.getInstance().getSqlSessionFactory().openSession();
            StudentMapper mapper = session.getMapper(StudentMapper.class);
            Student student1 = mapper.getStudent("20171509");
            log.info(student1);
            student1.setName("秦红霞");
            student1.setAge(22);
            Student student2 = mapper.getStudent("20171509");
            log.info(student2);
            log.info(student2.equals(student1));
        } finally {
            if (session != null) {
                session.close();
            }
        }
    }
    private static void testSelect_2() {
        Logger log = Logger.getLogger(CacheDemo_Main.class);
        SqlSession session = null;
        try {
            session =

DBOperatorMgr.getInstance().getSqlSessionFactory().openSession();
            StudentMapper mapper = session.getMapper(StudentMapper.class);
            Student student1 = mapper.getStudent("20171509");
            //System.out.println(student);
            log.info(student1);
            student1.setName("秦红霞");
            student1.setAge(22);
            mapper.deleteStudent("20170058");
            session.commit();
            /*再次查询学号为"20171509"的Student对象，因为DML操作会
                清空SqlSession内存，所以会再次执行Select语句*/
            Student student2 = mapper.getStudent("20171509");
            log.info(student2);
            log.info(student2.equals(student1));
        } finally {
            if (session != null) {
                session.close();
            }
        }
    }
}
```

先测试testSelect_1()方法。运行测试类CacheDemo_Main的main方法，在控制台上可以得到以下结果：

```
DEBUG [main] - ==> Preparing: select * from student where sno = ?
DEBUG [main] - ==> Parameters: 20171509(String)
TRACE [main] - <==    Columns: sno, name, sex, age, dept_no
TRACE [main] - <==        Row: 20171509, 刘娟, 女, 19, 2602
```

```
    DEBUG [main] - <==      Total: 1
    INFO  [main]  -  Student  [sno=20171509,  name=刘娟 ,  sex=女 ,  age=19,
dept_no=2602]
    INFO  [main]  -  Student  [sno=20171509,  name=秦红霞 ,  sex=女 ,  age=22,
dept_no=2602]
    INFO [main] - true
```

在第 1 次查询学号（sno）为"20171509"的 Student 对象时执行了一条 Select 语句，但是第 2 次获取学号为"20171509"的 Student 对象时并没有执行 Select 语句。因为此时一级缓存，即 SqlSession 缓存中已经缓存了学号为"20171509"的 Student 对象，MyBatis 直接从缓存中将对象取出来，并没有再次去数据库中查询。

注释掉 testSelect_1()的调用，并取消 texr Select_1()的注释，来测试 testSelect_2()方法。

先在 stu 数据库的 student 表中添加一条数据，

```
  INSERT INTO 'student' VALUES ('20170058', '白发魔女', '女', '56', '2603');
```

student 表中的数据如图 13-7 所示。

sno	name	sex	age	dept_no
20171508	李勇	男	20	2601
20171509	刘娟	女	19	2602
20171699	周小莉	女	20	2629
20171510	余小男	女	22	2602
20170058	白发魔女	女	56	2603

图 13-7　student 表中的数据

运行测试类 CacheDemo_Main 的 main 方法，控制台将得到如图 13-8 所示的结果。

```
DEBUG [main] - ==>  Preparing: select * from student where sno = ?
DEBUG [main] - ==> Parameters: 20171509(String)
TRACE [main] - <==    Columns: sno, name, sex, age, dept_no
TRACE [main] - <==        Row: 20171509, 刘娟, 女, 19, 2602
DEBUG [main] - <==      Total: 1
 INFO [main] - Student [sno=20171509, name=刘娟, sex=女, age=19, dept_no=2602]
DEBUG [main] - ==>  Preparing: delete from student where sno = ?
DEBUG [main] - ==> Parameters: 20170058(String)
DEBUG [main] - <==    Updates: 1
DEBUG [main] - Committing JDBC Connection [com.mysql.jdbc.JDBC4Connection@127bd23]
DEBUG [main] - ==>  Preparing: select * from student where sno = ?
DEBUG [main] - ==> Parameters: 20171509(String)
TRACE [main] - <==    Columns: sno, name, sex, age, dept_no
TRACE [main] - <==        Row: 20171509, 刘娟, 女, 19, 2602
DEBUG [main] - <==      Total: 1
 INFO [main] - Student [sno=20171509, name=刘娟, sex=女, age=19, dept_no=2602]
 INFO [main] - false
```

图 13-8　测试 testSelect_2()方法

第 1 次查询 sno 为"20171509"的 Student 对象时执行了一条 select 语句，接下来执行了一个 delete 操作，MyBatis 为了保证缓存中存储的是最新信息，会清空 SqlSession 缓存。当第 2 次获取 sno 为"20171509"的 Student 对象时，一级缓存已清空，所以 MyBatis 会再次执行 Select 语句去查询 sno 为"20171509"的 Student 对象。

13.2.2　二级缓存（SqlSessionFactory 层面）

MyBatis 的二级缓存是 Application 级别的缓存，它可以提高对数据库查询的效率，以提高应用的性能。当开始一个会话时，一个 SqlSession 对象会使用一个 Executor 对象来完成会话操作，MyBatis 的二级缓存机制的关键就是对 Executor 对象进行操作。

如果用户配置了"cacheEnabled=true"，那么 MyBatis 在为 SqlSession 对象创建 Executor 对象时，会对 Executor 对象加上一个装饰者——CachingExecutor，此时 SqlSession 使用

CachingExecutor 对象来完成操作请求。

对于查询请求，CachingExecutor 会先判断该查询请求在 Application 级别的二级缓存中是否有缓存结果，如果有查询结果，则直接返回缓存结果；如果缓存中没有结果，则交给真正的 Executor 对象来完成查询操作，之后 CachingExecutor 会将真正的 Executor 返回的查询结果放置到缓存中，再返回给用户。

CachingExecutor 是 Executor 的装饰者，以增强 Executor 的功能，使其具有缓存查询的功能，这里用到了设计模式中的装饰者模式。

MyBatis 对二级缓存的支持粒度很细，它会指定某一条查询语句是否使用二级缓存。虽然在 Mapper 中配置了<cache>，并且为此 Mapper 分配了 Cache 对象，但是并不表示使用 Mapper 中定义的查询语句查到的结果都会被放置到 Cache 对象之中，必须指定 Mapper 中的某条选择语句是否支持缓存，即在<select>节点中配置 useCache="true"，Mapper 才会对此 Select 的查询支持缓存特性。

总之，要想使某条 Select 查询支持二级缓存，需要保证以下几点。

（1）MyBatis 支持二级缓存的总开关：全局配置变量参数 cacheEnabled=true。

（2）该 Select 语句所在的 Mapper 配置了<cache>或<cached-ref>节点，并且有效。

（3）该 Select 语句的参数 useCache=true。

缓存的使用顺序：二级缓存→一级缓存→数据库。

【示例】MyBatis 的二级缓存。

1．创建工程 CacheDemo_2

在 Eclipse 中创建 Maven Project，选择 maven→archetype→quickstart，在“Group ID”文本框中输入“com.mialab”，在“Artifact ID”文本框中输入“CacheDemo_2”。最终完成的 CacheDemo_2 工程目录结构如图 13-9 所示，student 初始表中的数据如图 13-10 所示。

```
CacheDemo_2
  src/main/java
    com.mialab.CacheDemo_2.domain
      Student.java
    com.mialab.CacheDemo_2.main
      CacheDemo_Main.java
    com.mialab.CacheDemo_2.mapper
      StudentMapper.java
      StudentMapper.xml
    com.mialab.CacheDemo_2.utils
      DBOperatorMgr.java
  src/main/resources
    log4j.properties
    mybatis-config.xml
```

图 13-9 CacheDemo_2 工程目录结构

sno	name	sex	age	dept_no
20171508	李勇	男	20	2601
20171509	刘娟	女	19	2602
20171699	周小莉	女	20	2629
20171510	余小男	女	22	2602

图 13-10 student 初始表中的数据

pom.xml、mybatis-config.xml、数据库操作工具类 DBOperatorMgr.java、log4j.properties 创建及配置、POJO 类的创建及实现代码都类似以前示例，这里不再赘述。详见本书教学资源包 ch13 目录下的 CacheDemo_2 工程。

2．开启二级缓存

首先，保证二级缓存的全局配置开启，可以在 MyBatis 全局配置文件 mybatis-config.xml 中进行 settings 的设置，示例代码如下：

```
<settings>
 <setting name="cacheEnabled" value="true" />
```

```
</settings>
```

其次，需要在 Mapper.xml 中配置二级缓存，只需要添加<cache/>元素即可。例如，在 StudentMapper.xml 文件中添加<cache/>元素，示例代码如下：

```
<mapper namespace="com.mialab.CacheDemo_2.mapper.StudentMapper">
  <cache/>
  <!-- 其他配置 -->
</mapper>
```

默认的二级缓存会有如下效果。

（1）映射语句文件中的所有 SELECT 语句将会被缓存。

（2）映射语句文件中的所有 INSERT、UPDATE、DELETE 语句会刷新缓存。

（3）缓存会使用最近最少使用的（Least Recently Used，LRU）算法来进行收回。

（4）根据时间表（如 no Flush Interval，没有刷新间隔），缓存不会以任何时间顺序来刷新，且缓存会被视为可读/可写的，意味着其是线程安全的。

（5）缓存会存储集合或对象（无论查询方法返回什么类型的值）的 1024 个引用。

3. 测试

测试类 CacheDemo_Main.java 的主要代码如下：

```
public class CacheDemo_Main {
 public static void main(String[] args) {
     testSelect_1();
 }
 private static void testSelect_1() {
     Logger log = Logger.getLogger(CacheDemo_Main.class);
     SqlSession session = null;
     Student student1, student2;
     try {
         session =
DBOperatorMgr.getInstance().getSqlSessionFactory().openSession();
System.out.println(DBOperatorMgr.getInstance().getSqlSessionFactory());
         StudentMapper mapper = session.getMapper(StudentMapper.class);
         student1 = mapper.getStudent("20171509");  //第 1 次查询
         log.info(student1);
         student1.setName("秦红霞");
         student1.setAge(22);
         Student student3 = mapper.getStudent("20171509");  //第 2 次查询
         log.info(student3);
         log.info(student3.equals(student1));
     } finally {
         if (session != null) {
             session.close();
         }
     }
     //以下是第二部分的测试代码，前面是第一部分的测试代码
     log.info("开启新的 sqlSession：");
     try {
         session =
DBOperatorMgr.getInstance().getSqlSessionFactory().openSession();
```

```
System.out.println(DBOperatorMgr.getInstance().getSqlSessionFactory());
            StudentMapper mapper = session.getMapper(StudentMapper.class);
            student2 = mapper.getStudent("20171509");  //第 3 次查询
            log.info(student2);
            log.info(student2.equals(student1));
        } finally {
            if (session != null) {
                session.close();
            }
        }
    }
}
```

运行测试类 CacheDemo_Main.java 的 main 方法，控制台输出如图 13-11 所示的结果。

```
org.apache.ibatis.session.defaults.DefaultSqlSessionFactory@f67b76
DEBUG [main] - Cache Hit Ratio [com.mialab.CacheDemo_2.mapper.StudentMapper]: 0.0
DEBUG [main] - Opening JDBC Connection
DEBUG [main] - Created connection 3213500.
DEBUG [main] - Setting autocommit to false on JDBC Connection [com.mysql.jdbc.JDBC4Connection@3108bc]
DEBUG [main] - ==>  Preparing: select * from student where sno = ?
DEBUG [main] - ==> Parameters: 20171509(String)
TRACE [main] - <==    Columns: sno, name, sex, age, dept_no
TRACE [main] - <==        Row: 20171509, 刘娟, 女, 19, 2602
DEBUG [main] - <==      Total: 1
 INFO [main] - Student [sno=20171509, name=刘娟, sex=女, age=19, dept_no=2602]
DEBUG [main] - Cache Hit Ratio [com.mialab.CacheDemo_2.mapper.StudentMapper]: 0.0
 INFO [main] - Student [sno=20171509, name=秦红霞, sex=女, age=22, dept_no=2602]
 INFO [main] - true
DEBUG [main] - Resetting autocommit to true on JDBC Connection [com.mysql.jdbc.JDBC4Connection@3108bc]
DEBUG [main] - Closing JDBC Connection [com.mysql.jdbc.JDBC4Connection@3108bc]
DEBUG [main] - Returned connection 3213500 to pool.
 INFO [main] - 开启新的sqlSession:
org.apache.ibatis.session.defaults.DefaultSqlSessionFactory@f67b76
DEBUG [main] - Cache Hit Ratio [com.mialab.CacheDemo_2.mapper.StudentMapper]: 0.3333333333333333
 INFO [main] - Student [sno=20171509, name=秦红霞, sex=女, age=22, dept_no=2602]
 INFO [main] - false
```

图 13-11　测试 MyBatis 的二级缓存

从上述代码中可以看到有 3 次查询。第 1 次查询获取 student1 信息的时候，由于没有缓存，所以执行了数据库查询。在第 2 次查询获取 student3 信息的时候，student3 和 student1 是完全相同的实例，这里使用的是一级缓存，所以返回同一个实例。

当调用 close 方法关闭 SqlSession 时，SqlSession 才会保存查询数据到二级缓存中。在此之后，二级缓存才有了缓存数据，所以可以看到在前面的两次查询中，命中率都是 0。

当开启新的 SqlSession 以后，第 3 次查询获取 student2 信息的时候，日志中并没有输出数据库查询，而是输出了命中率为 0.333…（即 1/3），并且得到了缓存的值。因为是可读写缓存，student2 是反序列化得到的结果，所以 student1 和 student2 不是相同的实例。

13.3 本章小结

本章主要介绍了 MyBatis 注解方式和 MyBatis 缓存配置。

MyBatis 注解方式就是将 SQL 语句直接写在接口上，其优点在于，对于需求比较简单的系统，效率较高；其缺点在于，当 SQL 有变化时需要重新编译代码。

使用缓存可以使应用更快地获取数据，避免频繁的数据库交互。MyBatis 作为持久化框架，提供了非常强大的查询缓存特性，可以非常方便地配置和定制使用。

习题 13

1．如何使用 MyBatis 注解方式实现表数据的增、删、改、查？给出应用实例。

2．MyBatis 一级缓存配置是怎样的？试编程加以说明。

3．MyBatis 二级缓存配置是怎样的？试编程加以说明。

第14章 MyBatis应用

本章导读

"移动商城"是一款针对XX网络运营商开发的电子商务平台手机应用程序，通过该App，用户可以购买手机、流量及定制宽带等。本示例是"移动商城"App的服务器端接口编程实现，旨在说明如何使用MyBatis框架。本章主要内容有：（1）示例总体介绍；（2）典型代码及技术要点。

14.1 示例总体介绍

14.1.1 任务说明和准备数据

任务说明：使用MyBatis框架，编程实现"移动商城"App的后台（服务器端）接口。

本示例使用了PostgreSQL数据库，客户端使用Navicat Premium和pgAdmin，创建数据库mall，在mall中创建宽带信息表market_broadband_info，数据表结构如表14-1所示。

表14-1 数据表结构

字　段	类　型	说　明
id	integer	主键，商品id
subject	character(50)	标题
body	character(1000)	内容
provider_id	integer	供应商id
buy_price	real	购买价
sell_price	real	卖价
discount_price	real	折扣价
stock	integer	库存
img	character(255)	商品图片
package_category	integer	宽带产品分类（0—单产品包月；1—单产品包年，2—融合宽带）
owner	integer	收款方
type	integer	产品类型
start_time	timestamp	产品有效开始时间
end_time	timestamp	产品有效终止时间
create_time	timestamp	产品创建时间
last_modify_time	timestamp	产品最新修改时间
tag	smallint	备用
img_url	character (255)	列表小图标
city_id	integer	城市id

教学资源包ch14目录下有数据库脚本文件market_broadband_info.sql和数据库备份文件

mall.backup。有以下两种方法可以获得到宽带信息表 market_broadband_info 及其表数据。

（1）通过 Navicat Premium 客户端与 PostgreSQL 数据库服务器相连，在 Navicat Premium 中建立 PostgreSQL 数据库“mall”，再在 mall 中执行脚本 market_broadband_info.sql，便可以生成宽带信息表 market_broadband_info 及其表数据。

（2）通过 pgAdmin 客户端与 PostgreSQL 数据库服务器相连，在 pgAdmin 中建立 PostgreSQL 数据库“mall”，再在 mall 中使用 mall.backup 备份文件还原数据即可。具体操作：右击 mall 数据库，在弹出的快捷菜单中选择“恢复...”选项，在弹出的“Restore database mall”对话框中，找到 mall.backup 所在的路径，单击“恢复”按钮即可得到宽带信息表及其表数据。

14.1.2 总体框架

在 Eclipse 中创建 Maven Project，选择 maven→archetype→quickstart，最终完成的 mobile-mall-demo 工程目录及文件如图 14-1 所示。（详见本书教学资源 ch14 目录下的 mobile-mall-demo 工程。）

mobile-mall-demo 工程中各个包的作用如下。

（1）org.suda.app.common：存放通用的功能类文件。

（2）org.suda.app.db：存放配置文件以及数据库访问类。

（3）org.suda.app.market.persistence：数据持久层，放置数据访问接口及映射文件。

（4）org.suda.app.market.vo：POJO 实体类。

（5）org.suda.app. market、org.suda.app.common.mgr 和 org.suda.app. mgr：业务层。

```
mobile-mall-demo
  src/main/java
    org.suda.app
      Command.java
      MarketServlet.java
      Params.java
    org.suda.app.common
      CommonConstants.java
      ResponseErrorMessage.java
      ResponseMessage.java
      ResponseSuccessMessage.java
    org.suda.app.common.mgr
      AbstractMgr.java
      IMgr.java
    org.suda.app.db
    org.suda.app.market
      MarketMgr.java
    org.suda.app.market.persistence
      MarketMapper.java
      MarketMapper.xml
    org.suda.app.market.vo
      MarketProductInfo.java
    org.suda.app.mgr
      RequestManager.java
```

图 14-1 mobile-mall-demo 工程目录及文件

14.1.3 程序主要流程

客户端的请求都要由 MarketServlet 来做处理，所以程序的入口是 MarketServlet 中的 doGetPost()方法。首先，判断客户端请求是否包含 cmd 参数，若无则打印“OK”；若有则调用 RequestManager 的 getManager()方法获取 MarketMgr 对象。getManager()方法是通过调用 MarketMgr 的 getInstance()方法来获取 MarketMgr 对象实例的。其次，MarketMgr 继承了 AbstractMgr，而 AbstractMgr 实现了接口 IMgr，因此，MarketMgr 实现了 IMgr 中定义的接口方法 processRequest()。MarketServlet 获取到 IMgr 接口对象（MarketMgr 对象实例）后，便执行 IMgr 接口对象的 processRequest()方法来处理请求。

processRequest()方法主要根据请求参数 cmd 的值执行相应的业务，本例中执行的是调用 writeResponse()方法在浏览器中打印 showProductInfo()方法返回的 JSON 字符串。writeResponse()方法是在 AbstractMgr 中定义的，用于利用服务器返回的响应对象向页面打印响应信息。showProductInfo()使用 MyBatis 框架获取查询得到的商品信息。

14.2 典型代码及技术要点

14.2.1 通用功能包的类实现

org.suda.app.common 包主要有以下几个功能类。

（1）CommonConstants.java：该类用于存储常量。

（2）ResponseMessage、ResponseSuccessMessage、ResponseErrorMessage：这 3 个类均是用于操作结果消息的类。其中，ResponseMessage 封装了一个操作消息所需要的数据成员，而 ResponseSuccessMessage、ResponseErrorMessage 继承了 ResponseMessage，并分别定义了用于返回操作成功和操作失败的方法。

14.2.2 控制层

MarketServlet 实际上是控制器。由代码可以知道，客户端无论是采用 GET 请求还是 POST 请求，均调用控制器 MarketServlet 的 doGetPost()方法。该方法根据 HTTP 请求，判断是否有请求参数 CommonConstants.PARAMS_CMD（即“cmd”）。若无，则向页面打印“OK”。如果存在，则通过 RequestManager 类中的 getManager()方法获取 MarketMgr 对象实例。若返回的是空对象，则打印错误日志；若返回的是非空对象，则调用 processRequest ()方法，将 JSON 字符串打印到返回页面。MarketServlet.java 主要代码如下：

```
public class MarketServlet extends HttpServlet {
…
    public void doGet(HttpServletRequest request, HttpServletResponse response)
    throws ServletException, IOException {
        doGetPost(request, response);
    }
    public void doPost(HttpServletRequest request, HttpServletResponse response)
    throws ServletException, IOException {
        doGetPost(request, response);
    }
    private void doGetPost(HttpServletRequest request, HttpServletResponse response)
    throws ServletException,IOException {
        response.setContentType("text/html; charset=UTF-8");
        String command = request.getParameter(CommonConstants.PARAMS_CMD);
        PrintWriter out = response.getWriter();
        if (command == null) {
            out.println("ok");
            out.flush();
            out.close();
        } else {
            IMgr mgr = RequestManager.getManager(command);
            if (mgr == null) {…
            } else {
                try {
```

```
                    mgr.processRequest(request, response);
                } catch (Exception e) {…
                }
            }
        }
    }
}
```

14.2.3 业务层及使用 FastJson

业务层负责只要接收到业务请求就调用数据处理层来完成具体的数据处理。该层主要由4个类组成：RequestManager.java、MarketMgr.java、IMgr.java 和 AbstractMgr.java。

IMgr 为一个接口，定义了 processRequest()方法。AbstractMgr 是实现 IMgr 接口的抽象类，该类定义了 writeResponse()方法。

RequestManager 获取并返回 MarketMgr 实例，实际上也是 IMgr 接口对象。

```
public class RequestManager {
    public static IMgr getManager(String command) {
        if(command.startsWith(Command.CMD_MARKET_PRODUCT_INFO)) {
            return MarketMgr.getInstance();
        }
        return null;
    }
    …
}
```

MarketMgr 继承了 AbstractMgr，而 AbstractMgr 实现了 IMgr 接口，因此，在 MarketMgr 中需要实现 IMgr 接口定义的 processRequest()方法。

```
    public class MarketMgr extends AbstractMgr {
     private static MarketMgr  mgr;
     private SqlSessionFactory  sqlSessionFactory;
     private MarketMgr() {
         sqlSessionFactory  =  DBOperatorMgr.getInstance().getSessionFactory
();
     }
     public static synchronized MarketMgr getInstance() {
         if (mgr == null) {
             mgr = new MarketMgr();
         }
         return mgr;
     }
     @Override
     public void processRequest(HttpServletRequest request, HttpServletResponse
response)
     {
         String command = request.getParameter(CommonConstants.PARAMS_CMD);
         if (command.equalsIgnoreCase(Command.CMD_MARKET_PRODUCT_INFO))
         {
             writeResponse(response, showProductInfo(request, response));
         }
     }
     private String showProductInfo(HttpServletRequest request,
```

```
        HttpServletResponse response) {
        String json_str = new String();
        SqlSession session = null;
        String strTypeId = request.getParameter(Params.PARAM_TYPEID);
        String strId = request.getParameter(Params.PARAM_ID);
        String strPhoneId = strId;
        try {
            session = sqlSessionFactory.openSession();
            MarketMapper  marketMapper  =  session.getMapper(MarketMapper.
class);
            if (strTypeId.equals("1000")) {
                List<MarketProductInfo> list1 =
    marketMapper.getMarketBroadbandInfo(Integer.parseInt(strId));
                Map<String,List<MarketProductInfo>> map = new HashMap<String,
                    List<MarketProductInfo>>();
                map.put("BroadbandInfo", list1);
                json_str = JSON.toJSONString(map); //这里使用了 FastJson
                String jsonFormatStr = JSON.toJSONString(map, true);
                System.out.println(jsonFormatStr);
            }
        } catch (Exception e) {
            e.printStackTrace();
            logger.error("=======>showProductInfo:" + e.toString());

        } finally {
            if (session != null) {
                session.close();
                session = null;
            }
        }
        return json_str;
    }
}
```

MarketMgr 定义了 getInstance()方法来实例化一个 MarketMgr 对象，从而获取 DBOperatorMgr 类定义的 SqlSessionFactory 实例。processRequest()方法主要实现根据请求进行数据处理，从而输出 JSON 字符串的功能。

processRequest()方法判断请求的业务与实际相应业务是否一致，如果一致，则调用父类 AbstractMgr 中定义的 writeResponse()方法，实现 JSON 字符串的输出。writeResponse()方法的第二个参数即为待输出的 JSON 字符串，该字符串通过调用 showProductInfo()得到。

在 showProductInfo()方法中创建 SqlSession，调用 SqlSession 的 getMapper()方法，该方法根据配置文件生成 Mapper 对象，通过 Mapper 对象可以访问 MyBatis。Mapper 对象访问 MyBatis 是通过接口来实现的，因此，也可以说 getMapper()方法获取了接口对象。此处调用 MarketMapper 接口中定义的 getMarketBroadbandInfo()方法来实现商品详情信息。另外，在 showProductInfo()方法中还引用了 Params 类定义的两个静态变量——PARAM_TYPEID（值为 typeId）和 PARAM_ID（值为 id）。

要说明的是，Java 对象和 JSON 数据之间的数据转换通常使用第三方插件来协助完成，如 JSON-Lib、Jackson、FastJson 和 Gson 等。这里使用了 FastJson。

因为使用了 fastjson 包，所以在 pom.xml 文件中必须添加 FastJson 依赖。

```
<!-- FastJson 依赖 -->
<dependency>
 <groupId>com.alibaba</groupId>
 <artifactId>fastjson</artifactId>
 <version>1.2.47</version>
</dependency>
```

14.2.4 数据层及 JNDI 数据源

（1）MarketProductInfo.java：根据数据表 market_broadband_info 建立对应的实体类文件 MarketProductInfo.java，并存放于 vo 包中。该类根据数据表中的字段，建立了一一对应的对象属性以及生成属性的 get 和 set 方法。

（2）建立接口文件 MarketMapper.java：定义了显示宽带详情的操作。

```
public interface MarketMapper {
 List<MarketProductInfo> getMarketBroadbandInfo(int par);  //显示宽带详情
}
```

（3）建立与 MarketMapper.java 对应的映射文件 MarketMapper.xml：

```
<!——接口的实现，namespace 的路径必须与其对应的接口的类路径一致-->
<mapper namespace="org.suda.app.market.persistence.MarketMapper">
 <!——定义查询映射时的返回类型-->
 <resultMap id="MarketProductInfoMap" type="MarketProductInfo">
     <result property="id" column="id" />
     <result property="subject" column="subject" />
     …
 </resultMap>
 <select id="getMarketBroadbandInfo" parameterType="int"
  resultMap="MarketProductInfoMap">
     SELECT  id,subject,body,provider_id,buy_price,
sell_price,discount_price,
            stock,img,package_category,owner,type,start_time,
        end_time,create_time,last_modify_time,tag,img_url
     FROM
        market_broadband_info
     WHERE
        id=#{id}
     AND
        tag = 1
 </select>
</mapper>
```

（4）建立配置文件 Configuration.xml：配置文件包含对 MyBatis 系统的核心设置，包含获取数据库连接实例的数据源以及决定事务范围和控制的事务管理器。

Configuration.xml 文件中数据源的配置代码如下：

```
<environments default="development">
 <environment id="development">
     <transactionManager type="JDBC" />
     <dataSource type="JNDI">
        <property name="data_source" value="java:comp/env/jdbc/MallDB"
/>
```

```
    </dataSource>
  </environment>
</environments>
```

必须注意的是，这里使用的是 **JNDI 数据源**，还需要在 Tomcat 服务器的 context.xml 文件中的<Context>节点下添加 JNDI 数据源（MallDB）的说明。配置代码如下：

```
<Resource auth="Container" driverClassName="org.postgresql.Driver"
maxActive="100" maxIdle="30" maxWait="10000" name="jdbc/MallDB"
password="1" type="javax.sql.DataSource"
url="jdbc:postgresql://localhost:5432/mall?characterEncoding=UTF8"
username="postgres" />
```

（这是全局数据源，可供所有 Web 应用程序使用。）

开发时，可在 Eclipse 的 Servers 下的 context.xml 文件中添加以上 JNDI 数据源的配置代码并保存。部署时，要保证 Tomcat 安装路径的 conf 目录下的 context.xml 文件中有以上 JNDI 数据源的配置代码，否则应用程序（通过 Configuration.xml 文件）无法连接数据库。

<Resource>节点基本属性及含义如下。

① name：表示 JNDI 名称，供应用程序调用。

② auth：表示管理数据源的方式，有两个值——Container 和 Application。通常设置为 Container，表示由容器进行创建和管理数据源，而 Application 则表示由 Web 应用来创建和管理数据源。

③ type：表示数据源所属的类型，使用了标准的 javax.sql.DataSource。

④ factory：表示生成数据源的工厂类名。

⑤ driverClassName：表示 JDBC 数据库驱动器。

⑥ url：表示数据库 URL 地址。

⑦ username：表示访问数据库所需的用户名。

⑧ password：表示数据库用户的密码。

（5）DBOperatorMgr.java：数据库工具类。DBOperatorMgr.java 的主要代码如下：

```
public class DBOperatorMgr {
 private static DBOperatorMgr mgr;
 private SqlSessionFactory sqlSessionFactory;
 private DBOperatorMgr() {
     String resource = "org/suda/app/db/Configuration.xml";
     Reader reader;
     try {
         reader = Resources.getResourceAsReader(resource);
         sqlSessionFactory = new SqlSessionFactoryBuilder().build(reader);
     } catch (Exception e) {
     }
 }
 public static DBOperatorMgr getInstance() {
     if (mgr == null) {
         mgr = new DBOperatorMgr();
     }
     return mgr;
 }
 public SqlSessionFactory getSessionFactory() {
     return sqlSessionFactory;
 }
}
```

① DBOperatorMgr 的 getInstance()方法通过调用 DBOperatorMgr 构造方法，而获得 DBOperatorMgr 实例。

② 在 DBOperatorMgr 构造方法中，利用 Resources.getResourceAsReader()方法读取 MyBatis 配置文件 Configuration.xml，并获得 Reader 对象；并通过 SqlSessionFactoryBuilder 的 build()方法来创建 SqlSessionFactory 实例。

③ DBOperatorMgr 也提供了 getSessionFactory ()方法以得到 SqlSessionFactory 实例，供外部类使用。

④ 在 MarketMgr.java 中，通过调用 SessionFactory 实例的 openSession()方法得到数据库会话——SqlSession 对象，从而建立数据库连接。

14.2.5 部署发布

在 Eclipse 中，右击 mobile-mall-demo 工程中 Webapp 文件夹中的 index.jsp，在弹出的快捷菜单中选择“Run As→Run on Server”选项，即可将 mobile-mall-demo 工程自动部署发布到 Tomcat 服务器上。打开火狐浏览器，在浏览器中输入如下 URL：http://localhost:8080/mobile-mall-demo/mall?cmd=market.product.info&typeId=1000&id=10001001。页面显示效果如图 14-2 所示。

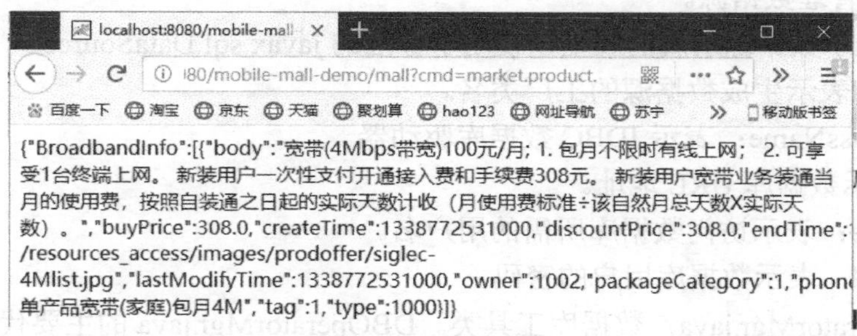

图 14-2 页面显示效果

14.2.6 使用 Jackson 和手工拼凑 JSON

源码包 ch14 目录中的 mobile-mall-demo2 工程是使用 Jackson 开源包的示例，源码包 ch14 目录中的 mialab-3gmarket-demo 工程未使用第三方关于 Java 对象和 JSON 数据转换的开源包，服务器端接口返回的 JSON 字符串是手工拼凑的，可作为参考。

14.3 本章小结

本章主要讲解了如何使用 MyBatis 框架编程实现“移动商城”App 的后台（服务器端）JSON 接口。其中，包括控制器 MarketServlet、处理业务逻辑组件 MarketMgr、映射接口 MarketMapper、映射文件 MarketMapper.xml、JNDI 数据源配置和使用 FastJson 等内容。

源码包 ch14 目录中提供了使用 Jackson 开源包的示例——mobile-mall-demo2 工程，以及手工拼凑 JSON 的示例——mialab-3gmarket-demo 工程。

习题 14

1．如何使用 MyBatis 编写服务器端接口（或 Mobile API）？试编程说明。
2．如何通过使用 JNDI 数据源访问数据库？试编程说明。
3．什么是 JNDI 数据源？为什么要使用 JNDI 数据源？
4．如何使用 FastJson 开源包实现 Java 对象和 JSON 数据的转换？试编程说明。
5．如何使用 Jackson 开源包实现 Java 对象和 JSON 数据的转换？试编程说明。

第三部分

Spring MVC

第 15 章　Spring 基础

本章导读

Spring 致力于 J2EE 应用的各层的解决方案，而不仅仅专注于某一层的方案。可以说 Spring 是企业应用开发的“一站式”选择，并贯穿表现层、业务层及持久层。然而，Spring 并不是取代那些已有的框架，而是与它们无缝地整合。本章主要内容有：（1）Spring 概述；（2）使用 Spring 容器；（3）依赖注入；（4）Spring 容器中的 Bean；（5）Bean 的生命周期；（6）两种后处理器；（7）装配 Spring Bean；（8）Spring 的 AOP。

15.1　Spring 入门

15.1.1　Spring 概述

1．简介

Spring 是于 2003 年兴起的一个轻量级的 Java 开发框架，由 Rod Johnson 创建。Spring 是一个开放源代码的设计层面框架，它解决的是业务逻辑层和其他各层的松耦合问题，因此，它将面向接口的编程思想贯穿于整个系统应用。

Spring 的源码设计精妙、结构清晰、匠心独运，处处体现着大师对 Java 设计模式的灵活运用以及在 Java 技术方面的高深造诣。**Spring 框架源码无疑是 Java 技术的最佳实践范例**。如果想在短时间内迅速提高自己的 Java 技术水平和应用开发水平，学习和研究 Spring 源码将会收到意想不到的效果。

下面列出的是使用 Spring 框架的主要好处。

（1）Spring 可以使开发人员使用 POJOs 开发企业级的应用程序。只使用 POJOs 的好处是不需要 EJB 容器产品，如一个应用程序服务器，但是可以选择使用一个健壮的 Servlet 容器，如 Tomcat 或者其他的商业产品。

（2）编写代码易于测试。使用依赖注入，不仅可以为 Bean 注入普通的属性值，还可以注入其他 Bean 的引用。依赖注入是一种优秀的解耦方式，其可以让 Bean 以配置文件的方式组织在一起，而不是以硬编码的方式耦合在一起。

（3）Spring 的 Web 框架是一个设计良好的 Web MVC 框架，它为诸如 Struts 或者其他工程上的 Web 框架，提供了一个很好的供替代的选择。

（4）Spring 框架是一个轻量级的 IoC 容器，特别是当与 EJB 容器相比的时候。这有利于在内存和 CPU 资源有限的计算机上开发和部署应用程序。

（5）Spring 提供了一个一致的事务管理界面，该界面可以缩小成一个本地事务（例如，使用一个单一的数据库）或扩展成一个全局事务（如使用 Java 事务 API）。

2．Spring 的整体架构

Spring 是一个分层架构，它包含一系列的功能要素，分为大约 20 个模块，这些模块大体上分为 Core Container、Data Access/Integration、Web、AOP（Aspect Oriented Programming，

即面向切面的编程）Instrumentation、Messaging 和 Test，其结构如图 15-1 所示。

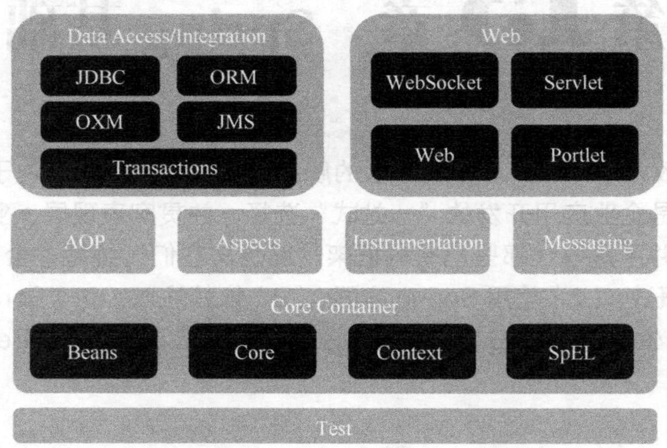

图 15-1 Spring 整体架构

组成 Spring 框架的每个模块（或组件）都可以单独存在，或者与其他一个或多个模块联合实现。这些模块被总结为以下几部分。

（1）核心容器。

核心容器由核心（Core）模块、Beans 模块、上下文（Context）模块和表达式语言（SpEL）模块组成。

① **核心**模块提供了框架的基本组成部分，包括 IoC 和依赖注入功能。

② **Beans** 模块提供了 BeanFactory，是一个工厂模式的复杂实现。

③ 上下文模块建立在 Core 和 Beans 模块的基础之上，它是访问定义和配置的任何对象的媒介。ApplicationContext 接口是上下文模块的重点。

④ 表达式语言模块是运行时查询和操作对象的强大的表达式语言。

（2）数据访问/集成。

数据访问/集成层包括 JDBC、ORM、OXM、JMS 和事务处理模块。

① **JDBC** 模块提供了删除冗余的 JDBC 相关编码的 JDBC 抽象层。

② **ORM** 模块为流行的对象关系映射 API（包括 JPA、JDO、Hibernate 和 iBatis），提供了集成层。

③ **OXM** 模块提供了一个支持对象/XML 映射的抽象层实现，如 JAXB、Castor、XMLBeans、JiBX 和 XStream。

④ Java 消息服务（**JMS** 模块）包含生产和消费的信息的功能。

⑤ 事务模块支持对实现特殊接口以及所有 POJO 类的编程和声明式的事务管理。

（3）Web。

Web 层由 Web、Web-MVC、Web-Socket 和 Web-Portlet 组成。

① **Web** 模块提供了基本的 Web 开发集成特性，例如，多文件上传、使用 Servlet 监听器初始化 IoC 容器以及 Web 应用上下文。

② **Web-MVC** 模块包含 Spring 的模型－视图－控制器（MVC），提供了 Web 应用的 MVC 实现。Spring 的 MVC 框架使得模型范围内的代码和 Web Forms 之间能够清楚地分离开来，并与 Spring 框架的其他特性集成在一起。

③ **Web-Socket** 模块为 WebSocket-based 提供了支持，而且在 Web 应用程序中提供了客户

端和服务器端之间通信的两种方式。

④ **Web-Portlet** 模块提供了在 portlet 环境中实现 MVC 并反映 Web-Servlet 模块的功能。

（4）其他。

还有其他较重要的模块，如 AOP、Aspects、Instrumentation 及测试模块。

① **AOP** 模块提供了面向切面编程实现，允许定义方法拦截器和切入点对代码进行干净的解耦。

② **Aspects** 模块提供了与 **AspectJ** 的集成功能，**AspectJ** 是一个功能强大且成熟的面向切面编程框架。

③ **Instrumentation** 模块提供了类工具的支持和类加载器的实现，可以在特定的应用服务器上使用。

④ **测试**模块支持使用 JUnit 或 TestNG 框架对 Spring 组件进行测试。

3．Spring 环境配置

这里所说的 Spring 环境配置，主要是指 Spring 框架的依赖 JAR 包的下载和使用，因为使用了 Maven，故 Spring 相关 JAR 包的下载和使用变得比较简单。

例如，后面的示例 SpringBasicDemo1 工程中使用了如下 pom.xml 后即可使用 Spring 框架（Spring 框架的依赖 JAR 包会自动下载）。

```
<project xmlns="http://maven.apache.org/POM/4.0.0"
 xmlns:xsi="http://www.w3.org/2001/XMLSchema-instance"
 xsi:schemaLocation="http://maven.apache.org/POM/4.0.0
 http://maven.apache.org/xsd/maven-4.0.0.xsd">
 <modelVersion>4.0.0</modelVersion>
 <groupId>com.mialab</groupId>
 <artifactId>SpringBasicDemo1</artifactId>
 <version>0.0.1-SNAPSHOT</version>
 <packaging>jar</packaging>
 <name>SpringBasicDemo1</name>
 <url>http://maven.apache.org</url>
 <properties>
     <java.version>1.8</java.version>
     <!-- Spring 版本号 -->
     <spring.version>4.3.6.RELEASE</spring.version>
     <project.build.sourceEncoding>UTF-8</project.build.sourceEncoding>
 </properties>
 <dependencies>
     <!-- 以下是 Spring 依赖 -->
     <dependency>
         <groupId>org.springframework</groupId>
         <artifactId>spring-context</artifactId>
         <version>${spring.version}</version>
     </dependency>
     <dependency>
         <groupId>junit</groupId>
         <artifactId>junit</artifactId>
         <version>4.12</version>
         <scope>test</scope>
     </dependency>
 </dependencies>
```

```
    <build>
        <plugins>
            <plugin>
                <artifactId>maven-compiler-plugin</artifactId>
                <version>3.6.2</version>
                <configuration>
                    <source>${java.version}</source>
                    <target>${java.version}</target>
                </configuration>
            </plugin>
        </plugins>
    </build>
</project>
```

推荐采用 Maven 或 Gradle 工具来下载 Spring 模块，具体操作步骤可以参见 Spring 官网：http://projects.spring.io/spring-framework/，如图 15-2 所示。

采用类似 Maven 或 Gradle 的工具有一个好处，即下载一个 Spring 模块时，会自动下载其所依赖的模块。

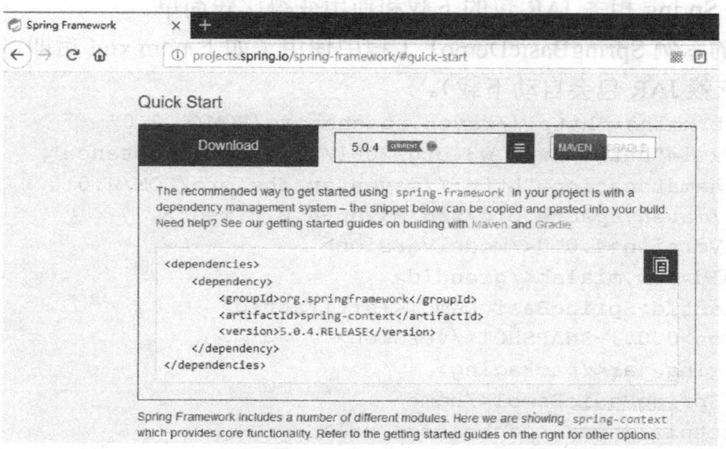

图 15-2　通过 Spring 官网来下载 Spring 核心模块

15.1.2　使用 Spring 容器

1. BeanFactory

BeanFactory 可视为 Spring 的 BeanFactory 容器，它的主要功能是为依赖注入（Dependency Injection，DI）提供支持。这个容器接口在 org.springframework.beans.factory.BeanFactor 中被定义。

BeanFactory 是用于访问 Spring Bean 容器的根接口，是一个单纯的 Bean 工厂，也就是常说的 IoC 容器的顶层定义，各种 IoC 容器是在其基础上为了满足不同需求而扩展的，包括经常使用的 ApplicationContext。

在 Spring 中，有大量对 BeanFactory 接口的实现。其中，最常被使用的是 XmlBeanFactory 类。这个容器从一个 XML 文件中读取配置元数据，由这些元数据来生成一个被配置化的系统或者应用。例如：

```
public class MainApp {
    public static void main(String[] args) {
```

```
        XmlBeanFactory factory = new XmlBeanFactory
                        (new ClassPathResource("Beans.xml"));
        HelloWorld obj = (HelloWorld) factory.getBean("helloWorld");
        obj.getMessage();
    }
}
```

2. ApplicationContext

ApplicationContext 是 BeanFactory 的子接口，使用它作为 Spring 容器更方便。它可以加载配置文件中定义的 Bean，将所有的 Bean 集中在一起，当有请求的时候分配 Bean。

ApplicationContext 包含 BeanFactory 所有的功能，一般情况下，相对于 BeanFactory，ApplicationContext 会被推荐使用。BeanFactory 仍然可以在轻量级应用中使用，如移动设备或者基于 Applet 的应用程序。

最常被使用的 ApplicationContext 接口实现如下。

① FileSystemXmlApplicationContext：该容器从 XML 配置文件中加载已被定义的 Bean。在这里，用户需要提供 XML 文件的完整路径给构造器。

② ClassPathXmlApplicationContext：以类加载路径下的 XML 配置文件创建 ApplicationContext 实例。（从类加载路径下搜索配置文件，并根据配置文件来创建 Spring 容器。）

③ WebXmlApplicationContext：该容器会在一个 Web 应用程序的范围内加载在 XML 文件中已被定义的 Bean。

一般不会使用 BeanFactory 实例作为 Spring 容器，而是使用 ApplicationContext 实例作为容器，因此也把 Spring 容器称为 Spring 上下文。必须注意的是，在使用 Spring 框架的 Web 项目中，ApplicationContext 容器的实例化工作会交由 Web 服务器来完成。Web 服务器实例化 ApplicationContext 容器时，通常会使用基于 ContextLoaderListener 的方式来实现。

3. Spring 入门示例

【示例】Spring 入门。

在 SpringBasicDemo1 工程中，演示了使用 Spring 容器（这里是 ApplicationContext）得到 bean 的入门示例。（代码详见本书源码包 ch15 目录中的 SpringBasicDemo1 工程。）

PersonService.java 主要代码如下：

```
public interface PersonService {
 public void say();
}
```

Person.java 主要代码如下：

```
public class Person implements PersonService {
 @Override
 public void say() {
     System.out.println("Hello, Spring Framework!");
 }
}
```

Spring 配置文件 applicationContext.xml 的内容如下：

```
<?xml version="1.0" encoding="UTF-8"?>
<beans xmlns="http://www.springframework.org/schema/beans"
 xmlns:xsi="http://www.w3.org/2001/XMLSchema-instance"
 xsi:schemaLocation="http://www.springframework.org/schema/beans
```

```
    http://www.springframework.org/schema/beans/spring-beans-4.3.xsd">

    <beanid="person"class="com.mialab.SpringBasicDemo1.service.impl.Person
" />
   </beans>
```

在配置文件中，一个普通的 Bean 只需要定义 id（或 name）和 class 两个属性即可。

测试类 TestBean.java 主要代码如下：

```
   public class TestBean {
    public static void main(String[] args) {
      // 创建 Spring 容器
      ApplicationContext ctx = new
          ClassPathXmlApplicationContext("applicationContext.xml");
      Person p = ctx.getBean("person", Person.class);   //获取 id 为 person
的 Bean
      p.say();  // 调用 person 对象的 say()方法
    }
   }
```

在测试类 TestBean-java 的 main 方法中，创建并初始化 Spring 容器（通过 Spring 配置文件 applicationContext.xml 实现），通过 Spring 容器获取 Person 实例（Java 对象），并调用 Person 实例的 say()方法。控制台输出结果如下：

```
   Hello, Spring Framework!
```

15.2 依赖注入

Spring 框架的核心功能有两个：Spring 容器作为超级工厂，负责创建、管理所有的 Java 对象，这些 Java 对象被称为 Bean；Spring 容器管理容器中 Bean 之间的依赖关系，Spring 使用一种被称为“依赖注入”的方式来管理 Bean 之间的依赖关系。

当某个 Java 实例（调用者）需要另一个 Java 实例（被调用者）时，在传统的程序设计过程中，通常由调用者来创建被调用者的实例。

在依赖注入的模式下，创建被调用者的工作不再由调用者完成，因此称为控制反转（Inversion of Control，IoC）；创建被调用者实例的工作通常由 Spring 容器来完成，然后注入调用者，因此也称为依赖注入。

1．理解依赖注入

当编写一个复杂的 Java 应用程序时，应用程序类应该尽可能独立于其他 Java 类来增加这些类重用的可能性，且在做单元测试时，测试独立于其他类的独立性。依赖注入有助于把这些类黏合在一起，同时保持它们的独立性。

假设有一个包含文本编辑器组件的应用程序，并要提供拼写检查功能。其代码如下：

```
   public class TextEditor {
     private SpellChecker spellChecker;
     public TextEditor() {
       spellChecker = new SpellChecker();
     }
   }
```

在这里需要创建 TextEditor 和 SpellChecker 之间的依赖关系。**在控制反转的场景中，反而会做这样的事情：**

```
public class TextEditor {
   private SpellChecker spellChecker;
   public TextEditor(SpellChecker spellChecker) {
      this.spellChecker = spellChecker;
   }
}
```

TextEditor 不应该担心 SpellChecker 的实现。SpellChecker 将会独立实现，并且在 TextEditor 实例化的时候将提供给 TextEditor，整个过程是由 Spring 框架控制的。

我们已经从 TextEditor 中删除了全面控制，并且把它保存到其他地方（即 XML 配置文件中），且依赖关系（即 SpellChecker 类）通过类构造函数被注入到 TextEditor 类中。因此，**控制流通过依赖注入已经“反转”，因为已经有效地委托了依赖关系到一些外部系统。**

依赖注入一般可以分为 3 种方式：构造器注入、setter 注入和接口注入。构造器注入和 setter 注入是主要方式，而接口注入则意味着注入的内容来自外界，例如，在 Web 应用中，配置的数据源是在 Tomcat 服务器的 context.xml 文件中配置的，并以 JNDI 的形式通过接口将它注入 Spring IoC 容器中。

setter 注入（设值注入）是指 IoC 容器通过成员变量的 setter 方法来注入被依赖对象。这种注入方式简单、直观，因而在 Spring 的依赖注入中被大量使用。

利用构造器来设置依赖关系的方式，被称为构造器注入。通俗来说，就是驱动 Spring 在底层以反射方式执行带指定参数的构造器，当执行带参数的构造器时，即可利用构造器参数对成员变量执行初始化，这就是构造器注入的本质。

下面将重点讲述构造器注入方式和 setter 注入方式。

2. Spring 基于设值函数的依赖注入

【示例】设值注入。

在 Eclipse 中选择“New Project→Maven Project”选项，在弹出的“New Maven Project”对话框中，取消勾选“Create a simple project（skip archetype selection）”复选框，单击“Next”按钮后，选择 maven-archetype-quickstart，在“Group ID”文本框中输入“com.mialab”，在“Artifact ID”文本框中输入“SpringBasicDemo1”，即可成功创建 Maven 项目“SpringBasicDemo1”工程。

而完成设值注入 Demo（包括前面的 Spring 入门示例 Demo）的 SpringBasicDemo1 工程目录层次及相应源程序文件如图 15-3 所示。（详见本书源码包 ch15 目录中的 SpringBasicDemo1 工程。）

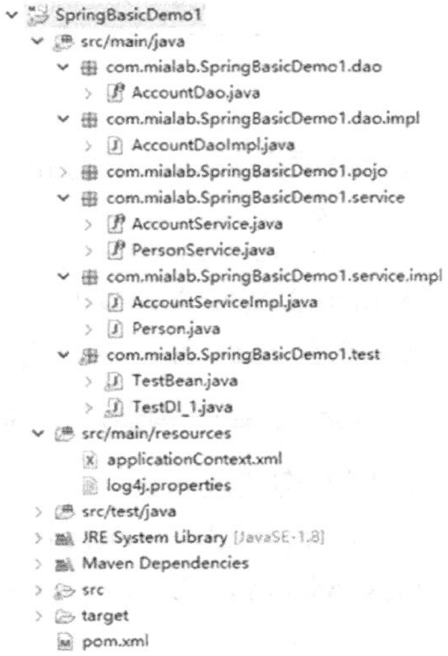

SpringBasicDemo1
src/main/java
com.mialab.SpringBasicDemo1.dao
AccountDao.java
com.mialab.SpringBasicDemo1.dao.impl
AccountDaoImpl.java
com.mialab.SpringBasicDemo1.pojo
com.mialab.SpringBasicDemo1.service
AccountService.java
PersonService.java
com.mialab.SpringBasicDemo1.service.impl
AccountServiceImpl.java
Person.java
com.mialab.SpringBasicDemo1.test
TestBean.java
TestDI_1.java
src/main/resources
applicationContext.xml
log4j.properties
src/test/java
JRE System Library [JavaSE-1.8]
Maven Dependencies
src
target
pom.xml

图 15-3 完成设值注入的 SpringBasicDemo1 工程目录层次及文件

（1）AccountDaoImpl.java 的主要代码如下：

```
    public class AccountDaoImpl implements
AccountDao {
     @Override
     public void add() {
         System.out.println("save account...");
     }
  }
```

（2）AccountServiceImpl.java 的主要代码如下：

```
public class AccountServiceImpl implements AccountService {
 private AccountDao accountDao;  // 声明 AccountDao 属性
 // 添加 AccountDao 属性的 setter 方法，用于实现依赖注入
 public void setAccountDao(AccountDao accountDao) {
     this.accountDao = accountDao;
 }
 @Override
 public void addAccount() {
     this.accountDao.add();
     System.out.println("account added!");
 }
}
```

（3）在 Spring 配置文件 applicationContext.xml 中添加如下内容：

```
<bean id="accountDaoInstance"
    class="com.mialab.SpringBasicDemo1.dao.impl.AccountDaoImpl" />
<bean id="accountService"

class="com.mialab.SpringBasicDemo1.service.impl.AccountServiceImpl">
        <!-- 将 id 为 accountDao 的 Bean 实例注入 accountService 实例中 -->
        <property name="accountDao" ref="accountDaoInstance" />
</bean>
```

（4）测试类 TestDI_1.java 的主要代码如下：

```
public class TestDI_1 {
 public static void main(String[] args) {
     ApplicationContext applicationContext = new
         ClassPathXmlApplicationContext("applicationContext.xml");
     AccountService accountService =
         (AccountService) applicationContext.getBean("accountService");
     accountService.addAccount();
 }
}
```

运行测试类 TestDI_1 的 main()方法，控制台上将有以下输出：

```
save account...
account added!
```

（pom.xml 文件的具体内容，以及 AccountDao.java、AccountService.java 等文件的代码详见本书源码包 ch15 目录中的 SpringBasicDemo1 工程。）

3. Spring 基于构造函数的依赖注入

【示例】构造注入。

（1）在 SpringBasicDemo1 工程的包 com.mialab.SpringBasicDemo1.service.impl 中创建类 AccountServiceImpl_2.java，其主要代码如下（详见本书源码包 ch15 目录中的 SpringBasic Demo1 工程）：

```
public class AccountServiceImpl_2 implements AccountService {
 private AccountDao accountDao;
 public AccountServiceImpl_2(AccountDao accountDao) {
      System.out.println("Inside AccountServiceImpl_2 constructor." );
      this.accountDao = accountDao;
    }
 @Override
```

```
        public void addAccount() {
            this.accountDao.add();
            System.out.println("account added again!");
        }
    }
```

（2）新建 Spring 配置文件 applicationContext2.xml，其中必须添加以下内容：

```
    <!--基于构造函数的依赖注入（Constructor-based dependency injection） -->
    <bean id="accountDao2"
        class="com.mialab.SpringBasicDemo1.dao.impl.AccountDaoImpl" />
    <bean id="accountService2"
class="com.mialab.SpringBasicDemo1.service.impl.AccountServiceImpl_2">
        <constructor-arg ref="accountDao2"/>
    </bean>
```

（3）测试类 TestDI_2.java 的主要代码如下：

```
public class TestDI_2 {
    public static void main(String[] args) {
        ApplicationContext applicationContext = new
            ClassPathXmlApplicationContext("applicationContext2.xml");
        AccountService accountService2 =
            (AccountService) applicationContext.getBean("accountService2");
        accountService2.addAccount();
    }
}
```

（4）运行测试类 TestDI_2 的 main()方法，控制台上将有以下输出：

```
Inside AccountServiceImpl_2 constructor.
save account...
account added again!
```

4．两种注入方式的对比

设值注入有如下优点。

（1）与传统的 JavaBean 的写法更相似，程序开发人员更容易理解、接受。通过 setter 方法设定依赖关系显得更加直观、自然。

（2）对于复杂的依赖关系，如果采用构造注入，会导致构造器过于臃肿，难以阅读。Spring 在创建 Bean 实例时，需要同时实例化其依赖的全部实例，因而导致性能下降。而使用设值注入能避免这些问题。

（3）在某些成员变量可选的情况下，多参数的构造器更加笨重。

构造注入有如下优点。

（1）构造注入可以在构造器中决定依赖关系的注入顺序，优先依赖的优先注入。

（2）对于依赖关系无须变化的 Bean，构造注入更有用处。因为没有 setter 方法，所有的依赖关系全部在构造器内设定，无须担心后续的代码对依赖关系产生破坏。

（3）依赖关系只能在构造器中设定，则只有组件的创建者才能改变组件的依赖关系，对组件的调用者而言，组件内部的依赖关系完全透明，更符合高内聚的原则。

注意：建议采用设值注入为主、构造注入为辅的注入策略。对于依赖关系无须变化的注入，尽量采用构造注入；而其他依赖关系的注入，应考虑采用设值注入。

15.3 Spring 容器中的 Bean

1. Spring Bean 定义

被称为 Bean 的对象是构成应用程序的支柱。Bean 是一个被实例化、组装，并通过 Spring IoC 容器所管理的对象。这些 Bean 是由用容器提供的配置元数据创建的。Bean 定义包含称为配置元数据的信息，具体包括如何创建一个 Bean、Bean 生命周期、Bean 的依赖关系等。这些信息一般通过 Spring 配置文件告知 Spring 容器。

上述所有的配置元数据转换成一组构成每个 bean 定义的属性，如表 15-1 所示。

表 15-1 bean 元素的常用属性

属　性	描　述
class	这个属性是强制性的，并且指定用来创建 bean 实例的具体实现类
name	这个属性指定唯一的 bean 标识符。在基于 XML 的配置元数据中，可以使用 name 或 ID 属性来指定 bean 标识符
scope	这个属性指定 bean 对象的作用域
constructor-arg	<.bean>元素的子元素，可以使用此元素传入构造参数进行实例化
properties	<.bean>元素的子元素，用于调用 Bean 实例中的 setter 方法以完成属性赋值，从而完成依赖注入
ref	<property>、<constructor-arg>等元素的属性或子元素，可以用于指定某个 Bean 实例的引用
value	<property>、<constructor-arg>等元素的属性或子元素，可以用于直接指定一个常量值

对于开发者来说，使用 Spring 框架主要做两件事：开发 Bean；配置 Bean。

对于 Spring 框架来说，它要做的就是根据配置文件来创建 Bean 实例，并调用 Bean 实例的方法完成“依赖注入”，这就是所谓 IoC 的本质。

2. Spring Bean 实例化

Spring 容器中的 Bean 实例化有 3 种方式：构造器实例化、静态工厂方式实例化和实例工厂方式实例化。这 3 种方式中以构造器实例化最为常见。

（1）使用构造器创建 Bean 实例。

使用构造器来创建 Bean 实例是最常见的情况。如果不采用构造注入，Spring 底层会调用 Bean 类的无参数构造器来创建实例，因此要求该 Bean 类提供无参数的构造器。

采用默认的构造器创建 Bean 实例，Spring 对 Bean 实例的所有属性执行默认初始化，即所有的基本类型的值初始化为 0 或 false；所有的引用类型的值初始化为 null。

关于使用构造器实例化的 Spring 配置文件内容，示例如下：

```
<bean id="bean1" class="com.mialab.SpringBeanDemo.constructor.Bean1" />
```

（2）使用静态工厂方法创建 Bean。

使用静态工厂方法创建 Bean 实例时，class 属性必须指定，但此时 class 属性并不是指定 Bean 实例的实现类，而是静态工厂类，Spring 通过该属性知道由哪个工厂类来创建 Bean 实例。

除此之外，还需要使用 factory-method 属性来指定静态工厂方法，Spring 将调用静态工厂方法返回一个 Bean 实例，一旦获得了指定 Bean 实例，Spring 后面的处理步骤就与采用普通方法创建 Bean 实例完全一样。如果静态工厂方法需要参数，则使用<constructor-arg.../>元素指定静态工厂方法的参数。

关于使用静态工厂方式实例化的 Spring 配置文件内容，示例如下：

```
    <bean id="bean2" class="com.mialab.SpringBeanDemo.factory.StaticFactory"
        factory-method="createBean" />
```

工厂类 StaticFactory 中须有静态方法 createBean()，来得到相应的 Bean 对象。

（3）调用实例工厂方法创建 Bean。

实例工厂方法与静态工厂方法只有一处不同：调用静态工厂方法只需使用工厂类即可，而调用实例工厂方法则需要工厂实例。使用实例工厂方法时，配置 Bean 实例的<bean.../>元素无须 class 属性，配置实例工厂方法时使用 factory-bean 指定工厂实例。

采用实例工厂方法创建 Bean 的<bean.../>元素时需要指定如下两个属性。

① factory-bean：该属性的值为工厂 Bean 的 id。

② factory-method：该属性指定实例工厂的工厂方法。

关于使用实例工厂方式实例化的 Spring 配置文件内容，示例如下：

```
    <!-- 配置工厂 -->
    <bean id="beanFactory"
     class="com.mialab.SpringBeanDemo.instance_factory.BeanFactory" />
    <!-- 使用 factory-bean 属性指向配置的实例工厂，使用 factory-method 属性确定使用工
厂中的哪个方法 -->
    <bean id="bean3" factory-bean="beanFactory" factory-method="getInstance"
/>
```

工厂类 BeanFactory 中必须有实例方法 getInstance()，用于得到相应的 Bean 对象。

（代码详见本书源码包 ch15 目录中的 SpringBeanDemo 工程。）

3．Spring Bean 配置依赖

Java 应用中各组件的相互调用的实质可以归纳为依赖关系，根据注入方式的不同，Bean 的依赖注入通常表现为如下两种形式。

（1）属性：通过< property >元素配置，对应设值注入。

（2）构造器参数：通过< constructor-arg >元素指定，对应构造注入。

不管是属性还是构造器参数，都视为 Bean 的依赖，受 Spring 容器管理。依赖关系的值要么是一个确定的值，要么是 Spring 容器中其他 Bean 的引用。例如，如果需要为 Bean 设置的属性值是容器中另一个 Bean 实例，则应该使用<ref>元素。

通常情况下，Spring 在实例化容器时，会校验 BeanFactory 中每一个 Bean 的配置，这些校验包括：Bean 引用的依赖 Bean 是否指向一个合法的 Bean；Bean 的普通属性值是否获得了一个有效值。

4．Spring Bean 设置属性值

< value >元素用于指定字符串类型、基本类型的属性值，Spring 使用 XML 解析器来解析出这些数据，然后利用 java.beans.PropertyEditor 完成类型转换，即从 java.lang.String 类型转换为所需的参数值类型。如果目标类型是基本数据类型，则通常可以正确转换。

5．Spring Bean 作用域

当通过 Spring 容器创建一个 Bean 实例时，不仅可以完成 Bean 实例的实例化，还可以为 Bean 指定特定的作用域。Spring 支持如下 5 种作用域。

① singleton：单例模式，在整个 Spring IoC 容器中，singleton 作用域的 Bean 将只生成一个实例。其为默认作用域。

② prototype：每次通过容器的 getBean()方法获取 prototype 作用域的 Bean 时，都将产生

一个新的 Bean 实例。

③ request：对于一次 HTTP 请求，request 作用域的 Bean 将只生成一个实例，这意味着，在同一次 HTTP 请求内，程序每次请求该 Bean，得到的总是同一个实例。只有在 Web 应用中使用 Spring 时，该作用域才真正有效。该作用域将 Bean 的定义限制为 HTTP 请求，只在 Web-aware Spring ApplicationContext 的上下文中有效。

④ session：该作用域将 Bean 的定义限制为 HTTP 会话，只在 Web-aware Spring ApplicationContext 的上下文中有效。

⑤ global session：每个全局的 HTTP Session 对应一个 Bean 实例。该作用域将 Bean 的定义限制为全局 HTTP 会话，只在 Web-aware Spring ApplicationContext 的上下文中有效。

（本章将讨论前两个作用域范围，即 singleton 作用域和 prototype 作用域。当讨论有关 Web-aware Spring ApplicationContext 时，将讨论其余 3 个作用域。）

如果不指定 Bean 的作用域，则 Spring 默认使用 singleton 作用域。prototype 作用域的 Bean 的创建、销毁代价比较大。而 singleton 作用域的 Bean 实例一旦创建成功，就可以重复使用。因此，应该尽量避免将 Bean 设置成 prototype 作用域。

【示例】singleton 作用域和 prototype 作用域

在 Eclipse 中选择“New Project→Maven Project”选项，在弹出的“New Maven Project”对话框中，取消勾选“Create a simple project（skip archetype selection）”复选框，单击“Next”按钮后，选择 maven-archetype-quickstart。在“Group ID”文本框中输入“com.mialab”，在“Artifact ID”文本框中输入“SpringBeanDemo”。成功创建 Maven 项目“SpringBeanDemo”工程。SpringBeanDemo 工程目录层次及相应源程序文件如图 15-4 所示。（详见本书源码包 ch15 目录中的 SpringBeanDemo 工程。）

图 15-4 SpringBeanDemo 工程目录层次及相应源程序文件

beans_scope.xml 文件的主要内容如下：

```
<?xml version="1.0" encoding="UTF-8"?>
<beans  xmlns="http://www.springframework.org/schema/beans"
    xmlns:xsi="http://www.w3.org/2001/XMLSchema-instance"
    xsi:schemaLocation="http://www.springframework.org/schema/beans
    http://www.springframework.org/schema/beans/spring-beans-4.3.xsd">
    <bean id="bean1" class="com.mialab.SpringBeanDemo.scope. Student"
        scope="singleton"/>
    <bean id="bean2" class="com.mialab.SpringBeanDemo.scope.Student"
        scope="prototype"/>
</beans>
```

Student.java 的主要代码如下：

```
public class Student {
 public String name;
 public String getName() {
     System.out.println("My name is : " + name);
     return name;
 }
```

```
    public void setName(String name) {
        this.name = name;
    }
}
```

TestBeanScope.java 的主要代码如下：

```
public class TestBeanScope {
  public static void main(String[] args) {
      ApplicationContext context = new
          ClassPathXmlApplicationContext("beans_scope.xml");
      Student stA = (Student) context.getBean("bean1");
      stA.setName("LiBai");
      stA.getName();
      Student stB = (Student) context.getBean("bean1");
      stB.getName();
      System.out.println(stA);
      System.out.println(stB);
  }
}
```

测试 TestBeanScope 的 main 方法，控制台上输出内容如下：

```
My name is : LiBai
My name is : LiBai
com.mialab.SpringBeanDemo.scope.Student@af6cff
com.mialab.SpringBeanDemo.scope.Student@af6cff
```

把上述代码 main 方法中的“bean1”改为“bean2”，重新运行 main 方法，控制台将得到以下输出内容。

```
My name is : LiBai
My name is : null
com.mialab.SpringBeanDemo.scope.Student@af6cff
com.mialab.SpringBeanDemo.scope.Student@135b478
```

15.4 容器中 Bean 的生命周期

Spring 可以管理 singleton 作用域 Bean 的生命周期，Spring 可以精确地知道该 Bean 何时被创建，何时被初始化完成，容器何时准备销毁该 Bean 实例。

对于 prototype 作用域的 Bean，Spring 容器仅仅负责创建。当容器创建了 Bean 实例后，Bean 实例完全交给客户端代码管理，容器不再跟踪其生命周期。

Spring Bean 的生命周期的整个执行过程描述如下。

（1）Spring 对 Bean 进行实例化，默认 Bean 是单例模式。

（2）Spring 对 Bean 进行依赖注入。

（3）如果 Bean 实现了 BeanNameAware 接口，则 Spring 会将当前 Bean 的 id 传给 setBeanName()方法。

（4）如果 Bean 实现了 BeanFactoryAware 接口，则 Spring 将调用 setBeanFactory()方法，将 BeanFactory 实例传进来。

（5）如果 Bean 实现了 ApplicationContextAware()接口，则 Spring 会将当前应用上下文（ApplicationContext 实例）的引用传入 setApplicationContext()方法。此时 Spring IoC 容器也必须是一个 ApplicationContext 接口的实现类。

（6）如果 Bean 实现了 BeanPostProcessor 接口，则 Spring 将调用该接口的预初始化方法 postProcessBeforeInitialization()对 Bean 进行加工操作。

（7）如果 Bean 实现了 InitializingBean 接口，则 Spring 将调用其 afterPropertiesSet 接口方法。类似的，如果 Bean 使用 init-method 属性声明了初始化方法，则该方法也会被调用。

（8）如果 Bean 实现了 BeanPostProcessor 接口，则 Spring 将调用该接口的初始化方法 postProcessAfterInitialization()。

（9）此时 Bean 已经准备就绪，可以被应用程序使用了。它们将一直驻留在应用上下文中，直到该应用上下文被销毁为止。如果 Bean 的作用域是 prototype，则 Spring 不再管理该 Bean。

（10）若 Bean 实现了 DisposableBean 接口，则 Spring 将调用它的 destroy()接口方法。同样，如果 Bean 使用 destroy-method 属性声明了销毁方法，则该方法将被调用。

上述生命周期的接口，大部分是针对单个 Bean 而言的，而 BeanPostProcessor 接口是针对所有 Bean 而言的，DisposableBean 接口则是针对 Spring IoC 容器自身的。

关于 Bean 自身方法 init-method 和 destroy-method，可以通过在配置文件 Bean 定义中添加相应属性来指定相应执行方法。例如：

```
<bean name="lifeBean" class="com.mialab.SpringBeanDemo.life.LifeBean"
    init-method="init" destroy-method="destory" />
```

如果在非 Web 应用程序环境中使用 Spring 的 IoC 容器，则可以在 JVM 中注册一个关闭钩子，以保证 Spring 容器被恰当关闭，且自动执行 singleton Bean 实例的析构回调方法。为了注册关闭钩子，只需要调用在 AbstractApplicationContext 中提供的 registerShutdownHook()方法即可。可以参考以下的示例代码。

```
ApplicationContext container = new ClassPathXmlApplicationContext("beans_
life.xml");
LifeBean bean = (LifeBean) container.getBean("lifeBean");
bean.say();
((AbstractApplicationContext) container).registerShutdownHook();
```

（关于容器中 Bean 生命周期的简单示例，可参考本书源码包 ch15 目录中的 Spring Bean Demo 工程的 beans_life.xml、LifeBean.java 和 TestBeanScope.java 等文件。）

15.5 两种后处理器

Spring 提供了两种常用的后处理器：Bean 后处理器，这种后处理器会对容器中的 Bean 进行后处理，对 Bean 进行额外加强；容器后处理器，这种后处理器会对 IoC 容器进行后处理，用于增强容器功能。

15.5.1 Bean 后处理器

Bean 后处理器是一种特殊的 Bean，这种特殊的 Bean 并不对外提供服务，它甚至可以无须 id 属性，它主要负责对容器中的其他 Bean 执行后处理，如为容器中的目标 Bean 生成代理等。

Bean 后处理器会在 Bean 实例创建成功之后，对 Bean 实例进行进一步的增强处理。

Bean 后处理器必须实现 BeanPostProcessor 接口，同时必须实现该接口的两个方法。

① Object postProcessBeforeInitialization (Object bean, String name) throws Beans-

Exception：该方法的第一个参数是系统即将进行后处理的 Bean 实例，第二个参数是该 Bean 的配置 id。

② Object postProcessAfterInitialization(Object bean, String name) throws BeansException：该方法的第一个参数是系统即将进行后处理的 Bean 实例，第二个参数是该 Bean 的配置 id。

容器中一旦注册了 Bean 后处理器，Bean 后处理器就会自动启动，在容器中，每个 Bean 创建时自动工作。Bean 后处理器两个方法的回调时机如图 15-5 所示。

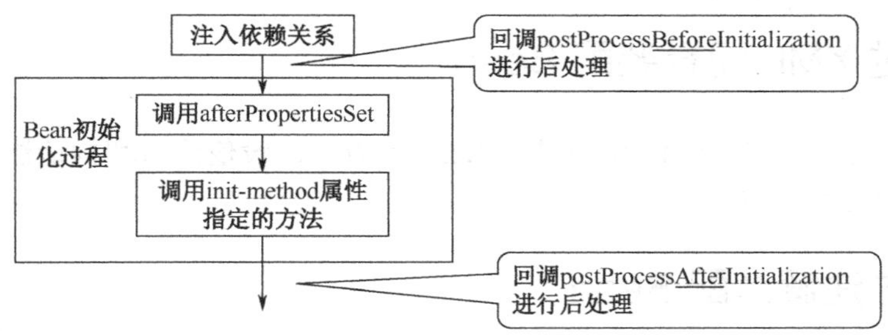

图 15-5 Bean 后处理器两个方法的回调时机

必须注意的是，如果使用 BeanFactory 作为 Spring 容器，则必须手动注册 Bean 后处理器，程序必须获取 Bean 后处理器实例，然后手动注册。以下是示例代码。

```
BeanPostProcessor bp = (BeanPostProcessor)beanFactory.getBean("bp");
beanFactory.addBeanPostProcessor(bp);
Person p = (Person)beanFactory.getBean("person");
```

15.5.2 容器后处理器

Bean 后处理器负责处理容器中的所有 Bean 实例，而容器后处理器则负责处理容器本身。容器后处理器必须实现 BeanFactoryPostProcessor 接口，并实现该接口的一个方法：

```
postProcessBeanFactory(ConfigurableListableBeanFactory beanFactory)
```

实现该方法的方法体就是对 Spring 容器进行的处理，这种处理可以对 Spring 容器进行自定义扩展，当然，也可以对 Spring 容器不进行任何处理。

类似于 BeanPostProcessor，ApplicationContext 可自动检测到容器中的容器后处理器，并且自动注册容器后处理器。但若使用 BeanFactory 作为 Spring 容器，则必须手动调用该容器后处理器来处理 BeanFactory 容器。

15.6 装配 Spring Bean

Bean 的装配可以理解为依赖关系注入，Bean 的装配方式即 Bean 依赖注入的方式。

Rod Johnson 是第一个高度重视以配置文件来管理 Java 实例的协作关系的人，他将这种方式称为控制反转。后来 Martine Fowler 将这种方式依赖注入。因此不管是依赖注入，还是控制反转，其含义完全相同。当某个 Java 对象（调用者）需要调用另一个 Java 对象（被依赖对象）的方法时，在传统模式下通常有以下两种做法。

（1）原始做法：调用者主动创建被依赖对象，再调用被依赖对象的方法。

（2）简单工厂模式：调用者先找到被依赖对象的工厂，再主动通过工厂去获取被依赖对象，最后调用被依赖对象的方法。

注意上面的“主动”二字，这必然会导致调用者与被依赖对象实现类的硬编码耦合，非常不利于项目的升级维护。使用 Spring 框架后，调用者无须主动获取被依赖对象，调用者只要被动接受 Spring 容器为调用者的成员变量赋值即可，由此可见，使用 Spring 后，调用者获取被依赖对象的方式由原来的主动获取变成了被动接受，所以 Rod Johnson 称之为控制反转。

另外，从 Spring 容器的角度来看，Spring 容器负责将被依赖对象赋值给调用者的成员变量，相当于为调用者注入其依赖的实例，因此 Martine Fowler 称之为依赖注入。

15.6.1 通过 XML 配置装配 Bean

Spring 提供了两种通过 XML 配置装配 Bean 的方式：设值注入和构造注入。前面的章节中已经介绍过，在此不再赘述。

15.6.2 通过注解装配 Bean

使用注解的方式可以减少 XML 的配置，注解功能更为强大。Spring 提供了如下几个 Annotation 来标注 Spring Bean：

（1）@Component：标注一个普通的 Spring Bean 类。

（2）@Controller：标注一个控制器组件类。

（3）@Service：标注一个业务逻辑组件类。

（4）@Repository：标注一个 DAO 组件类。

在 Spring 配置文件中做如下配置，指定自动扫描的包：

```
<context:component-scan base-package="…"/>
```

Spring 使用@Resource 配置依赖。

@Resource 位于 javax.annotation 包中，是来自 Java EE 规范的一个 Annotation，Spring 直接借鉴了该 Annotation，通过使用该 Annotation 为目标 Bean 指定协作者 Bean。使用@Resource 与<property.../>元素的 ref 属性有相同的效果。

@Resource 不仅可以修饰 setter 方法，还可以直接修饰实例变量，如果使用@Resource 修饰实例变量将会更加简单，此时 Spring 将会直接使用 Java EE 规范的 Field 注入，此时连 setter 方法都可以不使用。

【示例】通过注解装配 Bean。

在 Eclipse 中选择“New Project→Maven Project”选项，在弹出的“New Maven Project”对话框中，取消勾选“Create a simple project（skip archetype selection）”复选框，单击“Next”按钮后，选择 maven→archetype→quickstart。在“Group ID”文本框中输入“com.mialab”，在“Artifact ID”文本框中输入“SpringBasicDemo2”。成功创建 Maven 项目“SpringBasicDemo2”工程。SpringBasicDemo2 工程目录层次及相应源程序文件如图 15-6 所示。（详见本书源码包 ch15 目录中的 SpringBasicDemo2 工程。）

图 15-6 SpringBasicDemo2 工程目录层次及相应源程序文件

applicationContext.xml 文件主要内容如下：

```
<?xml version="1.0" encoding="UTF-8"?>
<beans xmlns="http://www.springframework.org/schema/beans"
 xmlns:xsi="http://www.w3.org/2001/XMLSchema-instance"
xmlns:context="http://www.springframework.org/schema/context"
 xsi:schemaLocation="http://www.springframework.org/schema/beans
 http://www.springframework.org/schema/beans/spring-beans-4.3.xsd
 http://www.springframework.org/schema/context

http://www.springframework.org/schema/context/spring-context-4.3.xsd">

    <!-- 指定需要扫描的包 -->
    <context:component-scan
base-package="com.mialab.SpringBasicDemo2.controller" />
    <context:component-scan
base-package="com.mialab.SpringBasicDemo2.service" />
    <context:component-scan  base-package="com.mialab.SpringBasicDemo2.dao"
/>
</beans>
```

BookController.java 的主要代码如下：

```
@Controller("bookController")
public class BookController {
  @Resource(name="bookService")
  private BookService bookService;
  public List<Book> findBookList(){
      System.out.println("-------调用 BookController 的 findBookList() 方
法------");
      return bookService.searchBooks();
  }
  public void setBookService(BookService bookService) {
      this.bookService = bookService;
  }
}
```

BookServiceImpl.java 的主要代码如下：

```
@Service("bookService")
public class BookServiceImpl implements BookService {
  @Resource(name="bookDao")
  private BookDao bookDao;
  @Override
  public List<Book> searchBooks() {
      System.out.println("-------调用 BookServiceImpl 的 searchBooks() 方
法------");
      return bookDao.findBookList();
  }
  public void setBookDao(BookDao bookDao) {
      this.bookDao = bookDao;
  }
}
```

BookDaoImpl.java 的主要代码如下：

```
@Repository("bookDao")
public class BookDaoImpl implements BookDao {
  public List<Book> findBookList() {
```

```
        System.out.println("-------调用 BookDaoImpl 的 findBookList() 方法
------");
        List<Book> bookList = new ArrayList<Book>();
        bookList.add(new Book("36201", "Android应用开发实践教程", 56.0f));
        bookList.add(new Book("36202", "IT项目管理", 42.0f));
        return bookList;
    }
  }
```

测试类 SpringAnnotationTest.java 的主要代码如下：

```
public class SpringAnnotationTest {
  public static void main(String[] args) {
        System.out.println("-------------调用 main 方法---------------");
        // 初始化Spring容器，加载配置文件
        ApplicationContext applicationContext = new
            ClassPathXmlApplicationContext("applicationContext.xml");
        // 获取BookController实例
        BookController bookController =
            (BookController) applicationContext.getBean("bookController");
        // 调用BookController中的findBookList()方法
        List<Book> bookList = bookController.findBookList();
        System.out.println(bookList);
    }
}
```

运行测试类 SpringAnnotationTest 的 main 方法，控制台输出内容如下：

```
-------------调用 main 方法---------------
-------调用 BookController 的 findBookList() 方法------
-------调用 BookServiceImpl 的 searchBooks() 方法------
-------调用 BookDaoImpl 的 findBookList() 方法------
[Book [bookId=36201, bookName=Android应用开发实践教程], Book [bookId=36202,
bookName=IT 项目管理]]
```

15.6.3 自动装配和精确装配

1. @Autowired 和@Qualifier 的使用

Spring 4.0 提供了增强的自动装配和精确装配功能。Spring 使用@Autowired 注解来指定自动装配，@Autowired 可以修饰 setter 方法、普通方法、实例变量和构造器等。

当使用@Autowired 标注 setter 方法时，默认采用 byType 自动装配策略。在这种策略下，符合自动装配类型的候选 Bean 实例常常有多个，此时就可能引起异常，为了实现精确的自动装配，Spring 提供了@Qualifier 注解，通过使用@Qualifier，会将@Autowired 默认的按 Bean 类型装配修改为根据 Bean 的 id（或实例名称）来执行自动装配，Bean 的实例名称由@Qualifier 注解的参数指定。

2. 使用自动装配注入合作者 Bean

Spring 能自动装配 Bean 与 Bean 之间的依赖关系，即无须使用 ref 显式指定依赖 Bean，而是由 Spring 容器检查 XML 配置文件内容，根据某种规则，为调用者 Bean 注入被依赖的 Bean。Spring 自动装配可通过<beans/>元素的 default-autowire 属性指定，该属性对配置文件中所有的 Bean 起作用；也可通过<bean/>元素的 autowire 属性指定，该属性只对该 Bean 起作用。autowire 和 default-autowire 可以接收如下值。

（1）no：不使用自动装配。Bean 依赖必须通过 ref 元素定义。这是默认配置，在较大的部署环境中不鼓励改变此配置，显式配置合作者能够得到更清晰的依赖关系。

（2）byName：根据 setter 方法名进行自动装配。Spring 容器查找容器中的全部 Bean，找出其 id 与 setter 方法名，并去掉 set 前缀和小写首字母后同名的 Bean 来完成注入。如果没有找到匹配的 Bean 实例，则 Spring 不会进行任何注入。

（3）byType：根据 setter 方法的形参类型来自动装配。Spring 容器查找容器中的全部 Bean，如果正好有一个 Bean 类型与 setter 方法的形参类型匹配，就自动注入这个 Bean；如果找到多个这样的 Bean，就抛出一个异常；如果没有找到这样的 Bean，则什么都不会发生，setter 方法不会被调用。

（4）constructor：与 byType 类似，区别是其是用于自动匹配构造器的参数。如果容器不能恰好找到一个与构造器参数类型匹配的 Bean，则会抛出一个异常。

（5）autodetect：Spring 容器根据 Bean 内部结构，自行决定使用 constructor 或 byType 策略。如果找到一个默认的构造函数，那么会应用 byType 策略。

当一个 Bean 既使用自动装配依赖，又使用 ref 显式指定依赖时，显式指定的依赖覆盖自动装配依赖；对于大型的应用，不鼓励使用自动装配。虽然使用自动装配可减少配置文件的工作量，但大大降低了依赖关系的清晰性和透明性。依赖关系的装配依赖于源文件的属性名和属性类型，导致 Bean 与 Bean 之间的耦合降低到代码层次，不利于高层次解耦。

通过设置可以将 Bean 排除在自动装配之外，如以下配置代码。

```
<bean id="…" autowire-candidate="false"/>
```

除此之外，还可以在 beans 元素中指定，支持模式字符串，如下所有以 abc 结尾的 Bean 都被排除在自动装配之外。

```
<beans default-autowire-candidates="*abc"/>
```

15.7 Spring 的 AOP

1. 为什么需要 AOP

面向切面编程 AOP（Aspect Orient Programming，AOP）作为面向对象编程的一种补充，已经成为一种比较成熟的编程方式。其实 AOP 问世的时间并不太长，AOP 和 OOP 互为补充，面向切面编程将程序运行过程分解成各个切面。

AOP 专门用于处理系统中分布于各个模块（不同方法）中的交叉关注点的问题，在 Java EE 应用中，常常通过 AOP 来处理一些具有横切性质的系统级服务，如事务管理、安全检查、缓存、对象池管理等，AOP 已经成为一种非常常用的解决方案。

2. 使用 AspectJ 实现 AOP

AspectJ 是一个基于 Java 语言的 AOP 框架，提供了强大的 AOP 功能，其他很多 AOP 框架都借鉴或采纳了其中的一些思想。其主要包括两部分：一部分定义了如何表达、定义 AOP 编程中的语法规范，通过这套语法规范，可以方便地用 AOP 来解决 Java 语言中存在的交叉关注点的问题；另一部分是工具部分，包括编译、调试工具等。

AOP 实现可分为以下两类。

（1）静态 AOP 实现。AOP 框架在编译阶段对程序进行修改，即实现对目标类的增强，生成静态的 AOP 代理类，以 AspectJ 为代表。

（2）动态 AOP 实现。AOP 框架在运行阶段动态生成 AOP 代理，以实现对目标对象的增强，以 Spring AOP 为代表。

一般来说，静态 AOP 实现具有较好的性能，但需要使用特殊的编译器。动态 AOP 实现是纯 Java 实现，因此无须特殊的编译器，但是通常性能略差。

3. AOP 的基本概念

关于面向切面编程的一些术语解释如下。

（1）切面（Aspect）：切面用于组织多个增强处理（Advice），Advice 放在切面中定义。

（2）连接点（Joinpoint）：程序执行过程中明确的点，它实际上是对象的一个操作，如方法的调用或者异常的抛出。在 Spring AOP 中，连接点就是方法的调用。

（3）增强处理：AOP 框架在特定的切入点执行的增强处理，即在定义好的切入点处所要执行的程序代码。

（4）切入点（Pointcut）：可以插入增强处理的连接点。简而言之，当某个连接点满足指定要求时，该连接点将被添加增强处理，该连接点也就变成了切入点。

15.8 本章小结

Spring 框架由 Rod Johnson 开发，2004 年发布了 Spring 框架的第一版。Spring 是一个从实际开发中抽取出来的框架，因此它完成了大量开发中的通用步骤，留给开发者的仅仅是与特定应用相关的部分，从而大大提高了企业应用的开发效率。

Spring 是低侵入式设计，代码的污染极低，独立于各种应用服务器，基于 Spring 框架的应用，可以真正实现“Write Once，Run Anywhere”的承诺。Spring 的 IoC 容器降低了业务对象替换的复杂性，提高了组件之间的解耦。

Spring 的 AOP 支持对一些通用任务（如安全、事务、日志等）进行集中式管理，从而提供了更好的复用性。Spring 的 ORM 和 DAO 提供了与第三方持久层框架的良好整合，并简化了底层的数据库访问。Spring 的高度开放性，并不强制应用完全依赖于 Spring，开发者可自由选用 Spring 框架的部分或全部。

习题 15

1. 如何理解 Spring 的依赖注入和控制反转？
2. 怎样在 Eclipse 中使用 Spring 框架创建 Bean？请编程加以说明。
3. 什么是设值注入？什么是构造注入？请编程加以说明。
4. Spring 容器中 Bean 的生命周期是怎样的？请编程举例加以说明。
5. 如何通过注解装配 Bean？请编程加以说明。
6. 如何通过注解自动装配 Bean？请编程加以说明。
7. 如何理解 Spring 的两种后处理器？请编程加以说明。
8. 使用 Spring 框架创建 Bean 有哪几种方式？请编程加以说明。
9. Spring 的 AOP 是什么？请编程加以说明。

第16章 Spring MVC入门

本章导读

Spring MVC 是围绕 DispatcherServlet 来设计的，DispatcherServlet 其实就是前端控制器，所有的前端请求都要经过它找到相对应的处理器。其工作原理是接收到用户的请求，找到相对应的处理器，也就是 Controller，通过处理器改变模型，调整视图的显示。本章包含的主要内容有：（1）如何创建基于 Maven 的 Web 应用；（2）Spring MVC 框架初使用；（3）使用 Spring MVC 框架进行表单提交；（4）编码过滤器；（5）创建控制器并添加注解；（6）Spring MVC 的工作流程。

16.1 Spring MVC 概述

Spring MVC 是基于 Spring 框架开发的，在 Spring 框架中加入了 MVC 框架。MVC 设计模式是 Web 开发中最常用到的，它将程序分为 3 层：Model 层（模型层）、View 层（视图层）和 Controller 层（控制层）。其中，Model 层是应用程序中处理数据逻辑的部分，通常操作的是数据库；View 层是将 Model 层的数据展示出来；Controller 层主要是接收用户的请求，然后将请求发送给相对应的 Model，控制哪个 View 显示相对应的数据。这样做可以降低耦合度，使得程序更加灵活，各个模块相互分离，而且多个视图可以共享同一个模型，使得代码更加易于管理。Spring MVC 框架的目的就是帮助开发者简化开发。

借助于注解，Spring MVC 提供了几乎是 POJO 的开发模式，使得控制器的开发和测试更加简单。这些控制器一般不直接处理请求，而是将其委托给 Spring 上下文中的其他 Bean，通过 Spring 的依赖注入功能，这些 Bean 将被注入控制器中。

16.2 Spring MVC 入门示例 1：Hello，Spring MVC！

16.2.1 创建 Maven 项目

1．创建工程 SpringMVC_Basic_Demo1

在 Eclipse 中选择“New Project→Maven Project”选项，在弹出的“New Maven Project”对话框中，取消勾选“Create a simple project（skip archetype selection）”复选框，单击“Next”按钮后，选择 maven→archetype→quickstart。在“Group ID”文本框中输入“com.mialab”，在 Artifact ID”文本框中输入“SpringMVC_Basic_Demo1”。（此处所用的 Eclipse 版本详见本书教学资源 tools 文件夹中的 eclipse-jee-oxygen-R-win32.zip。）

成功创建 Maven 项目“SpringMVC_Basic_Demo1”工程，如图 16-1 所示。需要注意的是，这是一个使用 Maven 的 Java Project，需要把它修改为 Maven Web 应用。

2．后续步骤说明

而完成后的 SpringMVC_Basic_Demo1 工程目录层次及相应源程序文件如图 16-2 所示。

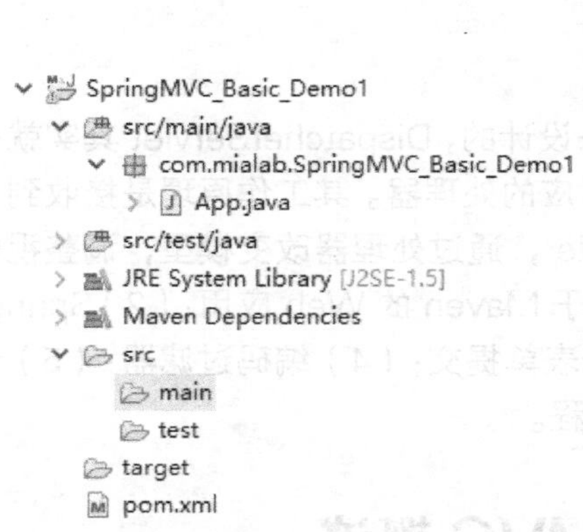

图16-1　SpringMVC_Basic_Demo1工程

SpringMVC_Basic_Demo1
src/main/java
com.mialab.SpringMVC_Basic_Demo1.controller
HelloController.java
src/main/resources
src/test/java
JRE System Library [JavaSE-1.8]
Maven Dependencies
Referenced Libraries
src
main
Webapp
Web-INF
config
springmvc-config.xml
jsp
hello.jsp
Web.xml
index.jsp
test
target
pom.xml

图16-2　完成后的SpringMVC_Basic_Demo1工程目录层次及相应源程序文件

这里需要做以下事情。

（1）在 src/main 中创建文件夹 Webapp 和 resources。

（2）在 src/main/Webapp 文件夹中创建文件夹 Web-INF；在 Web-INF 中创建子文件夹 config 和 jsp，以及 Web 应用部署描述文件 Web.xml。

（3）把 pom.xml 中的<packaging>的值 jar 改为 war，单击工具栏中的保存按钮（“Save”或“Save All”按钮），保存 pom.xml 后，须进行 Maven Project 的更新。具体操作如下：右击 SpringMVC_Basic_Demo1 工程，在弹出的快捷菜单中选择“Maven→Update Project...”选项，当 Web.xml 不存在时会弹出“Web.xml is missing...”错误信息的提示。

（4）修改 pom.xml 文件内容如后文所示或参见源码包，并进行更新操作。

（5）修改 Web.xml 文件内容如后文所示或参见源码包。

（6）在 src/main/Webapp/Web-INF/config 文件夹中，创建 Spring MVC 配置文件 springmvc-config.xml，修改 springmvc-config.xml 文件内容如后文所示或参见源码包。

（7）在 src/main/java 中创建包 com.mialab.SpringMVC_Basic_Demo1.controller，在 package 中创建类（控制器）HelloController.java，其内容如后文所示或参见源码包。

（8）在 src/main/Webapp/Web-INF/jsp 文件夹中，创建文件 hello.jsp，修改 hello.jsp 文件内容如后文所示或参见源码包。

（9）部署运行项目。

16.2.2　pom.xml

此处的 Maven 项目 SpringMVC_Basic_Demo1 中的 pom.xml 内容如下（详见本书教学资源包 ch16 目录中的 SpringMVC_Basic_Demo1 工程）：

```
<project xmlns="http://maven.apache.org/POM/4.0.0"
   xmlns:xsi="http://www.w3.org/2001/XMLSchema-instance"
  xsi:schemaLocation="http://maven.apache.org/POM/4.0.0
   http://maven.apache.org/xsd/maven-4.0.0.xsd">
```

```
    <modelVersion>4.0.0</modelVersion>
    <groupId>com.mialab</groupId>
    <artifactId>SpringMVC_Basic_Demo1</artifactId>
    <version>0.0.1-SNAPSHOT</version>
    <packaging>war</packaging>
    <name>SpringMVC_Basic_Demo1</name>
    <url>http://maven.apache.org</url>
    <properties>
        <java.version>1.8</java.version>
        <!-- Spring 版本号 -->
        <spring.version>4.3.6.RELEASE</spring.version>
        <project.build.sourceEncoding>UTF-8</project.build.sourceEncoding>
    </properties>
    <dependencies>
        <!-- 以下是 Spring 依赖 -->
        <!-- Spring 核心包 -->
        <dependency>
            <groupId>org.springframework</groupId>
            <artifactId>spring-core</artifactId>
            <version>${spring.version}</version>
        </dependency>
        <!-- Spring Web 支持包 -->
        <dependency>
            <groupId>org.springframework</groupId>
            <artifactId>spring-Web</artifactId>
            <version>${spring.version}</version>
        </dependency>
        <!-- Spring MVC 支持包 -->
        <dependency>
            <groupId>org.springframework</groupId>
            <artifactId>spring-Webmvc</artifactId>
            <version>${spring.version}</version>
        </dependency>
        <!-- Spring 面向切面编程支持包 -->
        <dependency>
            <groupId>org.springframework</groupId>
            <artifactId>spring-aop</artifactId>
            <version>${spring.version}</version>
        </dependency>
        <dependency>
            <groupId>javax.servlet</groupId>
            <artifactId>javax.servlet-api</artifactId>
            <version>4.0.0</version>
        </dependency>
        <dependency>
            <groupId>junit</groupId>
            <artifactId>junit</artifactId>
            <version>4.12</version>
            <scope>test</scope>
        </dependency>
    </dependencies>
    <build>
        <plugins>
```

```
            <plugin>
                <artifactId>maven-compiler-plugin</artifactId>
                <version>3.6.2</version>
                <configuration>
                    <source>${java.version}</source>
                    <target>${java.version}</target>
                </configuration>
            </plugin>
        </plugins>
    </build>
</project>
```

可右击 pom.xml，在弹出的快捷菜单中选择“Run As”→“Maven clean”或“Maven install”选项，来清理或者构建 Maven 项目。

16.2.3 Web 应用部署描述文件 Web.xml

Web.xml 的内容如下（详见本书教学资源包 ch16 目录中的 SpringMVC_Basic_Demo1 工程）：

```
<?xml version="1.0" encoding="UTF-8"?>
<Web-app xmlns:xsi="http://www.w3.org/2001/XMLSchema-instance"
      xmlns="http://java.sun.com/xml/ns/javaee"
      xsi:schemaLocation="http://java.sun.com/xml/ns/javaee
      http://java.sun.com/xml/ns/javaee/Web-app_3_0.xsd"
  id="WebApp_ID" version="3.0">
  <display-name>springmvc-basic-demo1</display-name>
  <welcome-file-list>
      <welcome-file>index.html</welcome-file>
      <welcome-file>index.htm</welcome-file>
      <welcome-file>index.jsp</welcome-file>
  </welcome-file-list>
  <servlet>
     <!-- 配置前端过滤器 -->
      <servlet-name>springmvc</servlet-name>
      <servlet-class>
            org.springframework.Web.servlet.DispatcherServlet
        </servlet-class>
      <!-- 初始化时加载配置文件 -->
      <init-param>
          <param-name>contextConfigLocation</param-name>

<param-value>/Web-INF/config/springmvc-config.xml</param-value>
      </init-param>
      <!-- 表示容器在启动时立即加载 Servlet -->
      <load-on-startup>1</load-on-startup>
  </servlet>
  <servlet-mapping>
      <servlet-name>springmvc</servlet-name>
      <url-pattern>/</url-pattern>
  </servlet-mapping>
</Web-app>
```

16.2.4　Spring MVC 配置文件

Spring MVC 配置文件为 springmvc-config.xml，在这个文件中需要配置控制器映射信息，springmvc-config.xml 内容如下（详见本书教学资源包 ch16 目录中的 SpringMVC_Basic_Demo1 工程）：

```
<?xml version="1.0" encoding="UTF-8"?>
<beans xmlns="http://www.springframework.org/schema/beans"
 xmlns:xsi="http://www.w3.org/2001/XMLSchema-instance"
 xsi:schemaLocation="http://www.springframework.org/schema/beans
 http://www.springframework.org/schema/beans/spring-beans-4.3.xsd">
 <!-- 配置处理器 Handle，映射"/helloController"请求 -->
 <bean name="/helloController"

class="com.mialab.SpringMVC_Basic_Demo1.controller.HelloController" />
 <!-- 处理器映射器，将处理器 Handle 的 name 作为 URl 进行查找 -->
 <bean class=

"org.springframework.Web.servlet.handler.BeanNameUrlHandlerMapping" />
 <!-- 处理器适配器，配置对处理器中 handleRequest()方法的调用-->
 <bean class=

"org.springframework.Web.servlet.mvc.SimpleControllerHandlerAdapter" />
 <!-- 视图解析器 -->
 <bean class=

"org.springframework.Web.servlet.view.InternalResourceViewResolver">
 </bean>
</beans>
```

16.2.5　基于 Controller 接口的控制器

在 Spring 2.5 以前，开发一个控制器的唯一方法是实现 org.springframework.Web.servlet.mvc.Controller 接口。Controller 接口必须实现 handleRequest 方法，下面是该方法的签名：

```
ModelAndView  handleRequest(HttpServletRequest  request,  HttpServlet
Response response)
```

Controller 接口的实现类可以通过 handleRequest 方法传递的参数访问对应请求的 HttpServletRequest 和 HttpServletResponse 对象，还必须返回一个包含视图路径或视图路径和模型的 ModelAndView 对象。

Controller 接口的实现类只能处理单一请求动作，而 Spring 2.5 之后新增的基于注解的控制器可以同时支持多个请求动作，并且无须实现任何接口。后文中将对此进行详细介绍。

HelloController.java 代码如下（详见本书教学资源包 ch16 目录中的 SpringMVC_Basic_Demo1 工程）：

```
public class HelloController implements Controller{
 @Override
 public ModelAndView handleRequest(HttpServletRequest request,
        HttpServletResponse response) {
    ModelAndView mav = new ModelAndView();        //创建 ModelAndView
对象
```

```
        mav.addObject("msg", "Hello,我的 Spring MVC!"); //向模型对象中添加数据
        mav.setViewName("/Web-INF/jsp/hello.jsp");          //设置逻辑视图名
        return mav; //返回 ModelAndView 对象
    }
}
```

16.2.6 视图

hello.jsp 代码如下（详见本书教学资源包 ch16 目录中的 SpringMVC_Basic_Demo1 工程）：

```
<%@ page language="java" contentType="text/html; charset=UTF-8"
   pageEncoding="UTF-8"%>
<!DOCTYPE html PUBLIC "-//W3C//DTD HTML 4.01 Transitional//EN"
"http://www.w3.org/TR/html4/loose.dtd">
<html>
<head>
<meta http-equiv="Content-Type" content="text/html; charset=UTF-8">
<title>Hello,Spring MVC</title>
</head>
<body> ${msg} </body>
</html>
```

16.2.7 部署发布项目

有两种方法部署发布项目。其中一种方法是在 Eclipse 中直接部署发布。在 Eclipse 的如图 16-3 所示的界面中，右击 Tomcat v8.0 Server at localhost，在弹出的快捷菜单中选择“Add and Remove..”选项，即可将项目部署发布到 Tomcat 上。此处添加的是 SpringMVC_Basic_Demo1。

图 16-3 在 Eclipse 中创建的 Tomcat Server

单击图 16-3 所示界面右上角的“Start the server”按钮，启动 Tomcat 服务器，在浏览器中输入网址 http://localhost:8080/SpringMVC_Basic_Demo1/helloController，便可进入如图 16-4 所示的界面。

图 16-4 运行 SpringMVC_Basic_Demo1

另一种方法是把工程中 target 文件夹下的 WAR 文件复制到 Tomcat 服务器下的 Webapps 文件夹中，启动 Tomcat 服务器，WAR 文件会自动解压缩成为 Web 应用。在浏览器上访问即可。

16.3　Spring MVC 入门示例 2：表单提交

16.3.1　创建 Maven 项目

同 Spring MVC 入门示例 1 一样，创建 Maven Project 名为 SpringMVC_Basic_Demo2。图 16-5 所示为完成后的 SpringMVC_Basic_Demo2 工程目录结构及文件。后续步骤也同 Spring MVC 入门示例 1 一样。

pom.xml 的内容，Book.java、BookForm.java、BookDetails.jsp 等文件的代码，详见本书源码包 ch16 目录中的 SpringMVC_Basic_Demo2 工程。以下主要对表单提交等相关内容进行讲解。

16.3.2　编码过滤器

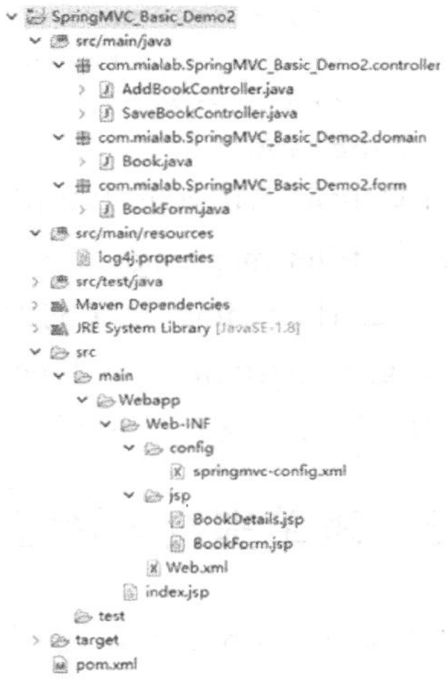

Web.xml 内容如下（详见本书源码包 ch16 目录中的 Spring- MVC_Basic_Demo2 工程）：

```
    <?xml version="1.0" encoding="UTF-8"?>
    <Web-app
xmlns:xsi="http://www.w3.org/2001/XML-
Schema- instance"

xmlns="http://java.sun.com/xml/ns/javaee"

xsi:schemaLocation="http://java.sun.com/xml
/ns/javaee

http://java.sun.com/xml/ns/javaee/Web-app_3
_0.xsd"
     id="WebApp_ID" version="3.0">
     <display-name>springmvc-basic-demo2</display-name>
     <welcome-file-list>
        <welcome-file>index.jsp</welcome-file>
     </welcome-file-list>
     <!-- 编码过滤器 -->
     <filter>
        <filter-name>encoding</filter-name>
        <filter-class>
           org.springframework.Web.filter.CharacterEncodingFilter
        </filter-class>
        <init-param>
           <param-name>encoding</param-name>
           <param-value>UTF-8</param-value>
        </init-param>
     </filter>
     <filter-mapping>
        <filter-name>encoding</filter-name>
        <url-pattern>/*</url-pattern>
     </filter-mapping>
     <servlet>
```

```
SpringMVC_Basic_Demo2
  src/main/java
    com.mialab.SpringMVC_Basic_Demo2.controller
      AddBookController.java
      SaveBookController.java
    com.mialab.SpringMVC_Basic_Demo2.domain
      Book.java
    com.mialab.SpringMVC_Basic_Demo2.form
      BookForm.java
  src/main/resources
    log4j.properties
  src/test/java
  Maven Dependencies
  JRE System Library [JavaSE-1.8]
  src
    main
      Webapp
        Web-INF
          config
            springmvc-config.xml
          jsp
            BookDetails.jsp
            BookForm.jsp
          Web.xml
        index.jsp
    test
  target
  pom.xml
```

图 16-5　完成后的 SpringMVC_Basic_ Demo2 工程目录结构及文件

```
        <servlet-name>springmvc</servlet-name>
        <servlet-class>
              org.springframework.Web.servlet.DispatcherServlet
          </servlet-class>
        <!-- 初始化时加载配置文件 -->
        <init-param>
            <param-name>contextConfigLocation</param-name>

    <param-value>/Web-INF/config/springmvc-config.xml</param-value>
        </init-param>
        <!-- 表示容器在启动时立即加载 Servlet -->
        <load-on-startup>1</load-on-startup>
    </servlet>
    <servlet-mapping>
        <servlet-name>springmvc</servlet-name>
        <url-pattern>/</url-pattern>
    </servlet-mapping>
  </Web-app>
```

为了防止前端传入的中文出现乱码，可在 Web.xml 中配置字符编码过滤器。在 Web.xml 中，通过<filter-mapping>元素的配置会拦截前端页面中的所有请求，并交由名称为 filter.CharacterEncodingFilter 的编码过滤器类进行处理。在<filter>元素中，首先配置了编码过滤器类 org.springframework.Web.filter.CharacterEncodingFilter，然后通过初始化参数设置统一的编码为 UTF-8。这样，所有的请求信息内容都会以 UTF-8 的编码格式进行解析。

16.3.3 表单提交及相应配置

AddBookController.java 的主要代码如下（详见本书源码包 ch16 目录中的 SpringMVC_Basic_ Demo2 工程）：

```
public class AddBookController implements Controller {
 @Override
 public ModelAndView handleRequest(HttpServletRequest request,
 HttpServletResponse response) throws Exception {
     return new ModelAndView("/Web-INF/jsp/BookForm.jsp");
 }
}
```

Spring MVC 配置文件为 springmvc-config.xml，在这个文件中需要配置控制器映射信息，其主要内容如下（详见本书源码包 ch16 目录中的 SpringMVC_Basic_Demo2 工程）：

```
<?xml version="1.0" encoding="UTF-8"?>
<beans xmlns="http://www.springframework.org/schema/beans"
 xmlns:xsi="http://www.w3.org/2001/XMLSchema-instance"
 xsi:schemaLocation="http://www.springframework.org/schema/beans
 http://www.springframework.org/schema/beans/spring-beans-4.3.xsd">
 <bean name="/book_add"

class="com.mialab.SpringMVC_Basic_Demo2.controller.AddBookController"/>
    <bean name="/book_save"

class="com.mialab.SpringMVC_Basic_Demo2.controller.SaveBookController"/>
</beans>
```

当访问 URL 为 http://localhost:8080/SpringMVC_Basic_Demo2/book_add 的网络资源时，实际上执行的是登录表单 BookForm.jsp 的操作，如图 16-6 和图 16-7 所示。

BookForm.jsp 主要内容如下（详见本书源码包 ch16 目录中的 SpringMVC_Basic_Demo2 工程）：

```
<form action="book_save" method="post">
 <table border="1">
     <tr> <td>Book Name:</td>
         <td><input type="text" id="name" name="name" tabindex="1"></td>
     </tr>
     <tr>    <td>Book Description:</td>
         <td><input type="text" id="description" name="description"
             tabindex="2"></td>
     </tr>
     <tr>    <td>Book Price:</td>
         <td><input  type="text"  id="price"  name="price"  tabindex="3">
</td>
     </tr>
     <tr>    <td><input id="reset" type="reset" tabindex="4"></td>
         <td><input id="submit" type="submit" tabindex="5"
             value="添加图书"></td>
     </tr>
 </table>
</form>
```

表单提交后处理的 action 是 book_save，由配置文件 springmvc-config.xml 可知，book_save 映射的是控制器 SaveBookController 的 handleRequest 方法。

SaveBookController.java 的主要代码如下：

```
public class SaveBookController implements Controller {
 @Override
 public ModelAndView handleRequest(HttpServletRequest request,
     HttpServletResponse response) throws Exception {
         BookForm bookForm = new BookForm();
        bookForm.setName(request.getParameter("name"));
        bookForm.setDescription(request.getParameter("description"));
        bookForm.setPrice(request.getParameter("price"));
        Book book = new Book();   // create model
        book.setName(bookForm.getName());
        book.setDescription(bookForm.getDescription());
        try {
         book.setPrice(Float.parseFloat(bookForm.getPrice()));
        } catch (NumberFormatException e) {
        }
        // insert code to save Book
        return new ModelAndView("/Web-INF/jsp/BookDetails.jsp", "book",
book);
 }
}
```

在 handleRequest 方法中，处理后的数据返回给 BookDetails.jsp 页面，进行视图渲染，如图 16-8 所示。（详见本书源码包 ch16 目录中的 SpringMVC_Basic_Demo2 工程。）

16.3.4 测试应用

将项目SpringMVC_Basic_Demo2部署发布到Tomcat服务器上，并启动Tomcat服务器。在浏览器地址栏中输入http://localhost:8080/SpringMVC_Basic_Demo2/book_add，进行测试。运行后效果如图16-6~图16-8所示。

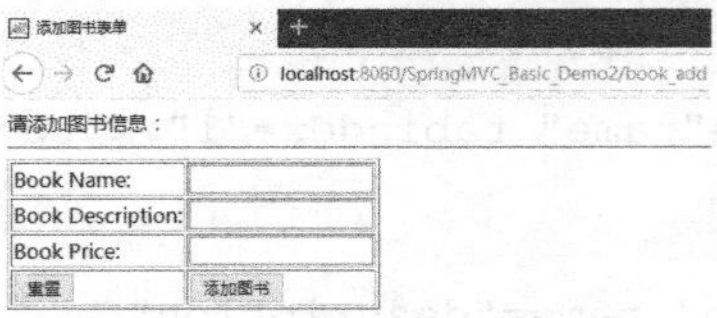

图16-6 添加图书表单① 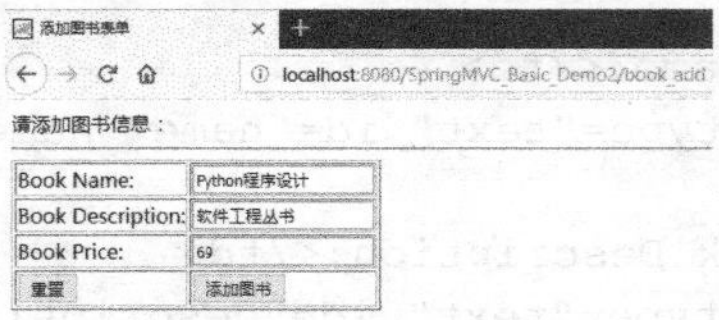　　图16-7 添加图书表单② 　　图16-8 图书保存页面

16.4 Spring MVC入门示例3：基于注解

Spring 2.5出现之前只能使用实现Controller接口的方式来开发一个控制器，从Spring 2.5开始，新增加了基于注解的控制器以及其他常用注解。注解的使用使得开发者的工作变得更为轻松。就目前的Spring MVC应用开发来说，一般会采用基于注解的开发方式。下面的示例是使用基于注解的控制器来改写SpringMVC_Basic_Demo1工程。

16.4.1 创建Maven项目

仍同前例一样，创建Maven Project，名为SpringMVC_Basic_Demo3。图16-9所示为完成后的SpringMVC_Basic_Demo3工程目录结构及文件。后续步骤也同前例一样。

本示例着重说明可通过在控制器中添加注解的方式来进行开发。pom.xml文件的内容，Web.xml、hello.jsp等文件的代码，可详见源码包ch16目录中的SpringMVC_Basic_Demo3工程。

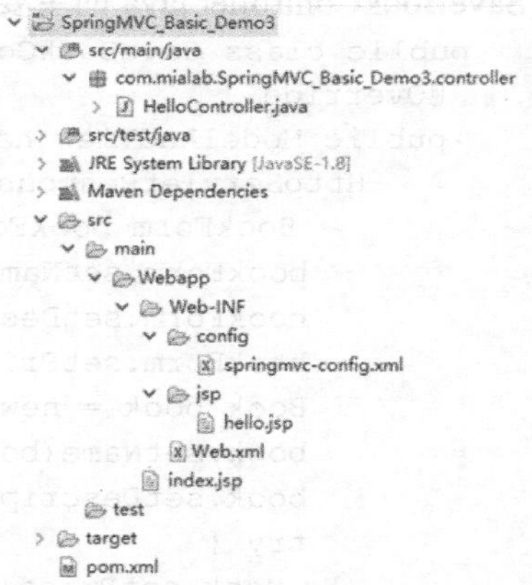

图16-9 完成后的SpringMVC_Basic_Demo3工程目录结构及文件

16.4.2 创建控制器并添加注解

控制器HelloController类不需要实现Controller接口，可改为使用注解类型来描述。Controller和RequestMapping注解类型是Spring MVC API最重要的两种注解类型。可在控制器HelloController类中添加这两种注解。

HelloController.java的主要代码如下（详见本书源码包ch16目录中的SpringMVC_Basic_Demo3工程）：

```
@Controller
@RequestMapping(value="/myboy")
public class HelloController {
 @RequestMapping(value="/hello")
```

```
        public String handleRequest(HttpServletRequest request,
                HttpServletResponse response, Model model) throws Exception {
            // 向模型对象中添加数据
            model.addAttribute("msg", "Hello myboy, 这是基于注解的 Spring MVC 示例!
");
            // 返回视图页面
            return "hello";
        }
    }
```

org.springframework.stereotype.Controller 注释类型表明该类的实例是一个控制器。Spring 通过@Controller 注解找到相应的控制器后，还需要知道控制器内部对每一个请求是如何处理的。org.springframework.Web.bind.annotation.RequestMapping 注释类型用来映射一个请求或一个方法，其注解形式为@RequestMapping，可以将该注解标注在一个方法或一个类上。

如上述代码在类和方法上都使用了@RequestMapping 注解，于是 handleRequest()方法可以通过网址 http://localhost:8080/SpringMVC_Basic_Demo3/myboy/hello 进行访问。

由于使用了注解类型，因此不需要再在配置文件中使用 XML 描述 Bean。需要在配置文件 springmvc-config.xml 中添加如下代码：

```
    <context:component-scan base-package="com.mialab.
    SpringMVC_Basic_Demo3.controller" />
```

上述代码指定了需要扫描的包。如果扫描到有 Spring 的相关注解的类，则把这些类注册为 Spring 的 bean。

16.4.3 视图解析器

Spring MVC 中的视图解析器（View Resolver）负责解析视图，可以通过在配置文件中定义一个 ViewResolver 来配置视图解析器。配置文件 springmvc-config.xml 内容如下（详见本书教学资源包 ch16 目录中的 SpringMVC_Basic_Demo1 工程）：

```
    <?xml version="1.0" encoding="UTF-8"?>
    <beans …>
     <!-- 指定需要扫描的包 -->
     <context:component-scan
         base-package="com.mialab.SpringMVC_Basic_Demo3.controller" />
     <!-- 定义视图解析器 -->
     <bean id="viewResolver"

class="org.springframework.Web.servlet.view.InternalResourceViewResolver">
          <!-- 设置前缀 -->
          <property name="prefix" value="/Web-INF/jsp/" />
          <!-- 设置后缀 -->
          <property name="suffix" value=".jsp" />
     </bean>
    </beans>
```

以上视图解析器设置了前缀和后缀两个属性。这样设置后，方法中所定义的 view 路径将可以简化。例如，本示例 handleRequest()方法返回的逻辑视图名只需设置为“hello”，而不必再设置视图路径为/Web-INF/jsp/hello.jsp，视图解析器将会自动增加前缀和后缀。

16.4.4 测试应用

使用 Eclipse 部署 SpringMVC_Basic_Demo3 Web 应用，在浏览器地址栏中输入 URL http://localhost:8080/SpringMVC_Basic_Demo3/myboy/hello，进入如图 16-10 所示的界面。

图 16-10 基于注解的 Spring MVC 入门示例

16.5 Spring MVC 的工作流程

Spring MVC 请求→响应的完整工作流程如下。

（1）用户向服务器发送请求，请求被 Spring MVC 前端控制器 DispatcherServlet 所拦截。

（2）前端控制器 DispatcherServlet 调用处理器映射器 HandlerMapping 查找 Handler 处理器，可以根据 XML 配置、注解进行查找。

（3）处理器映射器 HandlerMapping 根据请求 URL 找到具体的处理器，生成 Handler 对象及 Handler 对象对应的拦截器（如果有则生成），这些对象会被封装到一个 HandlerExecutionChain 对象中，一并返回给前端控制器 DispatcherServlet。

（4）DispatcherServlet 根据获得的 Handler，选择一个合适的 HandlerAdapter（处理器适配器）。（如果成功获得了 HandlerAdapter，将开始执行拦截器的 preHandler()方法。HandlerAdapter 会被用于处理多种 Handler，调用 Handler 实际处理请求的方法。）

（5）HandlerAdapter 调用并执行 Handler 处理器，这里的处理器就是程序中编写的 Controller 类。具体过程如下：提取 Request 中的模型数据，填充 Handler 入参，开始执行 Handler（Controller）。在填充 Handler 的入参过程中，根据用户的配置，Spring 将帮用户做一些额外工作。

① HttpMessageConveter：将请求消息（如 JSON、XML 等数据）转换成一个对象，将对象转换为指定的响应信息。

② 数据转换：对请求消息进行数据转换，如 String 转换成 Integer、Double 等。

③ 数据格式化：对请求消息进行数据格式化。如将字符串转换成格式化数字或格式化日期等。

④ 数据验证：验证数据的有效性（长度、格式等），验证结果存储到 BindingResult 或 Error 中。

（6）Handler（Controller）执行完成后给处理器适配器返回一个 ModelAndView 对象。处理器适配器向前端控制器 DispatcherServlet 返回 ModelAndView 对象。

（7）根据返回的 ModelAndView 对象，选择一个适合的 ViewResolver（必须是已经注册到 Spring 容器中的 ViewResolver）返回给 DispatcherServlet 。

（8）视图解析器进行视图解析，把逻辑视图名解析成真正的视图。视图解析器向前端控制器返回真正的视图对象。

（9）（此时前端控制器中既有视图又有 Model 对象数据。）前端控制器根据模型数据和视图对象进行视图渲染，并将渲染结果返回给客户端。[视图渲染可将模型数据（在 ModelAndView 对象中）填充到 request 域中]。

16.6 本章小结

Spring MVC 框架提供了 MVC 架构和用于开发灵活松散耦合的 Web 应用程序的组件。MVC 模式导致应用程序的不同方面（输入逻辑、业务逻辑和 UI 逻辑）分离，同时提供了这些元素之间的松散耦合。模型封装了应用程序数据，视图负责渲染模型数据，控制器负责处理用户请求及构建适当的模型，并将其传递给视图进行渲染。

本章开始讲解了 Spring MVC 框架，通过 3 个示例说明了控制器是开发的核心内容，需要知道如何获取请求参数，处理业务逻辑，并将得到的数据通过视图解析器和视图渲染出来展现给客户。这些内容在后续章节中会进一步深入学习。

习题 16

1. 在 Eclipse 中如何创建基于 Maven 的 Java Project，并把它修改为 Maven Web 应用？
2. 简述使用 Spring MVC 框架进行编程的步骤，并编程说明。
3. Spring MVC 配置文件一般有哪些内容？有什么含义？
4. 使用 Spring MVC 框架时，如何防止前端传入的中文出现乱码？
5. 如何使用 Spring MVC 框架进行表单提交？试编程说明。
6. 如何创建基于注解的控制器？试编程说明。
7. 简述 Spring MVC 的工作流程。
8. Spring MVC 是如何初始化 Spring IoC 容器上下文的？又是如何初始化映射请求上下文的？

第 17 章 基于注解的控制器

本章导读

从 Spring 2.5 开始引入注解，使用注解的方式可以减少 XML 的配置，其也提供自动装配的功能，使得开发工作变得更为轻松，这实际上是“约定优于配置”的开发原则。本章主要内容有：（1）Spring MVC 中的常用注解的使用；（2）Spring MVC 中如何处理模型数据；（3）应用@Autowired 和@Service 进行依赖注入；（4）通过 Flash 属性实现重定向传值。

17.1 Spring MVC 常用注解

使用基于注解的控制器有以下两个优点。

（1）一个使用基于注解的控制器可以处理多个动作，而一个实现了 Controller 接口的控制器只能处理一个动作。这就允许将相关的操作写在同一个控制器类中，从而减少了应用程序中类的数量。

（2）基于注解的控制器的请求映射不需要存储在配置文件中。使用 RequestMapping 注解类型，可以对一个方法进行请求处理。

17.1.1 @Controller 和@RequestMapping

1. @Controller

org.springframework.stereotype.Controller 注解类型用于指示 Spring 类的实例是一个控制器，其注解形式是@Controller，Spring MVC 使用扫描机制查找应用程序中所有基于注解的控制器类（使用@Controller 标记的类）。分发处理器会扫描使用了该注解的类的方法，并检测该方法是否使用了@RequestMapping 注解，而使用@RequestMapping 注解的方法才是真正处理请求的处理器。为了保证 Spring 能找到控制器，需要在 Spring MVC 的配置文件中添加相应的扫描配置信息。

（1）在配置文件的声明中引入 spring-context。

（2）使用<context:component-scan>元素指定需要扫描的类包。

整个配置文件应类似于如下所示：

```
<?xml version="1.0" encoding="UTF-8"?>
<beans xmlns="http://www.springframework.org/schema/beans"
 xmlns:xsi=http://www.w3.org/2001/XMLSchema-instance
 xmlns:context="http://www.springframework.org/schema/context"
 xsi:schemaLocation=
   "http://www.springframework.org/schema/beans
   http://www.springframework.org/schema/beans/spring-beans.xsd
   http://www.springframework.org/schema/context
```

```
http://www.springframework.org/schema/context/spring-context.xsd">
    <!-- 指定需要扫描的包 -->
    <context:component-scan base-package="com.mialab.example.controller" />
    …
</beans>
```

确保所有控制器类都在基本包下，并且指定扫描该包，即 com.mialab.example.controller，而不应该指定扫描范围更广泛的包，如假设指定扫描包 com.mialab.example，则会使得 Spring MVC 扫描了无关的包。

2. @RequestMapping

@RequestMapping 注解可以在控制器类的级别和（或）控制器类中的方法的级别上使用。类的级别上的注解会将一个特定请求或者请求模式映射到一个控制器之上。还可以另外添加方法级别的注解的方式来进一步指定到处理方法的映射关系。

下面是一个同时在类和方法上应用了@RequestMapping 注解的示例。正如下述代码所示，hostname:port/home/的请求会由 get()方法来处理，而 hostname:port/home/index/的请求会由 index()方法来处理。

```
@RestController
@RequestMapping("/home")
public class IndexController {
    @RequestMapping("/")
    String get() {
        //mapped to hostname:port/home/
        return "Hello from get";
    }
    @RequestMapping("/index")
    String index() {
        //mapped to hostname:port/home/index/
        return "Hello from index";
    }
}
```

@RequestMapping 注解支持的常用属性如下，且以下属性都是可选属性。

（1）value 属性：@RequestMapping 注解的默认属性。如果 value 属性是唯一的，则可以省略 value 属性名，如@RequestMapping("/index")。

（2）method 属性：用于指定该方法用于处理哪种类型的请求方式。例如，method=RequestMethod.POST 表示该方法只支持 POST 请求。如果没有指定 method 属性值，则请求处理方法可以处理任意的 HTTP 请求方式。

（3）params 属性：指定 request 中必须包含某些参数值时，才以该方法处理。

（4）headers 属性：指定 request 中必须包含某些指定的 header 值时，才以该方法处理。

（5）consumes 属性：用于指定处理请求的提交内容类型（Content-type），如 application/json、text/html 等。

（6）produces 属性：用于指定处理返回的内容类型，返回的内容类型必须是 request 请求头中所包含的类型。

3. 处理器的拦截器

Spring MVC 还定义了处理器的拦截器，当启动 Spring MVC 的时候，Spring MVC 就会解析@Controller 中的@RequestMapping 的配置，再结合所配置的拦截器，这样，它就会组成多个拦截器和一个控制器的形式，并存放到一个 HandlerMapping 中。当请求来到服务器时，首先会通过请求信息找到对应的 HandlerMapping，进而可以找到对应的拦截器和处理器，这样就能够运行对应的控制器和拦截器。

17.1.2 @Autowired 和@Service

使用 Spring 框架的一个好处是容易进行依赖注入。毕竟，Spring 框架一开始就是一个依赖注入容器。将依赖注入 Spring MVC 控制器中的最简单方法是通过注解@Autowired 到字段（属性）或方法来实现。Autowired 注解类型属于 org.springframework.beans.factory.annotation 包。被定位于实现业务逻辑功能的 Service，为了能被作为依赖注入，类必须要声明为@Service，该类型是 org.springframework.stereotype 的成员。Service 注解类型指示类是一个服务。此外，在 Spring MVC 配置文件中，还需要添加自动扫描控制器所在的包以及服务所在的包的代码。

本章后面的示例 AnnotationDemo2 工程进一步说明了 Spring MVC 如何应用@Autowired 和@Service 进行依赖注入。在 AnnotationDemo2 工程中，BookService 是 BookController 中的成员（或者字段），BookController 调用 BookService 的方法实现业务逻辑。于是，BookController.java 中有如下代码：

```
@Controller
public class BookController {
 @Autowired
 private BookService bookService;
 …
}
```

BookServiceImpl.java 中有如下代码：

```
@Service
public class BookServiceImpl implements BookService {
 …
}
```

Spring MVC 配置文件 springmvc-config.xml 中还需要加上这样的代码来自动扫描包，以在 Spring 容器中得到 BookController 对象和 BookServiceImpl 对象（服务实例）。

```
    <context:component-scan
base-package="com.mialab.AnnotationDemo2.controller" />
    <context:component-scan
base-package="com.mialab.AnnotationDemo2.service" />
```

17.1.3 @RequestParam 和@PathVariable

1. @RequestParam

@RequestParam 用于将请求参数区中的数据映射到功能处理方法的参数上。例如：

```
    @RequestMapping("/accounts/show")
    public void show(@RequestParam("num") String number, Map<String, Object>
model) {
        model.put("account", accountRepository.findAccount(number));
    }
```

这里@RequestParam 注解可以用来提取**请求参数区数据**名为"num"的 String 类型的参数，

并将之作为输入参数传给 number。这里**请求参数区数据**可以是 URL 请求中的参数，也可以是提交表单中的参数。例如，针对前一种可假设 URL 请求如下：

```
http://localhost:8080/context/accounts/show?num=168
```

针对后一种可假设提交表单如下：

```
<form action="accounts/show " method="get">
     inputNumber:<input type="text" name="num">
     …
    <input type="submit" value="提交">
</form>
```

@RequestParam 还有如下写法：

```
@RequestParam（value="loginname", required=true, defaultValue="admin"）
```

对传入参数指定为 loginname，如果前端不传入 loginname 参数名，则会报错。可以通过 required=false 或者 true 来要求@RequestParam 配置的前端参数是否一定要传入。

2. @PathVariable

@PathVariable 可以用来映射 URL 中的占位符到目标方法的参数中。例如，有以下代码：

```
@RequestMapping(value="/users/{userId}/topics/{topicId}")
 public String test(@PathVariable(value="userId") int userId,
                    @PathVariable(value="topicId") int topicId)
```

如请求的 URL 为“控制器 URL/users/123/topics/456”，则自动将 URL 中的模板变量{userId}和{topicId}绑定到通过@PathVariable 注解的同名参数上，即入参后 userId=123、topicId=456。

17.1.4 @CookieValue 和@RequestHeader

1. @CookieValue

考虑浏览器和服务器之间的交互。当用户第 1 次访问服务器时，服务器会在响应消息中增加 Set-Cookie 头字段，将用户信息以 Cookie 的形式发送给浏览器。一旦用户浏览器接收了服务器发送的 Cookie 信息，就会将它保存在浏览器的缓冲区中。这样，当浏览器后续访问该服务器时，都会在请求信息中将用户信息以 Cookie 的形式发送给 Web 服务器，从而使服务器端分辨出当前请求是由哪个用户发出的。

使用@Cookie 注解可以将请求的 Cookie 数据映射到功能处理方法的参数上。

参见以下示例代码（详见本书源码包 ch17 目录中的 CookieValueDemo 工程）。

index.jsp 的主要代码如下：

```
<a href="cookieValue/jsessionId">Test CookieValue</a>
```

CookieValueController.java 的主要代码如下：

```
@Controller
@RequestMapping("/cookieValue")
public class CookieValueController {
 @RequestMapping(value = "/jsessionId")
 public String jsessionId(
        @CookieValue(value = "JSESSIONID", required = true,
             defaultValue = "MyJsessionId") String jsessionId,  Model
model) {
     model.addAttribute("jsessionId", jsessionId);
     return "result";
```

```
    }
}
```

result.jsp 的主要代码如下：

```
jsessionId的值是：${requestScope.jsessionId}
```

测试应用。单击 index.jsp 页面中的超链接，得到 result.jsp 页面的输出结果：

```
jsessionId的值是：8F17DA624AEA04A9D0958BE7AF8655E0
```

2. @RequestHeader

org.springframework.Web.bind.annotation.RequestHeader 注解类型用于将请求的头信息区数据映射到功能处理方法的参数上。

在控制器的方法参数中使用@RequestHeader 注解，能够从 Http 请求头中提取指定的某个请求头，可以说其等价于 HttpServletRequest.getHeader(String)。

参见以下示例代码（详见本书源码包 ch17 目录中的 RequestHeaderDemo 工程）。

index.jsp 的主要代码如下：

```
<a href="requestHeader/accept">Test RequestHeader</a>
```

RequestHeaderController.java 的主要代码如下：

```
@Controller
@RequestMapping("/requestHeader")
public class CookieValueController {
 @RequestMapping(value = "/accept")
 public String accept(@RequestHeader(value = "Accept", required = true,
         defaultValue = "MyAccept") String accept, Model model) {
     model.addAttribute("accept", accept);
     return "result";
 }
}
```

result.jsp 的主要代码如下：

```
accept的值是：${requestScope.accept}
```

测试应用。单击 index.jsp 页面中的超链接，得到 result.jsp 页面的输出结果：

```
accept的值是：text/html,application/xhtml+xml,application/xml;q=0.9,*/*;
q=0.8
```

17.2 在 Spring MVC 中处理模型数据

17.2.1 数据模型

在讨论 Spring MVC 流程的时候，我们知道从控制器获取数据后，会装载数据到数据模型和视图中，一直用 ModelAndView 来定义视图类型，也用它来加载数据模型。

ModelAndView 有一个类型为 ModelMap 的属性 model，而 ModelMap 继承了 LinkedHashMap<String，Object>，因此它可以存放各种键值对。为了进一步定义数据模型功能，Spring 还创建了类 ExtendedModelMap，这个类实现了数据模型定义的 Model 接口（org.springframework.ui.Model），如图 17-1 所示，且在此基础上派生了关于数据绑定的类——BindingAwareModelMap。

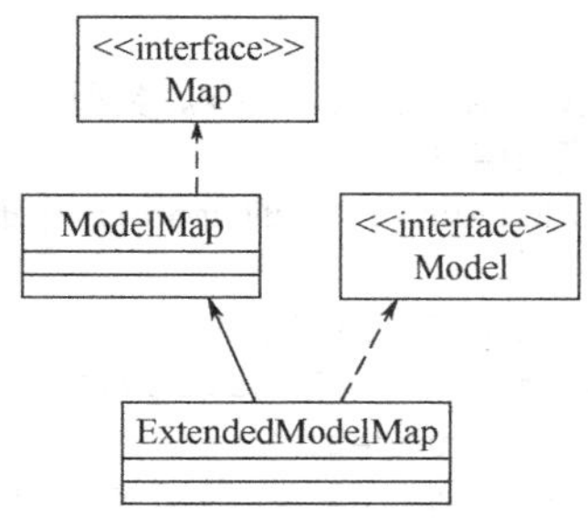

图 17-1　类 ExtendedModelMap 实现了 Model 接口

在控制器的方法中，可以把 ModelAndView、Model、ModelMap 作为参数。当 Spring MVC 运行的时候，会自动初始化它们，因此可以选择 ModelMap 或者 Model 作为数据模型。事实上，Spring MVC 创建的是一个 BindingAwareModelMap 实例。

ModelAndView 初始化后，model 属性为空，当调用它增加数据模型的方法后，会自动创建一个 ModelMap 实例，用于保存数据模型。

Spring MVC 提供了以下几种途径输出模型数据。

（1）ModelAndView：处理方法返回值类型为 ModelAndView 时，方法体即可通过该对象添加模型数据。

（2）Map 及 Model：入参为 java.util.Map、org.springframework.ui.ModelMap 或 org.springframework.ui.Model 时，处理方法返回时，Map 中的数据会自动添加到模型中。

（3）@SessionAttributes：将模型中的某个属性暂存到 HttpSession 中，以便多个请求之间可以共享这个属性。

（4）@ModelAttribute：方法入参标注该注解后，入参的对象就会放到数据模型中。

17.2.2 ModelAndView

1. 概念

可将控制器处理方法的返回值设为 ModelAndView，ModelAndView 中既可存放视图信息，又可存放模型数据信息。

Spring MVC 会把 ModelAndView 的 model 中的数据放入到 request 域对象中。

ModelAndView 设置视图的方法如下：

```
void setView(View view)
void setViewName(String viewName)
```

ModelAndView 添加模型数据的方法如下：

```
MoelAndView addObject(String attributeName, Object attributeValue)
ModelAndView addAllObject(Map<String, ?> modelMap)
```

2. 示例：testModelAndView 工程

（详见本书源码包 ch17 目录中的 testModelAndView 工程。）

index.jsp 的主要代码如下：

```
<a href=" mytest /testModelAndView">Test ModelAndView</a>
```

DateController.java 的主要代码如下：

```
@Controller
@RequestMapping("/mytest")
public class DateController {
```

```
/**
    * 目标方法的返回值可以是 ModelAndView 类型
    * 其中可以包含视图和模型信息
    * SpringMVC 会把 ModelAndView 的 model 中的数据放入 request 域对象中
    */
@RequestMapping("/testModelAndView")
  public ModelAndView testModelAndView(){
      String viewName = "success";
      ModelAndView modelAndView = new ModelAndView(viewName);
      //添加模型数据到 ModelAndView 中
      modelAndView.addObject("time", new Date());
      return modelAndView;
  }
}
```

success.jsp 的主要代码如下：

```
time : ${requestScope.time }
```

测试 testModelAndView 应用。部署 testModelAndView 应用后启动 Tomcat 服务器，在浏览器地址栏中输入 http://localhost:8080/testModelAndView/index.jsp，进入 index.jsp 页面，单击“testModelAndView”超链接，进入如图 17-2 所示的界面。

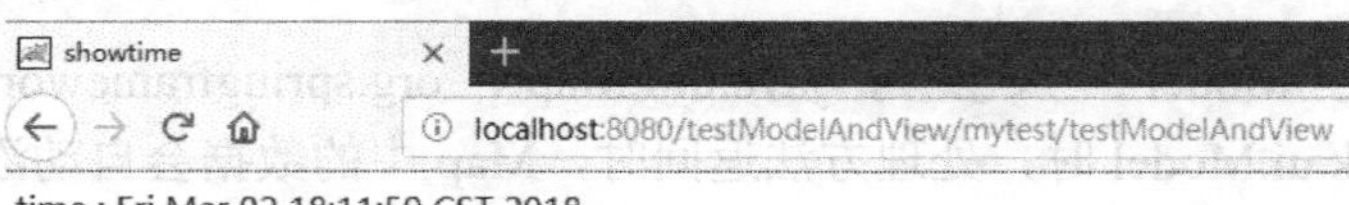

图 17-2　测试 testModelAndView 应用

17.2.3　Map 及 Model

1. 概念

Spring MVC 在调用方法前会创建一个隐含的模型对象作为模型数据的存储容器（事实上，这个隐含的模型对象是一个 BindingAwareModelMap 类型的对象，通过后面的例子可以验证），如果方法的入参为 Map、Model 或者 ModelMap 类型，则 Spring MVC 会将隐含模型的引用传递给这些入参（因为 BindingAwareModelMap 继承或实现了 Map、Model 或者 ModelMap）。在方法体内，开发者可以通过这个入参对象访问到模型中的所有数据，也可以向模型中添加新的属性数据，Spring MVC 也会把 Map 中的数据放入到 request 域对象中。　通常情况下使用的都是 Map 类型，下面以 Map 为例进行说明。

2. 示例：testMap 工程

（代码详见本书源码包 ch17 目录中的 testMap 工程。）

index.jsp 的主要代码如下：

```
<a href="mytest/testMap">Test Map</a>
```

NameController.java 的主要代码如下：

```
@Controller
@RequestMapping("/mytest")
public class NameController {
 @RequestMapping("/testMap")
   public String testMap(Map<String, Object> map) {
      System.out.println(map.getClass().getName());
```

```
        map.put("names", Arrays.asList("John", "Mary", "Marston"));
        return "success";
    }
}
```

success.jsp 的主要代码如下：

```
names: ${requestScope.names }
```

测试 testMap 应用。部署 testMap 应用后启动 Tomcat 服务器，在浏览器地址栏中输入 http://localhost:8080/testMap/index.jsp，进入 index.jsp 页面，单击“testMap”超链接，进入如图 17-3 所示的界面。

图 17-3　测试 testMap 应用

在控制台上输出了以下内容：

```
org.springframework.validation.support.BindingAwareModelMap
```

这说明 Spring MVC 在调用方法前创建的隐含的模型对象是 BindingAwareModelMap 类型。

17.2.4　@SessionAttributes

1．概念

在前面介绍的两种方式中，Spring MVC 都是将数据存放在 request 域对象中的，若希望在多个请求之间共用某个模型属性数据，则可以在控制器类上标注一个@SessionAttributes 注解（该注解只能放在类的上面，而不能修饰方法），Spring MVC 将把模型中对应的属性暂存到 HttpSession 中。

@SessionAttributes 除了可以通过属性名指定需要放到会话中的属性之外，还可以通过模型属性的对象类型指定哪些模型属性需要放到会话中。

2．示例

（代码详见本书源码包 ch17 目录中的 testSessionAttributes 工程。）

index.jsp 的主要代码如下：

```
<a href="mytest/testSessionAttributes">Test SessionAttributes</a>
```

UserController.java 的主要代码如下：

```
@Controller
@RequestMapping("/mytest")
@SessionAttributes(value="user",types={String.class})
public class UserController {
 @RequestMapping("/testSessionAttributes")
   public String testSessionAttributes(Map<String,Object> map){
      User user = new User("Mike","123456",20,"Mike@163.com");
      map.put("user", user);
      map.put("school", "ECNU");
      return "success";
   }
}
```

success.jsp 的主要代码如下：

```
request user : ${requestScope.user }
<br><br>
session user : ${sessionScope.user }
<br><br>
request school : ${requestScope.school }
<br><br>
session school : ${sessionScope.school }
<br><br>
```

测试 testSessionAttributes 应用。部署 testSessionAttributes 应用后启动 Tomcat 服务器，在浏览器地址栏中输入 http://localhost:8080/testSessionAttributes/index.jsp，进入 index.jsp 页面，单击“testSessionAttributes”超链接，进入如图 17-4 所示的界面。

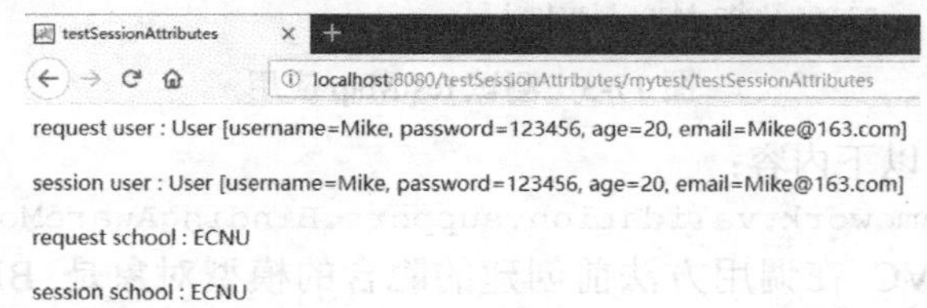

图 17-4 测试 testSessionAttributes 应用

17.2.5 @ModelAttribute

1. 概念

Spring MVC 在每次调用请求处理方法时，都会创建 Model 类型的一个实例。若打算使用该实例，则可以在方法中添加一个 Model 类型的参数。除此之外，还可以使用在方法中添加 org.springframework.Web.bind.annotation.ModelAttribute 注解类型的方法将请求参数绑定到 Model 对象上。

@ModelAttribute 注解只支持一个属性，即 value，类型为 String，表示绑定的属性名称。

可以用@ModelAttribute 来注解方法参数或方法。带@ModelAttribute 注解的方法（请求方法），会将其输入的或创建的参数对象添加到 Model 对象中（若方法中没有显式地增加）。

@ModelAttribute 还可以标注一个非请求的处理方法。被@ModelAttribute 注解的方法，会在每次调用该控制器类的请求处理方法时被调用。

Spring MVC 会在调用请求处理方法之前调用带有@ModelAttribute 注解的方法。带@ModelAttribute 注解的方法可以返回一个对象或一个 void 类型。如果返回一个对象，则返回对象会自动添加到 Model 中；若方法返回 void，则必须添加一个 Model 类型的参数，并自行将实例添加到 Model 中。

2. @ModelAttribute 注释 void 返回值的方法

（以下示例详见本书源码包 ch17 目录中的 ModelAttributeDemo1 工程。）

本示例的情景如下：有一个 User 类，其有 id、userName、email3 个属性。现在要完成一个修改操作，但是其中有一项属性不能被修改，如 id，那么只能修改两项属性——userName 和 email，所以从表单传递的信息只能有这两项。通常的处理方法是新建一个 User 对象 user，用表单传递的值为 user 赋值，然后用 user 作为参数去更新信息。现在由于表单无法传递 id 值，user.id 此时就为 null，因此无法满足我们要求的更新操作。

@ModelAttribute 注解为人们提供了这种解决方法的实现：使用@ModelAttribute 注解的方法，会在调用目标处理方法之前被调用。

index.jsp 的主要代码如下：

```
<!-- 模拟修改操作：①原始数据为 1，John，John@163.com；
    ②将 username 和 email 修改为 Smith、Smith@qq.com，id 不能被修改
-->
<form action="mytest/testModelAttribute" method="Post">
 username: <input type="text" name="username" value="John" /> <br>
 email: <input type="text" name="email" value="John@163.com" /> <br>
 <input type="submit" value="Submit" />
</form>
```

UserController.java 的主要代码如下：

```
@Controller
@RequestMapping("/mytest")
public class UserController {
 @ModelAttribute
    public void getUser(Map<String, Object> map) {
        System.out.println("modelAttribute method");
            // 模拟从数据库中获取对象
            User user = new User(1,"Jack", "Jack@163.com");
            System.out.println("从数据库中获取一个对象：" + user);
            map.put("user", user);
    }
    @RequestMapping("/testModelAttribute")
    public String testModelAttribute(User user){
        System.out.println("修改：" + user);
        return "success";
    }
}
```

测试 ModelAttributeDemo1 应用。部署 ModelAttributeDemo1 应用后启动 Tomcat 服务器，在浏览器地址栏中输入 http://localhost:8080/ModelAttributeDemo1/index.jsp，在表单提交页面（index.jsp）中将 username 和 email 修改为 Smith、Smith@qq.com（id 不能被修改），提交表单后，控制台输出内容如图 17-5 所示。

```
Problems  Javadoc  Declaration  Console  Servers
Tomcat v8.0 Server at localhost [Apache Tomcat] C:\Program Files (x86)\Java\jdk1.8.0_92\bin\javaw.exe (2018年3月4日 下午2:06:37)
modelAttribute method
从数据库中获取一个对象：User [id=1, username=Jack, email=Jack@163.com]
修改：User [id=1, username=Smith, email=Smith@163.com]
```

图 17-5　测试 ModelAttributeDemo1 应用时控制台的输出

3. @ModelAttribute 注释返回具体类的方法

（以下示例详见本书源码包 ch17 目录中的 ModelAttributeDemo2 工程。）

UserController.java 的主要代码如下：

```
@Controller
@RequestMapping("/mytest")
public class UserController {
 @ModelAttribute
 public User addUser() {
     return new User("John", "123", 36, "John@163.com");
 }
 @RequestMapping(value="/user")
```

```
    public String find() {
        return "newuser";
    }
}
```

在这种情况下，model 属性的名称没有指定，它由返回类型隐含表示，如这个方法返回 User 类型，那么 model 属性的名称是 user。

newuser.jsp 的主要代码如下：

```
request user : ${requestScope.user }
```

测试 ModelAttributeDemo2 应用。部署 ModelAttributeDemo2 应用后启动 Tomcat 服务器，在浏览器地址栏中输入 http://localhost:8080/ModelAttributeDemo2/mytest/user，进入如图 17-6 所示的界面。

图 17-6　测试 ModelAttributeDemo2 应用

4. @ModelAttribute(value="")注释返回具体类的方法

（以下示例详见本书源码包 ch17 目录中的 ModelAttributeDemo3 工程。）

UserController.java 的主要代码如下：

```
@Controller
@RequestMapping("/mytest")
public class UserController {
 @ModelAttribute(value="babyuser")
 public User addUser() {
     return new User("Baby", "9898", 12, "Baby@163.com");
 }
 @RequestMapping(value="/baby")
 public String find() {
     return "baby";
 }
}
```

此例中使用@ModelAttribute 注释的 value 属性来指定 model 属性的名称。model 属性对象就是方法的返回值。它无须特定的参数。

baby.jsp 的主要代码如下：

```
request user : ${requestScope.babyuser}
```

测试 ModelAttributeDemo3 应用。部署 ModelAttributeDemo3 应用后启动 Tomcat 服务器，在浏览器地址栏中输入 http://localhost:8080/ModelAttributeDemo3/mytest/baby，进入如图 17-7 所示的界面。

localhost:8080/ModelAttributeDemo3/mytest/baby

request user : User [username=Baby, password=9898, age=12, email=Baby@163.com]

图 17-7　测试 ModelAttributeDemo3 应用

5. @ModelAttribute 和@RequestMapping 同时注释一个方法

（以下示例详见本书源码包 ch17 目录中的 ModelAttributeDemo4 工程。）

LoveController.java 的主要代码如下：

```
@Controller
public class LoveController {
```

```
    @RequestMapping(value = "/baby")
    @ModelAttribute("love")
    public User bigsky() {
        return new User("Baby999", "9898", 12, "Baby999@163.com");
    }
}
```

在 LoveController 中，@ModelAttribute 和@RequestMapping 同时注释一个方法。

此时，这个方法的返回值并不表示一个视图名称，而是 model 属性的值，视图名称由 RequestToViewNameTranslator 根据请求“/baby”转换为逻辑视图 baby。

model 属性名称由@ModelAttribute(value="love")指定，相当于在 request 中封装了 key=love，value=User 对象。

baby.jsp 的主要代码如下：

```
request user : ${requestScope.love}
```

测试 ModelAttributeDemo4 应用。部署 ModelAttributeDemo4 应用后启动 Tomcat 服务器，在浏览器地址栏中输入 http://localhost:8080/ModelAttributeDemo4/baby，进入如图 17-8 所示的界面。

localhost:8080/ModelAttributeDemo4/baby
request user : User [username=Baby999, password=9898, age=12, email=Baby999@163.com]

图 17-8 测试 ModelAttributeDemo4 应用

6. @ModelAttribute 注释一个方法的参数

（以下示例详见本书源码包 ch17 目录中的 ModelAttributeDemo5 工程。）

UserController.java 的主要代码如下：

```
@Controller
@RequestMapping("/mytest")
public class UserController {
    @ModelAttribute("user")
    public User userModel() {
        return new User("李白", "9898", 1668, "LiBai@163.com");
    }
    @RequestMapping(value="/baby")
    public String user(@ModelAttribute("user") User user) {
        user.setUsername("假设是李白");
        return "babyli";
    }
}
```

关于 UserController 的 user 方法，它的参数 User 使用了@ModelAttribute("user")注解，表示参数 user 的值就是 userModel( )方法中的 model 属性。

babyli.jsp 的主要代码如下：

```
user : ${requestScope.user}
```

测试 ModelAttributeDemo5 应用。部署 ModelAttributeDemo5 应用后启动 Tomcat 服务器，在浏览器地址栏中输入 http://localhost:8080/ModelAttributeDemo5/mytest/baby，进入如图 17-9 所示的界面。

图 17-9　测试 ModelAttributeDemo5 应用

17.3　基于注解的控制器示例 1

17.3.1　创建 AnnotationDemo1 工程

同前面章节的示例一样，创建 Maven Project，名为 AnnotationDemo1。图 17-10 所示为完成后的 AnnotationDemo1 工程目录结构及文件。后续步骤同第 16 章中的示例。

```
AnnotationDemo1
  src/main/java
    com.mialab.AnnotationDemo1.controller
      BookController.java
    com.mialab.AnnotationDemo1.domain
      Book.java
    com.mialab.AnnotationDemo1.form
      BookForm.java
  src/main/resources
  src/test/java
  JRE System Library [JavaSE-1.8]
  Maven Dependencies
  src
    main
      Webapp
        Web-INF
          config
            springmvc-config.xml
          jsp
            BookDetails.jsp
            BookForm.jsp
          Web.xml
        index.jsp
    test
  target
  pom.xml
```

图 17-10　完成后的 AnnotationDemo1 工程目录结构及文件

本示例是对前面章节中 SpringMVC_Basic_Demo2 工程的改写。由于使用注解类型来开发，因此一个控制器类可以包含多个请求处理方法。原本在 SpringMVC_Basic_Demo2 工程中有两个控制器——AddBookController 和 SaveBookController，这里用一个控制器替代即可，这个控制器是 BookController。而控制器 BookController 包含了两个请求方法，其请求映射分别是/book_add 和/book_save。（代码可详见本书源码包 ch17 目录中的 AnnotationDemo1 工程。）

17.3.2　创建控制器并添加注解

BookController.java 的主要代码如下（详见本书源码包 ch17 目录中的 AnnotationDemo1 工程）：

```
@Controller
public class BookController {
 private static final Log logger = LogFactory.getLog(BookController.class);
 @RequestMapping(value="/book_add")
   public String add_Book() {
     logger.info("add_Book called");
     return "BookForm";
   }
```

```
        @RequestMapping(value="/book_save")
        public String saveBook(BookForm bookForm, Model model) {
            logger.info("saveBook called");
            Book book = new Book();  // create Book
            book.setName(bookForm.getName());
            book.setDescription(bookForm.getDescription());
            try {
                book.setPrice(Float.parseFloat(
                        bookForm.getPrice()));
            } catch (NumberFormatException e) {
            }
            model.addAttribute("book", book);  // add book
            return "BookDetails";
        }
    }
```

必须在 Spring MVC 配置文件中指定需要扫描的包，这样能自动生成控制器实例。

```
    <context:component-scan
base-package="com.mialab.AnnotationDemo1.controller" />
```

17.3.3　测试应用

将项目 AnnotationDemo1 部署发布到 Tomcat 服务器，并启动 Tomcat 服务器。在浏览器地址栏中输入 http://localhost:8080/ AnnotationDemo1/book_add，进行测试。运行效果类似于图 16-6~图 16-8。

17.4　基于注解的控制器示例 2

17.4.1　创建 AnnotationDemo2 工程

仍同前例一样，创建 Maven Project，名为 AnnotationDemo2。图 17-11 所示为完成后的 AnnotationDemo2 工程目录结构及文件。后续步骤也同第 16 章中的示例。

```
AnnotationDemo2
    src/main/java
        com.mialab.AnnotationDemo2.controller
            BookController.java
        com.mialab.AnnotationDemo2.domain
            Book.java
        com.mialab.AnnotationDemo2.form
            BookForm.java
        com.mialab.AnnotationDemo2.service
            BookService.java
        com.mialab.AnnotationDemo2.service.impl
            BookServiceImpl.java
    src/main/resources
    src/test/java
    JRE System Library [JavaSE-1.8]
    Maven Dependencies
    src
        main
            Webapp
                Web-INF
                    config
                        springmvc-config.xml
                    jsp
                        BookForm.jsp
                        BookView.jsp
                    Web.xml
                index.jsp
        test
    target
    pom.xml
```

图 17-11　完成后的 AnnotationDemo2 工程目录结构及文件

本示例着重说明：如何使用@Autowired 和@Service 进行依赖注入；Spring 如何通过 Flash 属性进行重定向传值；@PathVariable 的使用；Spring 组件自动扫描机制。

17.4.2 应用@Autowired 和@Service 进行依赖注入

BookController.java 的主要代码如下（详见本书源码包 ch17 目录中的 AnnotationDemo2 工程）：

```
@Controller
public class BookController {
 private static final Log logger = LogFactory.getLog(BookController.class);
 @Autowired
 private BookService bookService;
 @RequestMapping(value = "/book_add")
 public String addBook() {
     logger.info("addBook called");
     return "BookForm";
 }
 @RequestMapping(value = "/book_save", method = RequestMethod.POST)
 public String saveBook(BookForm bookForm, RedirectAttributes redirect
Attributes) {
     logger.info("saveBook called");
     Book book = new Book();
     book.setBookId(bookForm.getBookId());
     book.setName(bookForm.getName());
     book.setDescription(bookForm.getDescription());
     try {
         book.setPrice(Float.parseFloat(bookForm.getPrice()));
     } catch (NumberFormatException e) {
     }
     Book savedBook = bookService.add(book);  // add book
     redirectAttributes.addFlashAttribute("message",
         "The book was successfully added.");
     return "redirect:/book_view/" + savedBook.getBookId();
 }

 @RequestMapping(value = "/book_view/{bookId}")
 public String viewBook(@PathVariable String bookId, Model model) {
     Book book = bookService.getId(bookId);
     model.addAttribute("book", book);
     return "BookView";
 }
}
```

BookServiceImpl.java 的主要代码如下：

```
@Service
public class BookServiceImpl implements BookService {
 private Map<String, Book> books = new HashMap<String, Book>();
 public BookServiceImpl() {
     Book book = new Book();
     book.setName("Android 应用开发实践教程");
     book.setDescription("卓越工程师培养创新系列丛书");
     book.setPrice(68.99F);
```

```
            add(book);
        }
        @Override
        public Book add(Book book) {
            String newBookId = book.getBookId();
             books.put(newBookId, book);
             return book;
        }
        @Override
        public Book getId(String bookId) {
            return books.get(bookId);
        }
    }
```

BookView.jsp 的<body>部分的主要代码如下：

```
    <h3>${message}</h3>
    <p> <h3>Details:</h3>
        Book Id: ${book.bookId}<br/>
        Book Name: ${book.name}<br/>
        Book Description: ${book.description}<br/>
        Book Price: $${book.price}
    </p>
```

@Autowired 用于对 Bean 的属性变量、属性的 setter 方法及构造方法进行标注，配合对应的注解处理器完成 Bean 的自动装置工作。默认按照 Bean 的类型进行装配。

@Resource 的作用与@Autowired 一样。其区别在于@Autowired 默认按照 Bean 类型装配，而@Resource 默认按照 Bean 实例名称进行装配。@Resource 中有两个重要属性：name 和 type。Spring 将 name 属性解析为 Bean 实例名称，将 type 属性解析为 Bean 实例类型。如果指定 name 属性，则按实例名称进行装配；如果指定 type 属性，则按 Bean 类型进行装配；如果都无法匹配，则抛出 NoSuchBeanDefinitionException 异常。

@Qualifier 则与@Autowired 注解配合使用，会将默认的按 Bean 类型装配修改为按 Bean 的实例名称装配，Bean 的实例名称由@Qualifier 注解的参数指定。

如果 Web 应用程序采用了经典的三层分层结构，则最好在持久层、业务层和控制层分别采用@Repository、@Service 和@Controller 对分层中的类进行注释，而用@Component 对那些比较中立的类进行注释。

在一个稍大的项目中，通常会有上百个组件，如果这些组件采用 XML 的 Bean 定义来配置，显然会增加配置文件的体积，查找及维护起来也不太方便。

从 Spring 2.5 开始，Spring 引入了组件自动扫描机制，它可以在类路径下寻找标注了@Component、@Service、@Controller、@Repository 注解的类，并把这些类纳入 Spring 容器中进行管理。它的作用和在 XML 文件中使用 Bean 节点配置组件是一样的。

在本示例中，要使用自动扫描机制，Spring MVC 的配置文件 springmvc-config.xml 还需添加以下配置代码：

```
    <!-- 指定需要扫描的包 -->
    <context:component-scan
base-package="com.mialab.AnnotationDemo2.controller" />
    <context:component-scan
base-package="com.mialab.AnnotationDemo2.service" />
```

其中，base-package 为需要扫描的包（含所有子包），@Service 用于标注业务层组件，@Controller 用于标注控制层组件，@Repository 用于标注数据访问组件，即 DAO 组件，而

@Component 泛指组件，当组件不好归类的时候，可以使用此注解进行标注。

17.4.3 重定向

重定向是客户端行为，转发是服务器行为。重定向行为是做了两次请求，即产生了两个 Request 对象，重定向会导致 Request 对象信息丢失。

使用重定向的一个不便之处是，无法轻松地传值给目标页面。而采用转发则可以非常简单地将属性添加到 Model 中，使得目标视图可以轻松访问。由于重定向经过客户端，所以 Model 中的一切都在重定向时丢失。而 Spring 3.1 及更高版本通过 Flash 属性提供了一种重定向传值的方法。

要使用 Flash 属性，必须在 Spring MVC 配置文件中添加<mvc:annotation-driven />元素，此外，必须在方法上添加一个新的参数类型 org.springframework.Web.servlet.mvc. support. RedirectAttributes。例如：

```
    @RequestMapping(value = "/book_save", method = RequestMethod.POST)
    public String saveBook(BookForm bookForm, RedirectAttributes redirect
Attributes) {
     …
     redirectAttributes.addFlashAttribute("message","The book was success
fully added.");
     return "redirect:/book_view/" + savedBook.getBookId();
    }
```

带占位符的 URL 是 Spring 3.0 新增的功能，该功能在 Spring MVC 向 REST 目标发展过程中具有里程碑式的意义。通过@PathVariable 可以将 URL 中的占位符参数绑定到控制器处理方法的入参中：URL 中的{xxx}占位符可以通过@PathVariable("xxx") 绑定到操作方法的入参中。例如：

```
    //@PathVariable 可以用来映射 URL 中的占位符到目标方法的参数中
    @RequestMapping("/testPathVariable/{id}")
    public String testPathVariable(@PathVariable("id") Integer id){
       System.out.println("testPathVariable:"+id);
       return "someView";
    }
```

17.4.4 测试应用

将项目 AnnotationDemo2 部署发布到 Tomcat 服务器，并启动 Tomcat 服务器。在浏览器地址栏中输入 http://localhost:8080/AnnotationDemo2/book_add，进行测试。运行效果如图 17-12 和图 17-13 所示（详见本书源码包 ch17 目录中的 AnnotationDemo2 工程）。

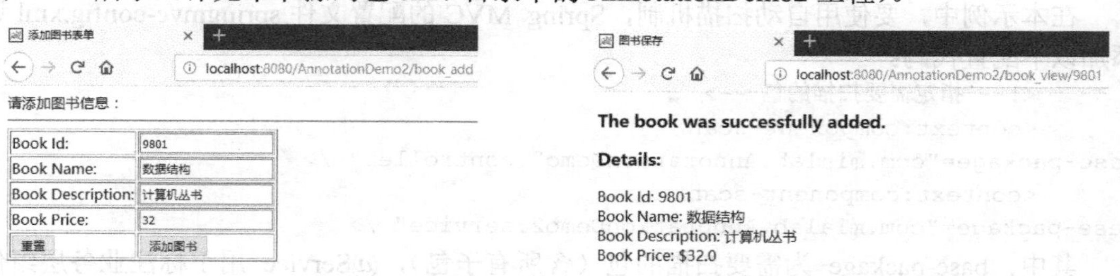

图 17-12 添加图书表单　　　　图 17-13 添加图书成功

17.5　本章小结

本章主要介绍了 Spring MVC 中的一些常用注解的使用，这里包括@Controller 和@RequestMapping、@Autowired 和@Service、@RequestParam 和@PathVariable、@CookieValue 和@RequestHeader、@SessionAttributes 和@ModelAttribute、@Resource 和@Qualifier、@Repository 和@Component，以及在 Spring MVC 中如何处理模型数据和重定向等内容。

习题 17

1．如何应用@Autowired、@Service 和@Repository 进行依赖注入？试编程加以说明。
2．@RequestParam 和@PathVariable 的区别是什么？试编程加以说明。
3．在 Spring MVC 中如何处理模型数据？试编程加以说明。
4．如何通过 Flash 属性实现重定向传值？试编程加以说明。

第18章 拦截器、数据转换和格式化

本章导读

Spring MVC 拦截器（Interceptor）的主要作用是拦截用户请求并进行相应的处理。例如，通过它来进行权限验证，或者判断用户是否登录，或者类似 12306 那样判断当前时间是否为购票时间。拦截器可以在进入处理器之前做一些操作，或者在处理器完成后进行操作，甚至是在渲染视图后进行操作。Spring 4 提供了转换器和格式化器，这样通过注解的信息和参数的类型，就能够把 HTTP 发送过来的各种消息转换成为控制器所需要的各类参数。本章主要内容有：（1）Spring MVC 的拦截器定义、注册及应用；（2）Spring MVC 的数据绑定；（3）Spring MVC 的数据转换和格式化；（3）JSON 格式的数据转换。

18.1 Spring MVC 的拦截器

Spring MVC 的处理器拦截器（如无特殊说明，下文所说的拦截器即处理器拦截器） 类似于 Servlet 开发中的过滤器，用于对处理器进行预处理和后处理。

Spring Web MVC 的拦截器常见应用场景如下。

（1）日志记录：记录请求信息的日志，以便进行信息监控、信息统计等。

（2）权限检查：如登录检测，进入处理器检测是否登录，如果没有则返回登录页面。

（3）性能监控：有时候系统在某段时间响应很慢，可以通过拦截器在进入处理器之前记录开始时间，在处理完成后，记录结束时间，从而得到该请求的处理时间。

18.1.1 拦截器的定义和注册

在 Spring 框架中，要实现拦截器功能，主要通过两种途径：第一种是实现 HandlerInterceptor 接口或继承 HandlerInterceptor 接口的实现类（如 HandlerInterceptorAdapter）；第二种是实现 WebRequestInterceptor 接口。下面主要对第一种实现方法进行讲解。

1. HandlerInterceptor 接口

org.springframework.Web.servlet. HandlerInterceptor 接口定义了以下 3 个方法。

```
    package org.springframework.Web.servlet;
    public interface HandlerInterceptor {
     boolean preHandle( HttpServletRequest   request,   HttpServletResponse
response,               Object handler) throws Exception;
     void   postHandle(HttpServletRequest   request,   HttpServletResponse
response,
         Object handler, ModelAndView modelAndView) throws Exception;
     void  afterCompletion(HttpServletRequest  request,  HttpServletResponse
response,               Object handler, Exception ex)  throws Exception;
    };
```

在实际应用中，一般是通过实现 HandlerInterceptor 接口或者继承 HandlerInterceptor Adapter 抽象类，覆写 preHandle()、postHandle()和 afterCompletion()这 3 个方法来对用户的请求进行拦截处理的。

（1）preHandle(HttpServletRequest request, HttpServletResponse response, Object handle)方法：该方法在请求处理之前进行调用。Spring MVC 中的 Interceptor 是链式调用的，在一个应用中或者说在一个请求中可以同时存在多个 Interceptor。每个 Interceptor 的调用会依据它的声明顺序依次执行，而且最先执行的都是 Interceptor 中的 preHandle 方法，所以可以在这个方法中进行一些前置初始化操作或者对当前请求做一个预处理，也可以在这个方法中进行一些判断来决定请求是否要继续进行下去。该方法的返回值是布尔值，当它返回 false 时，表示请求结束，后续的 Interceptor 和 Controller 都不会再执行；当返回值为 true 时，会继续调用下一个 Interceptor 的 preHandle 方法，如果已经是最后一个 Interceptor，就会调用当前请求的 Controller 中的方法。

（2）postHandle(HttpServletRequest request, HttpServletResponse response, Object handle, ModelAndView modelAndView)方法：这个方法包括后面的 afterCompletion 方法都只能在当前所属的 Interceptor 的 preHandle 方法的返回值为 true 的时候被调用。postHandle 方法在当前请求进行处理之后，即在 Controller 中的方法调用之后执行，但是它会在 DispatcherServlet 进行视图返回渲染之前被调用，所以可以在这个方法中对 Controller 处理之后的 ModelAndView 对象进行操作。postHandle 方法被调用的方向与 preHandle 是相反的，也就是说，先声明的 Interceptor 的 postHandle 方法反而会后执行。

（3）afterCompletion(HttpServletRequest request, HttpServletResponse response, Object handle, Exception ex)方法：也需要当前对应的 Interceptor 的 preHandle 方法的返回值为 true 时才会执行。因此，该方法将在整个请求结束之后，即在 DispatcherServlet 渲染了对应的视图之后执行，这个方法的主要作用是进行资源清理。

抽象类 HandlerInterceptorAdapter 主要代码如下：

```
    public abstract class HandlerInterceptorAdapter implements AsyncHandler
Interceptor {
        public  boolean  preHandle(HttpServletRequest  request,  HttpServlet
Response response,          Object handler) throws Exception {
            return true;
        }
        public void postHandle(HttpServletRequest request, HttpServletResponse
response,
          Object handler, ModelAndView modelAndView) throws Exception {
        }
        public  void  afterCompletion(HttpServletRequest  request,  HttpServlet
Response response,          Object handler, Exception ex) throws Exception {
        }
       public void afterConcurrentHandlingStarted(HttpServletRequest request,
          HttpServletResponse response, Object handler) throws Exception {
        }
    }
```

要使自定义的拦截器生效，还需要在 Spring MVC 的配置文件中进行注册。

```
<mvc:interceptors>
    <!-- 使用 Bean 定义一个 Interceptor，直接定义在 mvc:interceptors 下面-->
  <!--直接定义在 mvc:interceptors 下面的 Interceptor 将拦截所有的请求-->
```

```
        <bean class="com.hit.interceptor.WrongCodeInterceptor"/>
        <mvc:interceptor>
            <mvc:mapping path="/demo/hello.do"/>
            <!--定义在 mvc:interceptor 下面的 Interceptor，表示对特定请求进行拦截-->
            <bean class="com.hit.interceptor.LoginInterceptor"/>
        </mvc:interceptor>
    </mvc:interceptors>
```

2. WebRequestInterceptor 接口

在 WebRequestInterceptor 接口中也定义了 3 个方法，同 HandlerInterceptor 接口完全相同，可以通过覆写这 3 个方法来对用户的请求进行拦截处理。这 3 个方法都传递了同一个参数——WebRequest，WebRequest 是 Spring 中定义的一个接口，其中的方法定义和 HttpServlet Request 类似，在 WebRequestInterceptor 中对 WebRequest 进行的所有操作都将同步到 HttpServletRequest 中，并在当前请求中依次传递。

在 Spring 框架之中，还提供了一个和 WebRequestInterceptor 接口类似的抽象类，即 WebRequestInterceptorAdapter，它实现了 AsyncHandlerInterceptor 接口，并在内部调用了 WebRequestInterceptor 接口。

关于 WebRequestInterceptor 接口在这里不做过多讲述，有兴趣的读者可参阅相关文档。

18.1.2 拦截器的执行流程

下面的示例是单个拦截器的执行流程。

【示例】演示拦截器的执行流程。

在 Eclipse 中创建 Maven Project，名为 InterceptorDemo。InterceptorDemo 工程目录层次及文件如图 18-1 所示。创建的新文件有：TestController.java、TestInterceptor.java 和 hello.jsp 等。（代码详见本书源码包 ch18 目录中的 InterceptorDemo 工程。）

图 18-1 InterceptorDemo 工程目录层次及文件

（1）TestController.java 的主要代码如下：

```
@Controller
public class TestController {
 private int i = 0;
 // 拦截器测试
 @RequestMapping("/hello")
 public ModelAndView requestHello() {
     ModelAndView model = new ModelAndView("hello");
     model.addObject("count", ++i);
     System.out.println("==========处理请求中=======");
     System.out.println("........................");
     return model;
 }
}
```

（2）TestInterceptor.java 的主要代码如下：

```
public class TestInterceptor implements HandlerInterceptor {
 @Override
 public    void    afterCompletion(HttpServletRequest    request,
HttpServletResponse  response,  Object  handler,Exception  exception)  throws
Exception {
     System.out.println("=====视图渲染结束了，请求处理完毕====");
 }
 @Override
 public void postHandle(HttpServletRequest request, HttpServletResponse
response, Object handler,ModelAndView modelAndView) throws Exception {
     System.out.println("======处理请求后，渲染页面前======");
     modelAndView.addObject("post", "======处理请求后，渲染页面前====");
 }
 @Override
 public boolean preHandle(HttpServletRequest arg0, HttpServletResponse
arg1, Object arg2) throws Exception {
     System.out.println("======处理请求之前======");
     return true;
 }
}
```

（3）hello.jsp 的<body>部分的主要代码如下：

```
<h3>Hello World!</h3>
<h3>count=${count}</h3>
<h3>调用postHandle方法得到的：${post}</h3>
```

（4）springmvc-config.xml 的主要内容如下：

```
<?xml version="1.0" encoding="UTF-8"?>
<beans xmlns="http://www.springframework.org/schema/beans"
 xmlns:xsi="http://www.w3.org/2001/XMLSchema-instance"
 xmlns:mvc="http://www.springframework.org/schema/mvc"
 xmlns:context="http://www.springframework.org/schema/context"
 xsi:schemaLocation="http://www.springframework.org/schema/beans
```

```
        http://www.springframework.org/schema/beans/spring-beans-4.3.xsd
        http://www.springframework.org/schema/mvc
          http://www.springframework.org/schema/mvc/spring-mvc-4.3.xsd
        http://www.springframework.org/schema/context

http://www.springframework.org/schema/context/spring-context-4.3.xsd">
      <!-- 指定需要扫描的包 -->
      <context:component-scan
base-package="com.mialab.InterceptorDemo.controller" />
      <!-- 定义视图解析器 -->
      <bean id="viewResolver"

   class="org.springframework.Web.servlet.view.InternalResourceViewResolver">
         <!-- 设置前缀 -->
         <property name="prefix" value="/Web-INF/jsp/" />
         <!-- 设置后缀 -->
         <property name="suffix" value=".jsp" />
      </bean>
      <!-- 配置拦截器 -->
      <mvc:interceptors>
         <!--使用 bean 直接定义在<mvc:interceptors>中，拦截器将拦截所有请求 -->
         <bean
class="com.mialab.InterceptorDemo.interceptor.TestInterceptor" />
      </mvc:interceptors>
     </beans>
```

（5）部署发布并测试。在 InterceptorDemo 工程中右击 src/main/Webapp 目录下的 index.jsp 页面，在弹出的快捷菜单中选择“Run As→Run on Server”选项，在浏览器中访问 http://localhost:8080/InterceptorDemo/hello，Eclipse 的控制台输出内容如下：

```
      ======处理请求之前======
      …Invoking        request        handler        method:        public
org.springframework.Web.servlet.ModelAndView
com.mialab.InterceptorDemo.controller.TestController.requestHello()
      ==========处理请求中=======
      ........................
      ========处理请求后，渲染页面前======
      …Added model object 'count' of type [java.lang.Integer] to request in view
with name 'hello'
      …Added model object 'post' of type [java.lang.String] to request in view
with name 'hello'
      …Forwarding to resource [/Web-INF/jsp/hello.jsp] in InternalResourceView
'hello'
      ========视图渲染结束，请求处理完毕====
      … Successfully completed request
```

由以上可以看出，程序会先执行拦截器的 preHandle()方法，如果该方法的返回值为 true，则程序会继续向下执行处理器中的方法，否则将不再向下执行；在控制器（处理器）处理完请求后，会执行 postHandle()方法，并通过 DispatcherServlet 向客户端返回响应；在 DispatcherServlet 处理完请求后，才会执行 afterCompletion()方法。

18.1.3　多个拦截器执行的顺序

在使用 Spring MVC 的大型企业级项目中，往往会定义多个拦截器来实现不同的功能。多个拦截器会以一个怎样的顺序执行呢？

【示例】演示多个拦截器的执行流程。

在 Eclipse 中创建 Maven Project，名为 InterceptorDemo2。InterceptorDemo2 工程目录层次及文件如图 18-2 所示。创建的新文件有：Interceptor1.java、Interceptor2.java、Interceptor3.java、TestController.java 和 welcome.jsp 等。

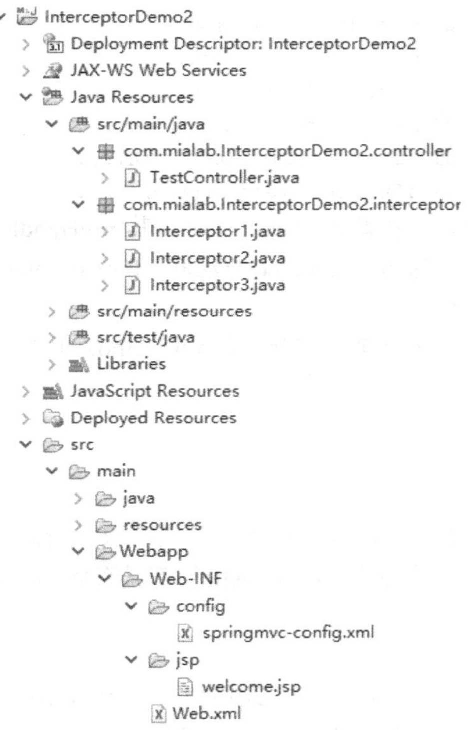

图 18-2　InterceptorDemo2 工程目录层次及文件

springmvc-config.xml 中关于拦截器的配置如下：

```
<mvc:interceptors>
  <!-- 拦截器 1 -->
  <mvc:interceptor>
      <!-- 配置拦截器作用的路径 -->
      <mvc:mapping path="/**" />
      <beanclass="com.mialab.InterceptorDemo2.interceptor.Interceptor1" />
  </mvc:interceptor>
  <!-- 拦截器 2 -->
  <mvc:interceptor>
      <mvc:mapping path="/hello" />
      <beanclass="com.mialab.InterceptorDemo2.interceptor.Interceptor2" />
  </mvc:interceptor>
  <!-- 拦截器 3 -->
  <mvc:interceptor>
```

```
        <mvc:mapping path="/**" />
        <beanclass="com.mialab.InterceptorDemo2.interceptor.Interceptor3" />
    </mvc:interceptor>
  </mvc:interceptors>
```

将 InterceptorDemo2 工程部署发布到 Tomcat 服务器上。在浏览器中访问以下 URL：http://localhost:8080/InterceptorDemo2/hello。Eclipse 的控制台输出内容如下：

```
调用 Intercepter1 的 preHandle() 方法...
调用 Intercepter2 的 preHandle() 方法...
调用 Intercepter3 的 preHandle() 方法...
访问 TestController 控制器的 Test() 方法
调用 Intercepter3 的 postHandle() 方法...
调用 Intercepter2 的 postHandle() 方法...
调用 Intercepter1 的 postHandle() 方法...
调用 Intercepter3 的 afterCompletion() 方法...
调用 Intercepter2 的 afterCompletion() 方法...
调用 Intercepter1 的 afterCompletion() 方法...
```

由以上可以看出，当有多个拦截器同时工作时，其 preHandle()方法会按照配置文件中拦截器的配置顺序执行，而它们的 postHandle()方法和 afterCompletion()方法会按照配置文件中拦截器配置顺序的反序来执行。

（代码详见本书源码包 ch18 目录中的 InterceptorDemo2 工程。）

18.1.4 拦截器应用

【示例】性能监控。

需求分析：要记录请求的处理时间，得到一些慢请求（如处理时间超过 100ms），从而进行性能改进，一般的反向代理服务器（如 Apache）都具有此功能，但此处演示使用拦截器怎样实现。

实现分析：

（1）在进入处理器之前记录开始时间，即在拦截器的 preHandle 中记录开始时间。

（2）在结束请求处理之后记录结束时间，即在拦截器的 afterCompletion 中记录结束实现，并用结束时间-开始时间得到这次请求的处理时间。

问题在于：拦截器是单例模式的，因此不管用户请求多少次都只有一个拦截器实现，即线程不安全。那么应该怎样记录时间呢？

解决方案：使用 ThreadLocal，它是线程绑定的变量，提供了线程局部变量（一个线程一个 ThreadLocal，A 线程的 ThreadLocal 只能看到 A 线程的 ThreadLocal，不能看到 B 线程的 ThreadLocal）。

在 Eclipse 中创建 Maven Project，名为 MonitorInterceptorDemo，其工程目录层次及文件如图 18-3 所示。创建的新文件有：MonitorInterceptor.java、TestController.java 和 finish.jsp 等。（代码详见本书源码包 ch18 目录中的 MonitorInterceptorDemo 工程。）

```
MonitorInterceptorDemo
  Deployment Descriptor: MonitorInterceptorDemo
  JAX-WS Web Services
  Java Resources
    src/main/java
      com.mialab.monitor.controller
        TestController.java
      com.mialab.monitor.interceptor
        MonitorInterceptor.java
      com.mialab.MonitorInterceptorDemo
    src/test/java
    src/main/resources
    Libraries
  JavaScript Resources
  Deployed Resources
  src
    main
      java
      resources
      Webapp
        Web-INF
          config
            springmvc-config.xml
          jsp
            finish.jsp
          Web.xml
        index.jsp
```

图 18-3　MonitorInterceptorDemo 工程目录层次及文件

MonitorInterceptor.java 的主要代码如下：

```
public class MonitorInterceptor extends HandlerInterceptorAdapter {
 private NamedThreadLocal<Long> startTimeThreadLocal =
     new NamedThreadLocal<Long>("StartTime");
 @Override
 public boolean preHandle(HttpServletRequest request, HttpServletResponse
response,
     Object handler) throws Exception {
     long beginTime = System.currentTimeMillis();     // ①开始时间
     // 线程绑定变量，该数据只有当前请求的线程可见
     startTimeThreadLocal.set(beginTime);
     return true;    // 继续流程
 }
 @Override
 public  void  afterCompletion(HttpServletRequest  request,  HttpServlet
Response response,          Object handler, Exception ex) throws Exception {
     long endTime = System.currentTimeMillis();     // ②结束时间
     // 得到线程绑定的局部变量（开始时间）
     long beginTime = startTimeThreadLocal.get();
     long consumeTime = endTime - beginTime;     // ③消耗的时间
     if (consumeTime > 1000) { //此处认为处理时间超过 1000ms 的请求为慢请求
     System.out.println(String.format("%s consume %d millis",
             request.getRequestURI(), consumeTime)); // 记录到日志文件中
     }
 }
}
```

NamedThreadLocal：Spring 提供的一个命名的 ThreadLocal 实现。

TestController.java 的主要代码如下：

```
@Controller
public class TestController {
 @RequestMapping("/mytask")
 public String mytask() {
```

```
        System.out.println("访问 TestController 控制器的 mytask() 方法");
        try {
            Thread.sleep(3000);
        } catch (InterruptedException e) {
            e.printStackTrace();
        }
        return "finish";
    }
}
```

访问 http://localhost:8080/MonitorInterceptorDemo/mytask，控制台上输出内容如下：

```
访问 TestController 控制器的 mytask() 方法
/MonitorInterceptorDemo/mytask consume 3006 millis
```

在测试时需要把 MonitorInterceptor 放在拦截器链的首位，这样得到的时间才是比较准确的。（代码详见本书源码包 ch18 目录中的 MonitorInterceptorDemo 工程。）

【示例】登录认证。

需求分析：在访问某些资源（如订单页面）时，需要用户登录后才能查看，因此需要进行登录检测。

实现分析：

（1）访问需要登录的资源时，由拦截器重定向到登录页面。

（2）如果访问的是登录页面，则拦截器不应该拦截。

（3）用户登录成功后，向 cookie/session 中添加登录成功的标识（如用户编号）。

（4）下次请求时，拦截器通过判断 cookie/session 中是否有该标识来决定继续流程还是返回到登录页面。

（5）此拦截器还应该允许游客访问其他资源。

在 Eclipse 中创建 Maven Project，名为 LoginInterceptorDemo。LoginInterceptorDemo 工程目录层次及文件如图 18-4 所示。创建的新文件有：LoginInterceptor.java、LoginController.java、info.jsp、login.jsp、User.java 等。

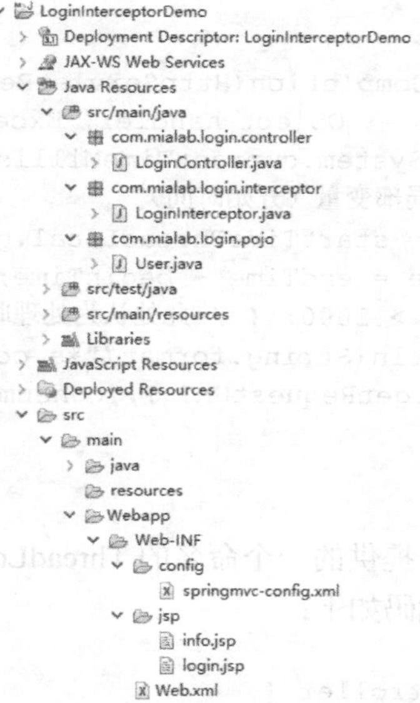

图 18-4 LoginInterceptorDemo 工程目录层次及文件

（代码详见本书源码包 ch18 目录中的 LoginInterceptorDemo 工程。）

LoginController.java 的主要代码如下：

```
@Controller
    public class LoginController {
     @RequestMapping(value = "/login", method = RequestMethod.GET)
     public String toLogin() {
         return "login";  //向用户登录页面跳转
     }
     @RequestMapping(value = "/login", method = RequestMethod.POST)
     public String login(User user, Model model, HttpSession session) {
         //获取用户名和密码
         String username = user.getUsername();
         String password = user.getPassword();
         // 此处模拟从数据库中获取用户名和密码后进行判断
         if (username != null && username.equals("snowgarden") && password !=
null
             && password.equals("168")) {
             session.setAttribute("USER_SESSION", user);  //将用户对象添加到
Session 中
             return "redirect:info";  //重定向到信息页面
         }
         model.addAttribute("msg", "用户名或密码错误，请重新登录！");
         return "login";
     }
     @RequestMapping(value = "/info")
     public String toMain() {
         return "info";  //向信息页面跳转
     }
     @RequestMapping(value = "/logout")
     public String logout(HttpSession session) {
         session.invalidate();   //清除 Session
         return "redirect:login";  //重定向到登录页面的跳转方法
     }
}
```

将 LoginInterceptorDemo 工程部署发布到 Tomcat 服务器，并启动服务器。

当通过 http://localhost:8080/LoginInterceptorDemo/info 试图访问 info.jsp 页面时，会被拦截器拦截，并跳转到 login.jsp 页面，对用户名和密码进行验证。用户名和密码验证正确后才能进入 info.jsp 页面。

18.2　Spring MVC 的数据转换和格式化

Spring MVC 会根据请求方法签名的不同，将请求消息中的信息以一定的方式转换并绑定到请求方法的参数中。在请求消息到达真正调用处理方法的这一段时间内，Spring MVC 还会完成很多其他工作，包括请求信息转换、数据转换、数据格式化及数据校验等。

18.2.1 Spring MVC 消息转换流程

Spring MVC 框架是围绕着 DispatcherServlet 而工作的。Spring MVC 还会给处理器加入拦截器，在处理器执行前后加入自己的代码，这样就构成了一个处理器的执行链。

当一个请求到来时，DispatcherServlet 首先根据请求和解析好的 HandlerMapping 配置，找到对应的处理器，并准备开始运行处理器和拦截器组成的执行链。而运行处理器需要有一个对应的环境，于是就需要使用处理器的适配器。通过这个适配器就能运行对应的处理器及其拦截器，这里的处理器包含了控制器的内容和其他增强的功能。

可以看到，这里的处理器和用户自己开发的控制器不是同一个概念。处理器是在控制器功能的基础上加上了一层包装，有了这层包装，在 HTTP 请求到达控制器之前它就能够对 HTTP 的各类消息进行处理。首先，当一个请求到达 DispatcherServlet 的时候，需要找到对应的 HandlerMapping；其次，根据 HandlerMapping 去找到对应的 HandlerAdapter 执行处理器。

处理器调用控制器之前，需要先获取 HTTP 发送过来的信息，并将其转变为控制器的不同类型的参数，这就是各类注解能够得到丰富类型参数的原因。它先用 HTTP 的消息转换器（HttpMessageConverter）对消息进行转换，但这是一个比较原始的转换，它是 String 类型和其他类型比较简易的转换，还需要进一步转换才能转换为 POJO 或者其他丰富的参数类型。为了拥有这样的能力，Spring 4 提供了转换器和格式化器，这样通过注解的信息和参数的类型，它就能够把 HTTP 发送过来的各种消息转换成为控制器所需要的各类参数。

当处理器处理完这些参数的转换后，就会进行验证，验证表单的方法将在后面的章节中进行讲述。完成了这些内容，下一步就是调用开发者所提供的控制器，将之前转换成功的参数传递进去，这样，开发的控制器就能够得到丰富的 Java 类型的支持，进而完成控制器的逻辑。

控制器完成了对应的逻辑，返回结果后，处理器如果可以找到对应处理结果类型的 HttpMessageConverter 的实现类，就会调用对应的 HttpMessageConverter 实现类方法，对控制器返回的结果进行 HTTP 转换。这一步不是必需的。可以转换的前提是能够找到对应的转换器，在讨论注解@ResponseBody 时会再次看到这个过程，完成这些处理器的功能即可。

18.2.2 Spring MVC 的数据绑定

1. 数据绑定流程

Spring MVC 通过反射机制对目标处理方法的签名进行分析，并将请求消息绑定到处理方法的参数中。数据绑定的核心组件是 DataBinder。

Spring MVC 框架将 ServletRequest 对象及处理方法的参数对象实例传递给 DataBinder，DataBinder 调用装配在 Spring Web 上下文中的 ConversionService 组件进行数据类型转换、数据格式化工作，并将 ServletRequest 中的消息填充到参数对象中。再调用 Validator 组件对已经绑定了请求消息数据的参数对象进行数据合法性校验，并最终生成数据绑定结果 BindingResult 对象。BindingResult 包含已完成数据绑定的参数对象，还包含相应的校验错误对象，Spring MVC 抽取 BindingResult 中的参数对象和校验错误对象，并将它们赋给处理方法的相应参数。

2．默认支持的类型的绑定

在 Spring MVC 中，有支持的默认类型的绑定。也就是说，直接在控制器方法形参上定义默认类型的对象即可使用这些对象。这些对象包括：HttpServletRequest 对象；HttpServletResponse 对象；HttpSession 对象；Model/ModelMap 对象。

在参数绑定过程中，如果遇到这些类型可直接进行绑定。也就是说，可以在控制器方法的形参中直接定义这些类型的参数，Spring MVC 会自动绑定。

这里要说明的是 Model/ModelMap 对象。Model 是一个接口，ModelMap 是一个接口实现，作用是将 Model 数据填充到 request 域中，和 ModelAndView 类似，关于它的使用，将在下面和简单类型参数绑定一起讲述。

3．简单类型的绑定

【示例】简单类型的数据绑定。

需求分析：根据商品的 id 来修改对应商品信息。所以，前台页面肯定要传入该商品的 id，交由 Spring MVC 的控制器进行处理，返回一个修改商品信息的页面。

前台页面通过 url 将参数传递过来，请求的是 editItems，代码如下所示：

```
<td><a>href="${pageContext.request.contextPath}/editItems?id=${item.id}"</a></td>
```

控制器的 editItems 方法的主要代码如下：

```
    @RequestMapping("/editItems")
    public String editItems(Model model, Integer id) throws Exception {
        //根据 id 查询对应的 Items
        ItemsCustom itemsCustom = itemsService.findItemsById(id);
        model.addAttribute("itemsCustom", itemsCustom);
        //通过形参中的 model 将 model 数据传到页面中
        //相当于 ModelAndView.addObject 方法
        return "/Web-INF/jsp/items/editItems.jsp";
    }
```

从前面的代码中可以看出 Model 对象可以直接作为参数，Spring MVC 默认会绑定它（前面章节中也讲述过），然后使用 Model 对象将查询到的数据放到 request 域中，这样即可在前台页面中取出该数据。要注意的是，简单类型的绑定中，方法形参中的参数名要和前台传入的参数名一样才能完成参数的绑定。如果定义的参数名不一样呢？此时可以使用注解 @RequestParam 对简单的类型进行参数绑定。例如：

```
    @RequestMapping("/editItems")
    public                          String                          editItems(Model
model,@RequestParam(value="/id",required=true)
    Integer items_id) throws Exception {
        //根据 id 查询对应的 Items
     ItemsCustom itemsCustom = itemsService.findItemsById(items_id);
        …
    }
```

通过@RequestParam 的 required 属性指定参数是否必须要传入，如果设置为 true，则没有传入参数就会报错。

4．POJO 类型的绑定

（1）普通 POJO 类型。

这里假设实体类 Items.java 有以下类型的属性。

```
    public class Items {
```

```
    private Integer id;
    private String name;
    private Float price;
    private String pic;
    private Date createtime;
    private String detail;
}
```

当提交表单后，要先将这些属性全部封装到一个 POJO 中，这里是 Items 对象，再去更新数据库。如何进行数据绑定呢？

绑定很简单，对于基本类型，只要前台页面中 input 标签的 name 属性值和控制器的 POJO 形参中的属性名称一致，即可将页面中数据绑定到 POJO。也就是说，前台页面传进来的 name 要和要封装的 POJO 属性名一致，然后可以将该 POJO 作为形参放到控制器的方法中。参考代码如下：

```
@RequestMapping("/editItemsSubmit")
public String editItemsSubmit(Model model, HttpServletRequest request,
Integer id,
Items items) throws Exception {
  //调用 service 更新商品信息，页面需要将商品信息传到此方法中
  itemsService.updateItems(id, items);
  return "/Web-INF/jsp/success.jsp";
}
```

这样即可将前台表单传入的不同属性值封装到 Items 中。但是运行后会报错，报错的信息是无法将 String 类型转换成 java.util.Date 类型，因为 Items 中有一个属性是 Date 类型的 createtime。这就需要用户自定义一个日期转换器。示例代码如下。

```
//需要实现 Converter 接口，这里是将 String 类型转换成 Date 类型
public class CustomDateConverter implements Converter<String, Date> {
  @Override
  public Date convert(String source) {
      //将日期串转换成日期类型(格式是 yyyy-MM-dd)
      SimpleDateFormat simpleDateFormat = new SimpleDateFormat("yyyy-
MM-dd);
      try {
          return simpleDateFormat.parse(source); //直接返回
      } catch (ParseException e) {
          e.printStackTrace();
      }
      return null; //如果参数绑定失败，则返回 null
  }
}
```

定义好转换器后，需要在 Spring MVC 配置文件中进行如下配置：

```
<mvc:annotation-driven conversion-service="conversionService" / >
<!-- 自定义的日期类型转换器 -->
<bean id="conversionService" class=
 "org.springframework.context.support.ConversionServiceFactoryBean">
 <!-- 转换器 -->
 <property name="converters">
     <list>
         <bean class="com.mialab.converter.CustomDateConverter"/>
     </list>
 </property>
```

```
</bean>
```

Spring MVC 能根据这个转换器将 String 类型正确转换成 Date 类型，然后封装到 Items 对象中。

（2）包装的 POJO 类型。

所谓的包装 POJO，是指在一个 POJO 中包含另一个简单 POJO。例如，在学生对象中包含班级对象。假定有一个学生类 Student，该类用于封装学生及其所在班级信息。

学生类 Student 的主要属性如下：

```
public class Student {
 private int studentId;
 private String studentName;
 private BanJi banJi;
}
```

班级类 BanJi 的主要属性如下：

```
public class BanJi {
 private int classId;
 private String className;
}
```

【示例】学生信息录入与保存。

在 Eclipse 中创建 Maven Project，名为 POJODemo。POJODemo 工程目录层次及文件如图 18-5 所示。创建的新文件有：StudentController.java、Student.java、BanJi.java、studentForm.jsp 和 result.jsp 等。（代码详见本书源码包 ch18 目录中的 POJODemo 工程。）

图 18-5　POJODemo 工程目录层次及文件

StudentController.java 的主要代码如下：

```
@Controller
public class StudentController {
 @RequestMapping("/student_input")
```

```
    public String inputStudent() {
        return "studentForm";
    }
    @RequestMapping("/student_save")
    public String saveStudent(Student student) {
        int studentId = student.getStudentId();
        String studentName = student.getStudentName();
        BanJi banJi = student.getBanJi();
        String className = banJi.getClassName();
        banJi.setClassId(302);
        System.out.println(student);
        System.out.println("studentId=" + studentId);
        System.out.println("studentName=" + studentName);
        System.out.println("className=" + className);
        return "result";
    }
}
```

由以上代码可看出，解决问题的思路如下：**在控制器（StudentController）请求方法的形参中使包装类型的 POJO**（这里是 Student 对象）**接收查询条件参数**。

studentForm.jsp 的<body>部分的主要代码如下：

```
<form   action="${pageContext.request.contextPath   }/student_save" method="POST">
      <p>学号：<input type="text" name="studentId"/></p>
    <p>姓名：<input type="text" name="studentName"/></p>
    <p>班级： <input type="text" name="banJi.className"/></p>
    <p><input type="submit" value="保存" /></p>
</form>
```

由以上内容可以看出，在使用包装 POJO 类型数据绑定时，前端请求的参数名编写必须符合以下两种情况。

① 如果查询条件参数是包装类的直接基本属性，则参数名直接使用对应的属性名，如 studentForm.jsp 代码中的 studentId 和 studentName。

② 如果查询条件参数是包装类中 POJO 的子属性，则参数名必须为**“对象.属性”**，其中“对象”要和包装 POJO 中的对象属性名称一致，“属性”要和包装 POJO 中的对象子属性名称一致，如 studentForm.jsp 代码中的 banJi.className。

result.jsp 的<body>部分的主要代码如下：

```
<p>学生学号:${requestScope.student.studentId}</p>
<p>学生姓名:${requestScope.student.studentName}</p>
<p>学生班级:${requestScope.student.banJi.className}</p>
<p>班级代码:${requestScope.student.banJi.classId}</p>
```

测试的结果如图 18-6 和图 18-7 所示。

图 18-6 学生信息录入

图 18-7 学生信息保存

（代码详见本书源码包 ch18 目录中的 POJODemo 工程。）

5. 集合类型的绑定

这里只讲述数组的绑定。数组的绑定指的是前台传来多个同一类型的数据（如勾选复选框），在控制器请求方法中使用数组形参来接收前台传来的数据。

【示例】批量删除图书（前台使用复选框）。

在前面创建的 POJODemo 工程中添加以下新文件：BookController.java、bookSelect.jsp 和 bookResult.jsp。在 bookSelect.jsp 页面中通过复选框选择欲删除的图书。必须注意的是，bookSelect.jsp 页面中复选框 name=bookChoose。

BookController.java 主要代码如下（代码详见本书源码包 ch18 目录中的 POJODemo 工程）：

```
@Controller
public class BookController {
 @RequestMapping("/selectBook")
 public String selectBooks() {
     return "bookSelect";
 }
 @RequestMapping("/deleteBooks")
 public String deleteUsers(Integer[] bookChoose) {
     if (bookChoose != null) {
         for (int id : bookChoose) {
             // 使用输出语句模拟已经删除了图书
             System.out.println("删除了 id 为" + id + "的图书！");
         }
     } else {
         System.out.println("bookChoose 的值为 null");
     }
     return "bookResult";
 }
}
```

测试的结果如图 18-8 和图 18-9 所示。

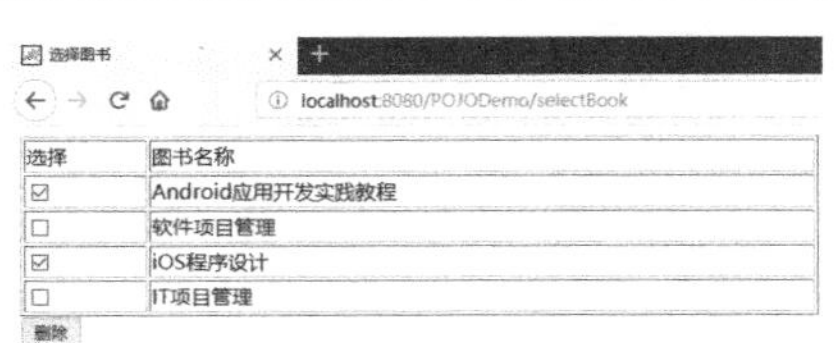

图18-8　选择图书页面

图18-9　图书删除成功

Eclipse 的控制台输出如下内容：

```
删除了 id 为 1 的图书！
删除了 id 为 3 的图书！
```

18.2.3　Spring MVC 的数据转换

在 Java 类型转换之前，Spring MVC 为了应对 HTTP 请求，定义了 HttpMessageConverter，它是一个总体的接口，通过它可以读入 HTTP 的请求内容。也就是说，在读取 HTTP 请求的参数和内容的时候会先用 HttpMessageConverter 读出，将其简单转换为 Java 类型，主要是字符串（String），然后即可用各类转换器进行转换。在逻辑业务处理完成后，还可以通过

HttpMessageConverter 把数据转换为响应给用户的内容。

在 Java 中，在 java.beans 包中提供了一个 PropertyEditor 接口来进行数据转换，PropertyEditor 的核心功能是将一个字符串转换为一个 Java 对象。但是 PropertyEditor 只能用于字符串和 Java 对象的转换，不适用于任意两个 Java 类型的互换。

Spring 从 3.0 开始添加了一个通用的类型转换模块，该类型转换模块位于 org.springframework.core.convert 包中。Spring 希望用这个类型转换体系替换 Java 标准的 PropertyEditor 接口。ConversionService 是 org.springframework.core.convert 包的核心组件，可以通过使用 ConversionServiceFactoryBean 在 Spring 的上下文中自定义一个 ConversionService，Spring 将自动识别这个 ConversionService，并在 Spring MVC 处理方法的参数绑定中使用它进行数据转换。可以通过 ConversionServiceFactoryBean 的 converts 属性注册自定义的类型转换器。

【示例】使用 ConversionService 转换数据。

在 Eclipse 中创建 Maven Project“ConverterDemo”。ConverterDemo 工程目录层次及文件如图 18-10 所示。创建的新文件有：StringToDateConverter.java、BookController.java、Book.java、bookForm.jsp 和 success.jsp。（代码详见本书源码包 ch18 目录中的 ConverterDemo 工程。）

实体类 Book 的属性有以下 3 个，注意图书登记日期 bookEntryDate 是 Date 类型的。

```
public class Book {
 private int bookId;
 private String bookName;
 private Date bookEntryDate;
}
```

```
ConverterDemo
  src/main/java
    com.mialab.ConverterDemo.controller
      BookController.java
    com.mialab.ConverterDemo.converter
      StringToDateConverter.java
    com.mialab.ConverterDemo.domain
      Book.java
  src/test/java
  JRE System Library [JavaSE-1.8]
  Maven Dependencies
  src/main/resources
  src
    main
      Webapp
        Web-INF
          config
            springmvc-config.xml
          jsp
            bookForm.jsp
            success.jsp
          Web.xml
```

图 18-10　ConverterDemo 工程目录层次及文件

BookController.java 的主要代码如下：

```
@Controller
public class BookController {
 private static final Log logger = LogFactory.getLog(BookController.class);
 @RequestMapping(value = "/{formName}")
```

```
    public String loginForm(@PathVariable String formName) {
        return formName;  // 动态跳转页面
    }
    @RequestMapping(value = "/bookEntry", method = RequestMethod.POST)
    public String bookEntry(Book book, Model model) {
        logger.info(book);
        model.addAttribute("book", book);
        return "success";
    }
}
```

StringToDateConverter.java 的主要代码如下：

```
public class StringToDateConverter implements Converter<String, Date> {
    private String datePattern;  // 日期类型模板：如 yyyy-MM-dd
    public void setDatePattern(String datePattern) {
        this.datePattern = datePattern;
    }
    // Converter<S,T>接口的类型转换方法
    @Override
    public Date convert(String date) {
        try {
            SimpleDateFormat dateFormat = new SimpleDateFormat(this.datePattern);
            return dateFormat.parse(date);  //将日期字符串转换成 Date 类型并返回
        } catch (Exception e) {
            e.printStackTrace();
            System.out.println("日期转换失败!");
            return null;
        }
    }
}
```

在 springmvc-config.xml 中添加关于自定义类型转换器的配置代码：

```
<!-- 装配自定义的类型转换器 -->
<mvc:annotation-driven conversion-service="conversionService" />
<!-- 自定义的类型转换器 -->
<bean id="conversionService"
 class="org.springframework.context.support.ConversionServiceFactoryBea
n">
 <property name="converters">
     <list>
         <bean  class="com.mialab.ConverterDemo.converter. StringToDate
Converter"
               p:datePattern="yyyy-MM-dd"></bean>
     </list>
 </property>
</bean>
```

success.jsp 中<body>部分的代码如下：

```
<p>图书编号：${requestScope.book.bookId}</p>
<p>图书名称：${requestScope.book.bookName}</p>
<p>登记日期：${requestScope.book.bookEntryDate}</p>
```

测试的结果如图 18-11 和图 18-12 所示，这说明 String 类型成功转换为了 Date 类型。

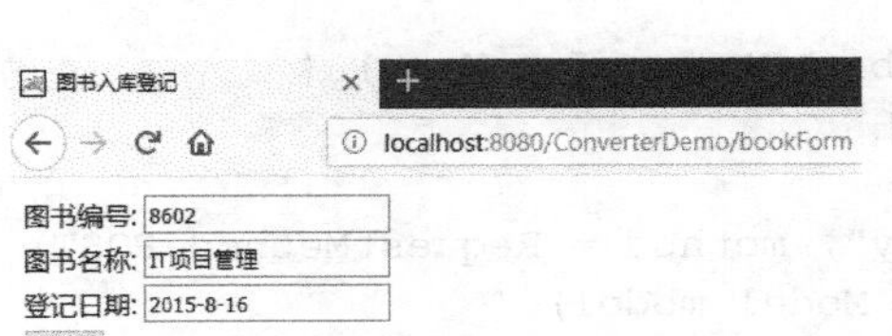

图18-11 图书入库登记页面

图18-12 图书登记成功

（代码详见本书源码包 ch18 目录中的 ConverterDemo 工程。）

18.2.4 Spring MVC 的数据格式化

1．使用 Formatter

Formatter 的作用也是将一种类型转换为另一种类型，但是 Formatter 的源类型必须是一个 String，而 Converter 则适用于任意的源类型。因此，Formatter 更适用于 Web 层的数据转换，而 Converter 可以用在任意层中。为了转换 Spring MVC 应用程序的表单中的用户输入，一般选用 Formatter，而不是 Converter。

【示例】使用 Formatter 格式化数据。

在 Eclipse 中创建 Maven Project“FormatterDemo”。FormatterDemo 工程目录层次及文件与 ConverterDemo 工程类似。创建的新文件有：DateFormatter.java、BookController.java、Book.java、bookForm.jsp 和 success.jsp。（代码详见本书源码包 ch18 目录中的 FormatterDemo 工程。）

自定义格式化器需要实现 Formatter 接口。DateFormatter.java 的主要代码如下：

```
public class DateFormatter implements Formatter<Date> {
  private String datePattern;  //日期类型模板：如 yyyy-MM-dd
  private SimpleDateFormat dateFormat;  //日期格式化对象
  //构造器，通过依赖注入的日期类型创建日期格式化对象
  public DateFormatter(String datePattern) {
      this.datePattern = datePattern;
      this.dateFormat = new SimpleDateFormat(datePattern);
      /*setLenient 用于设置 Calendar 是否宽松解析字符串，如果参数为 false，
      则严格解析；默认为 true，即宽松解析*/
      dateFormat.setLenient(false);
  }
  //显示 Formatter<T>的 T 类型对象
  @Override
  public String print(Date date, Locale locale) {
      return dateFormat.format(date);
  }
  //解析文本字符串，返回一个 Formatter<T>的 T 类型对象
  @Override
  public Date parse(String source, Locale locale) throws ParseException {
      try {
          return dateFormat.parse(source);
      } catch (Exception e) {
          throw new IllegalArgumentException("invalid date format");
      }
  }
```

```
    }
```

在 springmvc-config.xml 中添加关于自定义格式化器的配置代码：

```
    <!-- 装配自定义的格式化器 -->
    <mvc:annotation-driven conversion-service="conversionService" />
    <!-- 自定义的格式化器 -->
    <bean id="conversionService" class="org.springframework.format.support.
FormattingConversionServiceFactoryBean">
     <property name="formatters">
        <set>
           <bean class="com.mialab.FormatterDemo.formatter.DateFormatter">
           <constructor-arg type="java.lang.String" value="yyyy-MM-dd" />
           </bean>
        </set>
     </property>
    </bean>
```

测试的结果与前面所述类似。

2. 使用 FormatterRegistrar 注册 Formatter

【示例】使用 FormatterRegistrar 注册 Formatter。

在 Eclipse 中创建 Maven Project“FormatterRegistrarDemo”。FormatterRegistrarDemo 工程目录层次及文件与 FormatterDemo 工程类似。创建的新文件有：MyFormatterRegistrar.java、BookController.java、Book.java、bookForm.jsp 和 success.jsp。（代码详见本书源码包 ch18 目录中的 FormatterRegistrarDemo 工程。）

想实现 FormatterRegistrar 只需要实现一个方法，即 registerFormatters()方法，在该方法中添加需要注册的 Formatter。MyFormatterRegistrar.java 的主要代码如下：

```
    public class MyFormatterRegistrar implements FormatterRegistrar {
     private String datePattern;
     public MyFormatterRegistrar(String datePattern) {
        this.datePattern = datePattern;
     }
     @Override
     public void registerFormatters(FormatterRegistry registry) {
        registry.addFormatter(new DateFormatter(datePattern));
     }
    }
```

有了 Registrar，就不需要在 Spring MVC 配置文件中注册任何 Formatter 了，只在 Spring MVC 配置文件中注册 Registrar 即可。springmvc-config.xml 注册 Registrar 的代码如下：

```
    <mvc:annotation-driven conversion-service="conversionService" />
    <bean id="conversionService" class="org.springframework.format.support.
FormattingConversionServiceFactoryBean">
     <property name="formatterRegistrars">
        <set>
           <bean class="com.mialab.FormatterRegistrarDemo.
                                   formatter.MyFormatterRegistrar">
           <constructor-arg type="java.lang.String" value="yyyy-MM-dd" />
           </bean>
        </set>
     </property>
```

```
</bean>
```

测试的结果与前面所述类似。

18.2.5 JSON 格式的数据转换

JSON 数据格式在接口调用中、HTML 页面中比较常用。JSON 格式比较简单，解析也比较方便，所以使用很普遍。在 Spring MVC 中，也支持对 JSON 数据的解析和转换。

HttpMessageConveter<T>是 Spring 3.0 之后新增的一个重要接口。在 Spring MVC 的设计中，HttpMessageConverter 接口扮演着重要的角色，它提供了强大的数据转换功能，可以将请求数据转换为 Java 对象，也可以将 Java 对象转化为特定格式输出。Spring 提供了众多的 HttpMessageConveter 实现类，如 MappingJacksonHttpMessageConverter，可以利用 Jackson 开源包处理 JSON 格式的请求或响应消息。

转换器的装配方式有以下两种形式。

（1）注册 org.springframework.Web.servlet.mvc.annotation.AnnotationMethodHandlerAdapter 来装配 messageConverters，如下所示：

```
<bean class="org.springframework.Web.servlet.mvc.annotation.
                        AnnotationMethodHandlerAdapter">
    <property name="messageConverters"> <!-- 装配数据转换器 -->
        <list>
            <ref bean="jsonConverter" /> <!-- 指定装配 JSON 格式的数据转换器 -->
        </list>
    </property>
</bean>
<bean id="jsonConverter" class="……">
    <property name="supportedMediaTypes" value="application/json" />
    <!-- 设置转换的 media 类型为 application/json -->
</bean>
```

（2）启用注解<mvc:annotation-driven /> 。该注解会初始化一些转换器。例如：

① ByteArrayHttpMessageConverter：读写二进制数据。

② StringHttpMessageConverter：将请求信息转换为字符串。

③ ResourceHttpMessageConverter：读写 org.springframework.core.io.Resource 对象。

④ SourceHttpMessageConverter：读写 javax.xml.transform.Source 类型的数据。

⑤ XmlAwareFormHttpMessageConverter： 转换表单属性中可能的 XML 数据。

⑥ Jaxb2RootElementHttpMessageConverter：通过 JAXB2 读写 XML 消息。

⑦ MappingJacksonHttpMessageConverter：利用 Jackson 开源包读写 JSON 数据。

通过以上两种方法，即可完成转换器注册。

如果要在控制层完成数据的输入输出转换，则有以下两种途径。一种途径是，可使用 @RequestBody 和@ResponseBody 对处理方法进行标注；其中，@RequestBody 通过合适的 HttpMessageConverter 将 HTTP 请求正文转换为需要的对象内容，而@ResponseBody 则将对象内容通过合适的 HttpMessageConverter 转换后作为 HTTP 响应的正文输出。另外一种途径是，使用 HttpEntity、ResponseEntity 作为处理方法的入参或返回值，以来完成数据的输入输出转换。

1．使用@ResponseBody 返回 JSON 格式的数据

@ResponseBody 注解用于将控制器的方法返回的对象，通过 HttpMessageConverter 接口

转换为指定格式的数据，如 JSON、XML 等，通过 Response 响应给客户端。

【示例】返回 JSON 格式的数据。

在 Eclipse 中创建 Maven Project“JsonDemo1”。JsonDemo1 工程目录层次及文件如图 18-13 所示。创建的新文件有：BookController.java、BookService.java 和 Book.java。（代码详见本书源码包 ch18 目录中的 JsonDemo1 工程。）

BookController.java 的主要代码如下：

```
@Controller
@RequestMapping("/book")
public class BookController {
 @Autowired
   private BookService bookService;

 @RequestMapping(value="/bookList")
 // @ResponseBody 会将集合数据转换成 JSON 格式返回客户端
 @ResponseBody
 public List<Book> findBooks(){
     return bookService.getBookList();
 }
}
```

```
JsonDemo1
  src/main/java
    com.mialab.JsonDemo1.controller
      BookController.java
    com.mialab.JsonDemo1.domain
      Book.java
    com.mialab.JsonDemo1.service
      BookService.java
  src/main/resources
  src/test/java
  JRE System Library [JavaSE-1.8]
  Maven Dependencies
  src
    main
      Webapp
        Web-INF
          config
            springmvc-config.xml
          jsp
          Web.xml
```

图 18-13　JsonDemo1 工程目录层次及文件

BookService.java 的主要代码如下：

```
@Service("bookService")
public class BookService {
 public List<Book> getBookList(){
     List<Book> list = new ArrayList<Book>();
     list.add(new Book(8901, "Android 应用开发实践教程", 56));
     list.add(new Book(8902, "iOS 程序设计", 48));
     list.add(new Book(8903, "IT 项目管理", 89));
     list.add(new Book(8904, "信息系统项目管理", 36));
     return list;
 }
```

```
}
```

因为使用了 Jackson 开源包，所以在 pom.xml 文件中必须添加 Jackson 依赖。

```
<properties>
 <!-- Jackson 版本号 -->
 <jackson.version>2.8.5</jackson.version>
</properties>
<dependencies>
 <!-- Jackson 依赖 -->
 <dependency>
     <groupId>com.fasterxml.jackson.core</groupId>
     <artifactId>jackson-databind</artifactId>
     <version>${jackson.version}</version>
 </dependency>
 <dependency>
     <groupId>com.fasterxml.jackson.core</groupId>
     <artifactId>jackson-annotations</artifactId>
     <version>${jackson.version}</version>
 </dependency>
 <dependency>
     <groupId>com.fasterxml.jackson.core</groupId>
     <artifactId>jackson-core</artifactId>
     <version>${jackson.version}</version>
 </dependency>
</dependencies>
```

springmvc-config.xml 中必须添加以下配置代码，以得到 Jackson 开源包读写 JSON 的支持和用来构建数据的 BookService 实例。在实际开发中，BookService 是用来执行业务逻辑的。

```
<!-- 指定需要扫描的包 -->
<context:component-scan base-package="com.mialab.JsonDemo1.controller"/>
<context:component-scan base-package="com.mialab.JsonDemo1.service" />
<mvc:annotation-driven />
```

<mvc:annotation-driven/> 的作用是自动注册 DefaultAnnotationHandlerMapping 与 AnnotationMethodHandlerAdapter 两个 Bean，这是 Spring MVC 为@Controllers 分发请求所必需的。其提供了以下支持：数据绑定支持；@NumberFormatannotation 支持；@DateTimeFormat 支持；@Valid 支持；读写 XML 的支持（JAXB）；读写 JSON 的支持（使用 Jackson 开源包）。

测试的结果如图 18-14 所示。（代码详见本书源码包 ch18 目录中的 JsonDemo1 工程。）

localhost:8080/JsonDemo1/book/bookList

[{"bookId":8901,"bookName":"Android应用开发实践教程","bookPrice":56.0},{"bookId":8902,"bookName":"iOS程序设计","bookPrice":48.0},{"bookId":8903,"bookName":"IT项目管理","bookPrice":89.0},{"bookId":8904,"bookName":"信息系统项目管理","bookPrice":36.0}]

图 18-14　使用@ResponseBody 返回 JSON 格式的数据

倘若使用其他开源包来处理 JSON，又该如何配置 HttpMessageConverter 呢？

【示例】使用 FastJson 返回 JSON 格式的数据。

在 Eclipse 中创建 Maven Project"JsonDemo2"。JsonDemo2 工程目录层次及文件与图 18-13 类似。创建的新文件有：BookController.java、BookService.java 和 Book.java。（代码详见本书源码包 ch18 目录中的 JsonDemo2 工程。）

因为使用到了 FastJson 包，所以在 pom.xml 文件中必须添加 FastJson 依赖。

```
<!-- FastJson 依赖 -->
<dependency>
```

```
        <groupId>com.alibaba</groupId>
        <artifactId>fastjson</artifactId>
        <version>1.2.47</version>
    </dependency>
```

springmvc-config.xml 必须添加以下配置代码：

```
    <mvc:annotation-driven>
       <mvc:message-converters register-defaults="true">
       <!-- 配置 FastJson 中实现 HttpMessageConverter 接口的转换器 -->
       <bean id="fastJsonHttpMessageConverter" class="com.alibaba.fastjson.
support.spring.
                        FastJsonHttpMessageConverter">
          <!-- 加入支持的媒体类型：返回 contentType -->
             <property name="supportedMediaTypes">
                <list>
                 <!--这里顺序不能反，一定要先写 text/html，否则 IE 会弹出下载提示 -->
                    <value>text/html;charset=UTF-8</value>
                    <value>application/json;charset=UTF-8</value>
                 </list>
             </property>
       </bean>
       </mvc:message-converters>
    </mvc:annotation-driven>
```

测试的结果与图 18-14 类似。（代码详见本书源码包 ch18 目录中的 JsonDemo2 工程。）

2．使用@RequestBody 接收 JSON 格式的数据

@RequestBody 注解用于读取 HTTP 请求的内容（字符串），通过 Spring MVC 提供的 HttpMessageConverter 接口将读到的内容转换为 JSON、XML 等格式的数据，并绑定到控制器的方法参数上。例如，这里的参数是某种 Java 对象类型。

【示例】使用@RequestBody 接收传递的 JSON 参数。

在 Eclipse 中创建 Maven Project“JsonDemo3”。JsonDemo3 工程目录层次及文件如图 18-15 所示。创建的新文件有：bookForm.jsp、BookController.java、BookService.java 和 Book.java。（代码详见本书源码包 ch18 目录中的 JsonDemo3 工程。）

BookController.java 的主要代码如下：

```
    @Controller
    public class BookController {
     @Autowired
       private BookService bookService;

     @RequestMapping("/book_query")
     public String query_Book() {
         logger.info("query_Book called");
         return "bookForm";
     }

     //接收页面请求的 JSON 数据
     @RequestMapping("/book_find")
     @ResponseBody
     public List<Book> findBook(@RequestBody BookParams bookParams) {
         System.out.println(bookParams);
         //此处只是模拟：根据查询参数 BookParams 对象从数据库中取出数据
```

```
        List<Book> bookList = bookService.getBookList(bookParams);
        return bookList;
    }
}
```

```
JsonDemo3
  src/main/java
    com.mialab.JsonDemo3.controller
      BookController.java
    com.mialab.JsonDemo3.domain
      Book.java
      BookParams.java
    com.mialab.JsonDemo3.service
      BookService.java
  src/main/resources
  src/test/java
  JRE System Library [JavaSE-1.8]
  Maven Dependencies
  src
    main
      Webapp
        js
          jquery-3.3.1.min.js
        Web-INF
          config
            springmvc-config.xml
          jsp
            bookForm.jsp
```

图 18-15　JsonDemo3 工程目录层次及文件

bookForm.jsp 的主要代码如下：

```
<%@ page language="java" contentType="text/html; charset=UTF-8"
 pageEncoding="UTF-8"%>
<html>
<head>
<title>图书查询</title>
<meta http-equiv="Content-Type" content="text/html; charset=UTF-8">
<script type="text/javascript"
 src="${pageContext.request.contextPath }/js/jquery-3.3.1.min.js">
</script>
<script type="text/javascript">
 function findBook() {
     //获取输入的查询关键词、出版社名称
     var bookKey = $("#bookKey").val();
     var bookPublish = $("#bookPublish").val();
     $
             .ajax({
                 url : "${pageContext.request.contextPath }/book_find",
                 type : "post",
                 // data 表示发送的数据
                 data : JSON.stringify({
                     bookKey : bookKey,
                     bookPublish : bookPublish
                 }),
                 // 定义发送请求的数据格式为 JSON 字符串
                 contentType : "application/json;charset=UTF-8",
                 //定义回调响应的数据格式为 JSON 字符串，该属性可以省略
```

```
                    dataType : "json",
                    //成功响应的结果
                    success : function(result) {
                        if (result != null) {
                            alert("查询图书列表: \n" + result[0].bookName +
"---"
                                + result[0].author + "---"
                                + result[0].bookPublish + "\n"
                                + result[1].bookName + "---"
                                + result[1].author + "---"
                                + result[1].bookPublish);
                        }
                    }
                });
            }
        </script>
        </head>
        <body>
          <form>
            <table>
                <tr>
                    <td><label>关键词: </label></td>
                    <td><input type="text" id="bookKey" name="bookKey"></td>
                </tr>
                <tr>
                    <td><label>出版社: </label></td>
                    <td><input type="text" id="bookPublish" name="bookPublish">
</td>
                </tr>
                <tr>
                    <td><input type="button" value="查询图书"
                        onclick="findBook()"></td>
                </tr>
            </table>
          </form>
        </body>
        </html>
```

需要说明的是，因为要访问静态资源，所以在 springmvc-config.xml 中加入了以下代码：

```
    <!--配置静态资源的访问映射，此配置中的文件，将不被前端控制器拦截 -->
    <!-- <mvc:resources location="/js/" mapping="/js/**" /> -->
    <mvc:default-servlet-handler />
```

测试的结果如图 18-16 和图 18-17 所示。

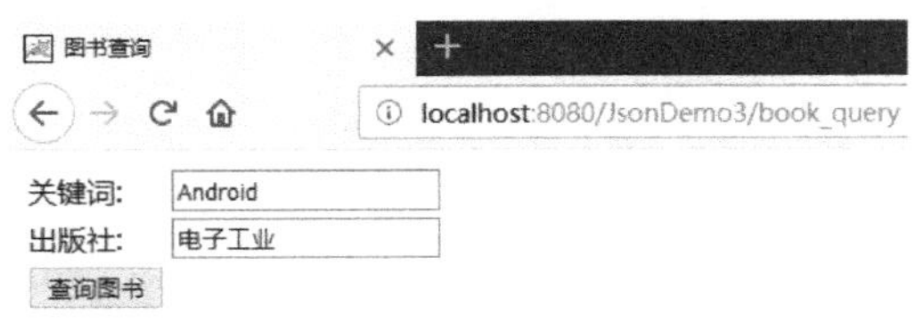

图18-16　输入图书查询参数

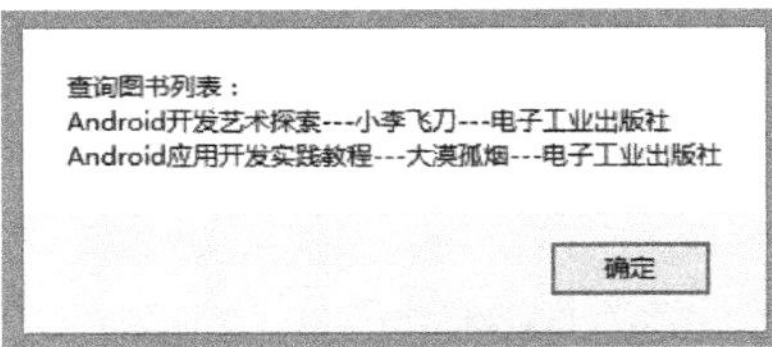

图18-17　查询图书列表

（代码详见本书源码包 ch18 目录中的 JsonDemo3 工程。）

18.3 本章小结

本章主要介绍了拦截器和数据转换及绑定。Spring MVC 中的拦截器类似于 Servlet 中的过滤器，它主要用于拦截用户请求并做相应处理。例如，通过拦截器可以进行权限验证、记录请求信息的日志、判断用户是否登录等。要使用 Spring MVC 中的拦截器，就需要对拦截器类进行定义和配置。通常，拦截器类可以通过两种方式来定义：通过实现 HandlerInterceptor 接口，或继承 HandlerInterceptor 接口的实现类（如 HandlerInterceptorAdapter）来定义；通过实现 WebRequestInterceptor 接口，或继承 WebRequestInterceptor 接口的实现类来定义。

Spring MVC 的数据转换是发生在数据绑定中的。Spring MVC 数据绑定流程如下。

① Spring MVC 框架将 ServletRequest 对象及目标方法的入参实例传递给 WebDataBinderFactory 实例，用以创建 DataBinder 实例对象，DataBinder 是数据绑定的核心组件。

② DataBinder 调用装配在 Spring MVC 上下文中的 ConversionService 组件进行数据类型转换、数据格式化工作，并将 Servlet 中的请求信息填充到入参对象中。

③ 调用 Validate 组件对已经绑定了请求消息的入参对象进行数据合法性校验，并最终生成数据绑定结果——BindingData 对象

④ Spring MVC 抽取 BindingResult 中的入参对象和校验错误对象，将其赋给处理方法的响应入参。

习题 18

1．Spring MVC 拦截器的执行流程是怎样的？试编程加以说明。

2．Spring MVC 框架中多个拦截器执行的顺序是怎样的？试编程加以说明。

3．如何使用 Spring MVC 拦截器实现用户登录权限验证？试编程加以说明。

4．Spring MVC 的数据绑定是怎样的？试编程加以说明。

5．Spring MVC 消息转换流程是怎样的？

6．Spring MVC 是如何实现数据转换和格式化的？试编程加以说明。

7．如何使用@ResponseBody 返回 JSON 格式的数据？如果是 Jackson 开源包又应该怎么办？如果使用的是 FastJson 开源包又应该怎么办？试编程加以说明。

8．如何使用@RequestBody 接收 JSON 格式的数据？试编程加以说明。

第19章　Spring MVC 其他

本章导读

数据绑定是将用户输入绑定到领域模型的一种特性。有了数据绑定，类型总是为 String 的 HTTP 请求参数，可用于填充不同类型的对象属性。为了高效地使用数据绑定，还需要 Spring 的表单标签库，这些标签都可以访问到 ModelMap 中的内容。本章主要内容有：（1）Spring MVC 的表单标签库；（2）表单验证之 Spring 验证和 JSR 303 验证；（3）Spring MVC 的文件上传和下载；（4）Spring MVC 的国际化。

19.1　Spring MVC 的表单标签库

在使用 Spring MVC 表单标签之前，需要先在 JSP 中声明使用的标签，具体做法是在 JSP 页面的开头处声明 taglib 指令。

```
<%@ taglib uri="http://www.springframework.org/tags/form" prefix="form" %>
```

1. form 标签

Spring 的 form 标签主要有两个作用：一是它会自动绑定 Model 中的一个属性值到当前 form 对应的实体对象上，默认是 command 属性，这样可以在 form 中方便地使用该对象的属性；二是它支持在提交表单的时候使用除 GET 和 POST 之外的其他方法进行提交，包括 DELETE 和 PUT 等。

Spring MVC 指定 form 标签默认绑定的是 Model 的 command 属性值，那么当 form 对象对应的属性名称不是 command 时，又如何呢？实际上，Spring 提供了 commandName 属性或者 modelAttribute 属性，通过该属性值来指定将使用 Model 中的哪个属性作为 form 标签需要绑定的 command 对象。

2. input 标签

Spring MVC 的 input 标签会被渲染为一个 type 为 text 的普通 HTML input 标签。使用 Spring MVC 的 input 标签的唯一作用就是它能绑定表单数据。当表单标签不需要绑定数据时，应该使用普通的 HTML 标签。

【示例】form 和 input 标签的使用。

在 Eclipse 中创建 Maven Project“FormSimpleDemo”。FormSimpleDemo 工程目录层次及文件如图 19-1 所示。创建的新文件有：bookForm.jsp、BookController.java、Book.java、success.jsp、bookForm2.jsp 和 success2.jsp。（代码详见本书源码包 ch19 目录中的 FormSimpleDemo 工程。）

```
∨ FormSimpleDemo
  ∨ src/main/java
    ∨ com.mialab.FormSimpleDemo.controller
      > BookController.java
    ∨ com.mialab.FormSimpleDemo.domain
      > Book.java
  > src/test/java
  > JRE System Library [JavaSE-1.8]
  > Maven Dependencies
  ∨ src/main/resources
      log4j.properties
  ∨ src
    ∨ main
      ∨ Webapp
        ∨ Web-INF
          ∨ config
              springmvc-config.xml
          ∨ jsp
              bookForm.jsp
              bookForm2.jsp
              success.jsp
              success2.jsp
            Web.xml
```

图 19-1　FormSimpleDemo 工程目录层次及文件

BookController.java 的主要代码如下：

```
@Controller
public class BookController {
 @RequestMapping(value = "/bookForm", method = RequestMethod.GET)
 public String loginForm(Model model) {
     Book bookDefault = new Book(9999, "Web 应用开发","2016-8-22");
     logger.info(bookDefault);
     //在 model 中添加属性 command，值是 bookDefault（Book 对象）
     model.addAttribute("command", bookDefault);
     return "bookForm";
 }
 @RequestMapping(value = "/bookEntry", method = RequestMethod.POST)
 public String bookEntry(Book book,Model model) {
     model.addAttribute("command", book);
     return "success";
 }
 @RequestMapping(value = "/bookForm2", method = RequestMethod.GET)
 public String loginForm2(Model model) {
     Book bookDefault = new Book(8888, "IT 项目管理","2018-6-12");
     model.addAttribute("bookRegister", bookDefault);
     return "bookForm2";
 }
 @RequestMapping(value = "/bookEntry2", method = RequestMethod.POST)
 public String bookEntry2(Book book,Model model) {
     model.addAttribute("bookRegister", book);
     return "success2";
 }
}
```

bookForm.jsp 的<body>部分的主要代码如下：

```
<form:form action="bookEntry" method="post">
    <table>
        <tr>
```

```
                <td>图书编号:</td>
                <td><form:input path="bookId" /></td>
            </tr>
            <tr>
                <td>图书名称:</td>
                <td><form:input path="bookName" /></td>
            </tr>
            <tr>
                <td>登记日期:</td>
                <td><form:input path="bookEntryDate" /></td>
            </tr>
            <tr>
                <td><input id="submit" type="submit" value="提交"></td>
            </tr>
        </table>
    </form:form>
```

而 bookForm2.jsp 的<body>部分的主要代码如下：

```
    <form:form modelAttribute="bookRegister" action="bookEntry" method="post">
        ...
    </form:form>
```

测试的结果如图 19-2 和图 19-3 所示。

图19-2　请求bookForm

图19-3　请求bookForm2

3．hidden 标签

hidden 标签会被渲染为一个 type 为 hidden 的普通 HTML input 标签。其用法与 input 标签一样，也能绑定表单数据，只是它生成的是一个隐藏域。

4．checkbox 标签

checkbox 标签会被渲染为一个 type 为 checkbox 的普通 HTML input 标签。checkbox 标签也是支持绑定数据的。我们知道 checkbox 就是一个复选框，有选中和不选中两种状态，那么在使用 checkbox 标签的时候是如何来设定它的状态的呢？checkbox 标签的选中与否状态是根据其绑定的值来判断的。

（1）绑定 boolean 数据。

当 checkbox 绑定的是一个 boolean 数据的时候，checkbox 的状态和该 boolean 数据的状态是一样的，即 true 对应选中，false 对应不选中。

```
    <form:form action="askAction" method="post" commandName="user">
        <table>
            <tr><td>Male:</td><td><form:checkbox path="male"/></td></tr>
            <tr><td colspan="2"><input type="submit" value="提交"/></td></tr>
        </table>
    </form:form>
```

根据以上代码，假设在渲染该视图之前向 ModelMap 中添加了一个 user 属性，并且该 user

对象有一个类型为 boolean 的属性 male，那么此时如果 male 属性为 true，则 Male 一栏的复选框将会被选中。

（2）绑定列表数据。

这里的列表数据包括数组、List 和 Set。下面将以 List 为例讲解 checkbox 是如何根据绑定的列表数据来设定选中状态的。现在假设有一个类 User，其有一个类型为 List 的属性 roles，如下所示：

```
public class User {
    private List<String> roles;
    public List<String> getRoles() {
       return roles;
    }
    public void setRoles(List<String> roles) {
       this.roles = roles;
    }
}
```

那么当需要展现该 User 是否拥有某一个 Role 的时候，可以使用 checkbox 标签来绑定 roles 数据进行展现。当 checkbox 标签的 value 在绑定的列表数据中存在的时候，该 checkbox 将为选中状态。来看下面一段代码：

```
<form:form action=" askAction " method="post" commandName="user">
    <table>
        <tr>
            <td>Roles:</td>
            <td>
               <form:checkbox path="roles" value="role1"/>Role1<br/>
               <form:checkbox path="roles" value="role2"/>Role2<br/>
               <form:checkbox path="roles" value="role3"/>Role3
            </td>
        </tr>
    </table>
</form:form>
```

就上面的代码而言，当 User 拥有 role1 的时候，对应的<form:checkbox path="roles" value="role1"/>为选中状态，也就是说，roles 列表中包含 role1 的时候，该 checkbox 为选中状态。

（3）绑定一个 Object 数据。

checkbox 还支持绑定数据类型为 Object 的数据，这种情况下，Spring 会以所绑定对象数据的 toString 结果与当前 checkbox 的 value 进行比较，如果能够进行匹配，则该 checkbox 将为选中状态。来看下面的例子，有一个 User 类的代码如下：

```
public class User {
    private Blog blog;
    public Blog getBlog() {
       return blog;
    }
    public void setBlog(Blog blog) {
       this.blog = blog;
    }
}
```

Blog 类的代码如下：

```
public class Blog {
```

```
        public String toString() {
            return "HelloWorld";
        }
    }
```

可以看到Blog类的toString方法已经被写为"HelloWorld"。此时，假设向ModelMap中放了一个user对象，而且给该user对象设定了blog属性，那么当使用该ModelMap对象渲染如下视图代码时，checkbox标签的选中状态是怎样的呢？根据前面描述的内容，当checkbox标签绑定的是一个Object对象的时候，会以该Object对象的toString和checkbox的value值进行比较，如果匹配，则当前checkbox为选中状态。这里的checkbox将为选中状态。

```
<form:form action=" askAction " method="post" commandName="user">
    <table>
        <tr>
            <td>HelloWorld:</td>
            <td>
                <form:checkbox path="blog" value="HelloWorld"/>
            </td>
        </tr>
        <tr>
            <td colspan="2"><input type="submit" value="提交"/></td>
        </tr>
    </table>
</form:form>
```

5. checkboxes 标签

相对于一个checkbox标签只能生成一个对应的复选框而言，一个checkboxes标签将根据其绑定的数据生成N个复选框。checkboxes绑定的数据可以是数组、集合和Map。在使用checkboxes时，有两个属性是必须指定的，一个是path，另一个是items。items表示当前要用来展现的项，而path所绑定的表单对象的属性表示当前表单对象拥有的项，即在items所展现的所有项中表单对象拥有的项会被设定为选中状态。

（1）使用List。先来看以下一段代码：

```
<form:form action=" askAction " method="post" commandName="user">
    <table>
        <tr>
           <td>Roles:</td>
            <td>
                <form:checkboxes path="roles" items="${roleList}"/>
            </td>
        </tr>
        <tr>
            <td colspan="2"><input type="submit" value="提交"/></td>
        </tr>
    </table>
</form:form>
```

前面的JSP视图对应着如下处理器方法：

```
@RequestMapping(value="form", method=RequestMethod.GET)
public String askAction (Map<String, Object> map) {
   User user = new User();
   List<String> roles = new ArrayList<String>();
   roles.add("role1");
```

```
        roles.add("role3");
        user.setRoles(roles);
        List<String> roleList = new ArrayList<String>();
        roleList.add("role1");
        roleList.add("role2");
        roleList.add("role3");
        map.put("user", user);
        map.put("roleList", roleList);
        return "someForm";
    }
```

从以上代码中可以看到放在 ModelMap 中的 roleList 对象有 3 个元素，分别是 role1、role2 和 role3，而表单对象 User 的 roles 属性只拥有两个元素，分别是 role1 和 role3，所以当访问该处理器方法返回如上所示的视图页面时，要展现的复选框项是 roleList，也就是 role1、role2 和 role3，而表单对象只拥有 role1 和 role3，所以在页面进行渲染的时候会展示 3 个复选框项，但只有 role1 和 role3 会被设定为选中状态，如图 19-4 所示。

Roles: ☑role1 ☐role2 ☑role3
提交

图 19-4　使用 List

（2）使用 Map。上面介绍的这种情况是使用 List 作为展现复选框项的数据源，这种情况中，其呈现出来的标签 Label 和它的值是一样的。使用 Array 和 Set 作为数据源也会是这种情况。如果要让 checkboxes 呈现出来的 Label 和实际上传入的 value 不同应该怎么做呢？此时可以使用 Map 作为数据源。使用 Map 作为 checkboxes 的 items 属性的数据源时，Key 将作为真正的复选框的 value，而 Map 的 value 将作为 Label 进行展示。当使用 Map 作为 checkboxes 的 items 属性的数据源时，绑定的表单对象属性的类型可以是 Array、集合和 Map，这种情况会通过判断 items Map 中是否含有对应的 key 来决定当前的复选框是否处于选中状态。

来看以下处理器方法及其对应的视图代码。

处理器方法：

```
    @RequestMapping(value="form", method=RequestMethod.GET)
    public String askAction (Map<String, Object> map) {
        User user = new User();
        List<String> roles = new ArrayList<String>();
        roles.add("role1");
        roles.add("role3");
        user.setRoles(roles);
        Map<String, String> roleMap = new HashMap<String, String>();
        roleMap.put("role1", "角色 1");
        roleMap.put("role2", "角色 2");
        roleMap.put("role3", "角色 3");
        map.put("user", user);
        map.put("roleMap", roleMap);
        return "someForm";
    }
```

其对应的 JSP 视图代码如下：

```
    <form:form action=" askAction " method="post" commandName="user">
        <table>
            <tr>
```

```
            <td>Roles:</td>
            <td>
                <form:checkboxes path="roles" items="${roleMap}"/>
            </td>
        </tr>
        <tr>
            <td colspan="2"><input type="submit" value="提交"/></td>
        </tr>
    </table>
</form:form>
```

此时会呈现出 3 个复选框，而 checkboxes 绑定的表单对象 user 的 roles 属性是一个集合对象，其包含的两个元素都能在 checkboxes 的 items 数据源中找到对应的 Key，所以以这两个元素为 value 的 checkbox 将处于选中状态。其效果如图 19-5 所示。

Roles: ☑角色3 ☐角色2 ☑角色1
提交

图 19-5 使用 Map

当使用 Array 或者集合作为数据源，且里面的元素都是 POJO 时，还可以使用 checkboxes 标签的 itemLabel 和 itemValue 属性来表示。

6．radiobutton 标签

radiobutton 标签会被渲染为一个 type 为 radio 的普通 HTML input 标签。radiobutton 标签也是可以绑定数据的。以下是一个 radiobutton 的简单应用示例：

```
<form:form action=" askAction " method="post" commandName="user">
    <table>
        <tr>
            <td>性别:</td>
            <td>
                <form:radiobutton path="sex" value="1"/>男
                <form:radiobutton path="sex" value="0"/>女
            </td>
        </tr>
        <tr>
            <td colspan="2"><input type="submit" value="提交"/></td>
        </tr>
    </table>
</form:form>
```

在上面的代码中，radiobutton 标签都是绑定了表单对象 user 的 sex 属性，当 sex 为 1 时就代表性别为男，性别为男的那一行就会被选中，当 sex 为 0 时就代表性别为女，性别为女的那一行就会被选中。

7．radiobuttons 标签

radiobuttons 标签和 radiobutton 标签的区别如同 checkboxes 标签对 checkbox 标签的区别。使用 radiobuttons 标签的时候将生成多个单选按钮。使用 radiobuttons 时有两个属性也是必须指定的：一个是 path 属性，表示绑定的表单对象对应的属性；另一个是 items 属性，表示用于生成单选按钮的数据源。与 checkboxes 一样，radiobuttons 的 items 属性和 path 属性都可以是 Array、集合或者 Map。

现在假设 user 在篮球、足球、乒乓球、羽毛球和排球这 5 种运动中选择一种作为自己最喜欢的球类运动。处理器方法和返回的对应的视图代码如下。

处理器方法：

```
@RequestMapping(value="form", method=RequestMethod.GET)
public String askAction (Map<String, Object> map) {
   User user = new User();
   user.setFavoriteBall(4);//设置最喜爱的球类运动是羽毛球
   Map<Integer, String> ballMap = new HashMap<Integer, String>();
   ballMap.put(1, "篮球");
   ballMap.put(2, "足球");
   ballMap.put(3, "乒乓球");
   ballMap.put(4, "羽毛球");
   ballMap.put(5, "排球");
   map.put("user", user);
   map.put("ballMap", ballMap);
   return "formTag/form";
}
```

视图代码：

```
<form:form action=" askAction " method="post" commandName="user">
   <table>
      <tr><td>最喜欢的球类:</td>
         <td>
            <form:radiobuttons path="favoriteBall" items="${ballMap}"
                        delimiter=" "/>
         </td>
      </tr>
      <tr> <td colspan="2"><input type="submit" value="提交"/></td></tr>
   </table>
</form:form>
```

在上述代码中，可以看到使用了 radiobuttons 的 delimiter 属性，该属性表示进行展示的 radiobutton 之间的分隔符。这里使用的是一个空格。radiobuttons 标签的数据绑定结果如图 19-6 所示。

最喜欢的球类：◎篮球 ◎足球 ◎乒乓球 ◉羽毛球 ◎排球
[提交]

图 19-6 radiobuttons 标签的数据绑定

8. password 标签

password 标签将会被渲染为一个 type 为 password 的普通 HTML input 标签。

9. select 标签

select 标签将会被渲染为一个普通的 HTML select 标签。这里以前面的 user 最喜欢的球类运动来做示例，有如下处理器方法和对应的视图页面。

处理器方法：

```
@RequestMapping(value="form", method=RequestMethod.GET)
public String askAction (Map<String, Object> map) {
   User user = new User();
   user.setFavoriteBall(4);//设置最喜爱的球类运动是羽毛球
   Map<Integer, String> ballMap = new HashMap<Integer, String>();
```

```
            ballMap.put(1, "篮球");
            ballMap.put(2, "足球");
            ballMap.put(3, "乒乓球");
            ballMap.put(4, "羽毛球");
            ballMap.put(5, "排球");
            map.put("user", user);
            map.put("ballMap", ballMap);
            return "someForm";
        }
```

视图页面代码：

```
<form:form action="formTag/form.do" method="post" commandName="user">
    <table>
        <tr>
            <td>最喜欢的运动:</td>
          <td><form:select path="favoriteBall" items="${ballMap}"/></td>
        </tr>
        <tr>
            <td colspan="2"><input type="submit" value="提交"/></td>
        </tr>
    </table>
</form:form>
```

此时会渲染出如图 19-7 所示的结果。

图 19-7　select 标签的 items 属性

从此示例可以看出，通过 items 属性给 select 标签指定了一个数据源，并且绑定了表单对象 user 的 favoriteBall 属性。items 属性是用于指定当前 select 的所有可选项的，但是它对于 select 标签而言不是必需的，因为还可以手动在 select 标签中间加上 option 标签来指定 select 可选的 option。

select 标签支持的 items 属性的数据类型可以是 Array、Collection 和 Map，当数据类型为 Array 或 Collection 且其中的元素为一个 POJO 时，可以通过属性 itemLabel 和 itemValue 来指定将用于呈现的 option Label 和 Value，其他情况下，Array 和 Collection 数据源中的元素将既作为可选项 option 的 Value 又作为它的 Label。

当 items 的数据类型为 Map 时，Map 的 key 将作为可选项 option 的 Value，而 Map 的 Value 将作为 option 的 Label 标签。

10．option 标签

option 标签会被渲染为一个普通的 HTML option 标签。当一个 Spring MVC select 标签没有通过 items 属性指定自己的数据源的时候，可以在 select 标签中通过普通 HTML option 标签或者 Spring MVC option 标签来指定可以选择的项。

Spring MVC option 标签和普通 HTML option 标签的区别就在于普通 HTML option 标签不具备数据绑定功能，而 SpringMVC option 标签具有数据绑定功能，它能把当前绑定的表单对象的属性对应的值对应的 option 置为选中状态。

11．options 标签

使用 options 标签的时候需要指定其 items 属性，它会根据其 items 属性生成一系列的普通 HTML option 标签。这里的 items 属性的可取数据类型及其对应的渲染规则和 select 的 items 属性是一样的。

```
<form:form action="formTag/form.do" method="post" commandName="user">
    <table>
        <tr>
            <td>最喜欢的运动:</td>
            <td>
                <form:select path="favoriteBall">
                    <option>请选择</option>
                    <form:options items="${ballMap}"/>
                </form:select>
            </td>
        </tr>
       <tr>
            <td  colspan="2"><input  type="submit" 
value="提交"/></td>
        </tr>
    </table>
</form:form>
```

此段代码将渲染出如图 19-8 所示的结果。

最喜欢的运动: 羽毛球
提交
请选择
篮球
足球
乒乓球
羽毛球
排球

图 19-8 使用 options 标签

12．textarea 标签

Spring MVC 的 textarea 标签将被渲染为普通 HTML textarea 标签。例如：

```
<form:textarea path="introduction" cols="20" rows="10"/>
```

13．errors 标签

Spring MVC 的 errors 标签是对应于 Spring MVC 的 Errors 对象的。其用于展现 Errors 对象中包含的错误信息。我们利用 errors 标签来展现 Errors 的时候是通过 errors 标签的 path 属性来绑定一个错误信息的。可以通过 path 属性来展现两种类型的错误信息。

（1）所有的错误信息，此时 path 的值应该置为“*”。

（2）当前对象的某一个属性的错误信息，此时 path 值应为所需展现的属性的名称。

例如：

```
<td><form:input path="bookName" /></td>
<td><font  color="red"><form:errors 
path="bookName" /></font></td>
```

【示例】使用 Spring MVC 的表单标签库。

在 Eclipse 中创建 Maven Project “FormTagDemo”。FormTagDemo 工程目录层次及文件如图 19-9 所示。创建的新文件有：UserController.java、User.java、user.jsp 和 success.jsp。（代码详见本书源码包 ch19 目录中的 FormTagDemo 工程。）

部署发布 FormTagDemo 应用。在浏览器地址栏中输入以下 URL：

```
http://localhost:8080/FormTagDemo/us
er
```

测试结果如图 19-10 所示。（代码详见本书源码包 ch19 目录中的 FormTagDemo 工程。）

图 19-9 FormTagDemo 工程目录层次及文件

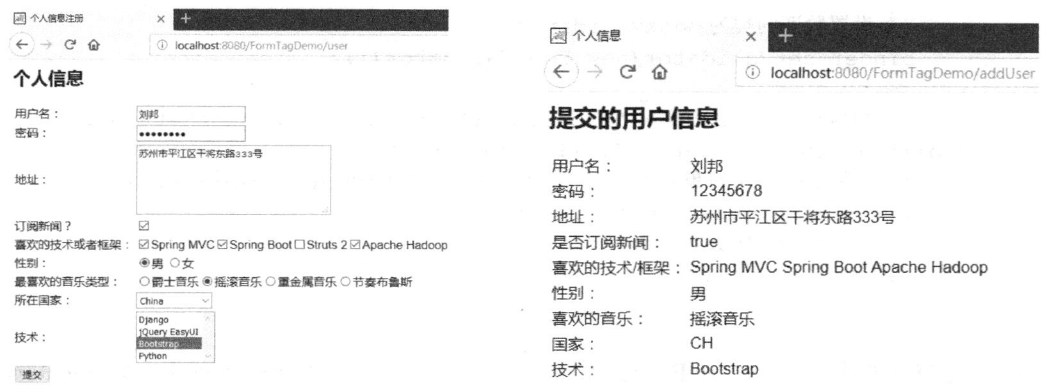

图 19-10　个人信息注册

【示例】使用 errors 标签。

在 Eclipse 中创建 Maven Project“FormErrorsDemo”。FormErrorsDemo 工程目录层次及文件如图 19-11 所示。创建的新文件有：BookController.java、BookValidator.java、Book.java 和 bookForm.jsp。（代码详见本书源码包 ch19 目录中的 FormErrorsDemo 工程。）

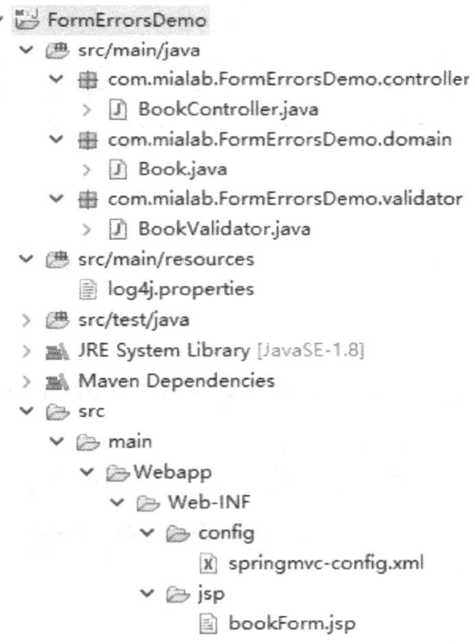

图 19-11　FormErrorsDemo 工程目录层次及文件

BookController.java 的主要代码如下：

```
@Controller
public class BookController {
 @RequestMapping(value = "/bookEntry", method = RequestMethod.GET)
 public String bookRegister(Model model) {
     Book book = new Book(9999,"","");
     //在 model 中添加属性 book，值是 book 对象
     model.addAttribute("book", book);
     return "bookForm";
 }
 @InitBinder
 public void initBinder(DataBinder binder) {
```

```
        // 设置验证的类为 BookValidator
        binder.setValidator(new BookValidator());
    }
    @RequestMapping(value = "/bookEntry", method = RequestMethod.POST)
    public String bookEntry(@Validated Book book, Errors errors) {
        // 如果 Errors 对象有 Field 错误，则重新跳回注册页面，否则正常提交
        if (errors.hasFieldErrors())
            return "bookForm";
        return "success";
    }
}
```

在上述控制器类中通过 DataBinder 对象给该类设定了一个用于验证的 BookValidator，这样当请求该控制器的时候，BookValidator 将生效。

BookValidator.java 的代码如下。

```
public class BookValidator implements Validator {
    @Override
    public boolean supports(Class<?> clazz) {
        return Book.class.equals(clazz);
    }
    @Override
    public void validate(Object object, Errors errors) {
        // 验证 bookId、bookName 和 bookEntryDate 是否为 null
        ValidationUtils.rejectIfEmpty(errors, "bookId", null, "图书编号不能为空");
        ValidationUtils.rejectIfEmpty(errors, "bookName", null, "图书名称不能为空");
        ValidationUtils.rejectIfEmpty(errors, "bookEntryDate", null, "登记日期不能为空");
    }
}
```

部署发布 FormErrorsDemo 应用。在浏览器地址栏中输入以下 URL：

http://localhost:8080/FormErrors Demo/bookEntry

测试结果如图 19-12 所示。（代码详见本书源码包 ch19 目录中的 FormErrorsDemo 工程。）

图 19-12　使用 Spring MVC 的 errors 标签

19.2　表单验证

Spring MVC 提供了强大的数据校验功能，有两种方法可以验证输入：一种是利用 Spring 自带的 Validation 校验框架；另一种是利用 JSR 303（Java 验证规范）实现校验功能。

Converter 和 Formatter 作用于 Field 级，它们将 String 转换或格式化成另一种 Java 类型，如 java.util.Date。验证器则作用于 Object 级，它决定了某一个对象中的所有 Field 是否均是有效的，以及是否遵循某些规则。

如果一个应用程序中既使用了 Formatter，又有 Validator（验证器），那么，它们的事件顺序是这样的：在调用 Controller 期间，将会有一个或多个 Formatter 试图将输入字符串转换成 Domain 对象中的 Field 值。一旦格式化成功，验证器就会介入。

19.2.1 Spring 验证

为了创建 Spring 验证器，需要实现 org.springframework.validation.Validator 接口。Validator 接口代码如下所示，其有 supports 和 validate 两个方法。

```
package org.springframework.validation;
public interface Validator {
 boolean supports(Class<?> clazz);
 void validate(Object target, Errors errors);
}
```

如果验证器可以处理指定的 Class，则 supports 将返回 true。validate 方法会验证目标对象，并将验证错误填入 Errors 对象。

Errors 对象是 org.springframework.validation. Errors 接口的一个实例。Errors 对象中的错误消息，可以利用表单标签库的 Errors 标签显示在 HTML 页面中。错误消息可以通过 Spring 支持的国际化特性进行本地化。

【示例】Spring 验证。

在 Eclipse 中创建 Maven Project“ValidatorDemo”。ValidatorDemo 工程目录层次及文件与 FormErrorsDemo 工程类似。创建的新文件有：BookController.java、BookValidator.java、Book.java、bookForm.jsp 和 success.jsp。（代码详见本书源码包 ch19 目录中的 ValidatorDemo 工程。）

BookController.java 的主要代码如下：

```
@Controller
public class BookController {
 private static final Log logger = LogFactory.getLog(BookController.class);
 // 注入 BookValidator 对象
 @Autowired
 @Qualifier("bookValidator")
 private BookValidator bookValidator;

 @RequestMapping(value = "/{formName}")
 public String bookForm(@PathVariable String formName, Model model) {
     Book book = new Book();
     model.addAttribute("book", book);
     return formName;
 }
 @RequestMapping(value = "/bookEntry", method = RequestMethod.POST)
 public  String  bookEntry(@ModelAttribute  Book  book,BindingResult
bindingResult, Model model) {
     logger.info(book);
     // 调用 bookValidator 的验证方法
     bookValidator.validate(book, bindingResult);
     // 如果验证不通过，则跳转到 bookForm 视图
```

```
        if (bindingResult.hasErrors()) {
            FieldError fieldError = bindingResult.getFieldError();
            logger.info("Code:" + fieldError.getCode() + ",field" + fieldError.
getField());
            return "bookForm";
        }
        model.addAttribute("book", book);
        return "success";
    }
}
```

BookValidator.java 的主要代码如下：

```
@Repository("bookValidator")
public class BookValidator implements Validator {
    @Override
    public boolean supports(Class<?> clazz) {
        return Book.class.isAssignableFrom(clazz);
    }
    @Override
    public void validate(Object target, Errors errors) {
        // 验证 bookId、bookName 和 bookPrice 是否为 null
        ValidationUtils.rejectIfEmpty(errors, "bookId", null, "图书编号不能为
空");
        ValidationUtils.rejectIfEmpty(errors, "bookName", null, "图书名称不能
为空");
        ValidationUtils.rejectIfEmpty(errors, "bookPrice", null, "图书价格不
能为空");
        Book book = (Book) target;
        Float bookPrice = book.getBookPrice();
        if (bookPrice != null && bookPrice < 0)
            errors.rejectValue("bookPrice", null, "图书价格不能小于 0！");
    }
}
```

Class.isAssignableFrom()用来判断一个类Class1和另一个类Class2是否相同或为另一个类的子类或接口。其格式如下：

```
Class1.isAssignableFrom(Class2)
```

bookForm.jsp 的<body>部分的主要代码如下：

```
<form:form modelAttribute="book" action="bookEntry" method="post">
    <table>
        <tr>
            <td>图书编号:</td>
            <td><form:input path="bookId" /></td>
            <td><font color="red"><form:errors path="bookId" /></font>
</td>
        </tr>
        …
        <tr>
            <td><input id="submit" type="submit" value="提交"></td>
        </tr>
    </table>
</form:form>
```

springmvc-config.xml 文件中必须添加以下配置代码。

```
    <context:component-scan
base-package="com.mialab.ValidatorDemo.controller" />
    <context:component-scan
base-package="com.mialab.ValidatorDemo.validator" />
```

部署发布 ValidatorDemo 应用。在浏览器地址栏中输入以下 URL：

```
    http://localhost:8080/FormErrors Demo/bookEntry
```

测试结果如图 19-13 所示。（代码详见本书源码包 ch19 目录中的 ValidatorDemo 工程。）

图 19-13　表单数据校验之 Spring 验证

19.2.2　JSR 303 验证

JSR 303 是 Java 为 Bean 数据合法性校验提供的标准框架，已经包含在 Java EE 6.0 中。JSR 303 通过在 Bean 属性中标注类似 @NotNull 、@Max 等标准的注解指定校验规则，并通过标准的验证接口对 Bean 进行验证。

可以在 http://jcp.org/en/jsr/detail?id=303 下载 JSR 303 规范。

Hibernate Validator 是 JSR 303 的一个参考实现。Hibernate Validator 提供了 JSR 303 规范中所有内置 constraint 的实现，除此之外，还有一些附加的 constraint。如果想了解更多有关 Hibernate Validator 的知识，可以查看 http://hibernate.org/validator/。

【示例】JSR 303 验证。

在 Eclipse 中创建 Maven Project“JSR303Demo”。JSR303Demo 工程目录层次及文件与 ValidatorDemo 工程类似，但不需要有 Validator。创建的新文件有：BookController.java、Book.java、bookForm.jsp 和 success.jsp。（代码详见本书源码包 ch19 目录中的 JSR303Demo 工程。）

BookController.java 的主要代码如下：

```
    @Controller
    public class BookController {
     @RequestMapping(value = "/{formName}")
     public String bookForm(@PathVariable String formName, Model model) {
         Book book = new Book();
         model.addAttribute("book", book);
         return formName;
     }

     // 数据校验使用@Valid，后面有 Errors 对象，以保存校验信息
     @RequestMapping(value = "/bookEntry", method = RequestMethod.POST)
     public String bookEntry(@Valid @ModelAttribute Book book, Errors errors,
Model model) {
         if (errors.hasErrors()) {
             return "bookForm";
         }
```

```
        model.addAttribute("book", book);
        return "success";
    }
}
```

Book.java 的主要代码如下：

```
public class Book {
  @NotEmpty(message = "图书编号不能为空")
  private String bookId;
  @NotEmpty(message = "图书名称不能为空")
  private String bookName;
  @NotNull(message = "图书价格不能为空")
  @Range(min =  (long) 0.0, message = "图书价格不能为负数")
  private Float bookPrice;
  public String getBookId() {
      return bookId;
  }
  …
}
```

Book 类使用 Hibernate Validator 的注解对前台提交的数据进行了验证。

pom.xml 中必须添加 hibernate-validator 依赖。

```
<dependency>
  <groupId>org.hibernate</groupId>
  <artifactId>hibernate-validator</artifactId>
  <version>6.0.9.Final</version>
</dependency>
<dependency>
  <groupId>javax.validation</groupId>
  <artifactId>validation-api</artifactId>
  <version>2.0.1.Final</version>
</dependency>
```

测试结果与 ValidatorDemo 工程类似。（代码详见本书源码包 ch19 目录中的 JSR303 Demo 工程。）

19.3 Spring MVC 的文件上传和下载

19.3.1 文件上传

负责上传文件的表单须满足 3 个条件：form 的 method 属性设置为 post；form 的 enctype 属性设置为 multipart/form-data；提供<input type="file" name="filename" >的文件上传输入框。

文件上传表单的示例代码如下：

```
<form action="uploadFile" enctype="multipart/form-data" method="post">
 Select File<input type="file" name="filename" multiple="mulpiple">
 <input type="submit" value="Upload">
</form>
```

【示例】Spring MVC 的文件上传。

在 Eclipse 中创建 Maven Project“FileUploadDemo”。FileUploadDemo 工程目录层次及文件如图 19-14 所示。创建的新文件有：FileUploadController.java、User.java 和 uploadForm.jsp、info.jsp、error.jsp。（代码详见本书源码包 ch19 目录中的 FileUploadDemo 工程。）

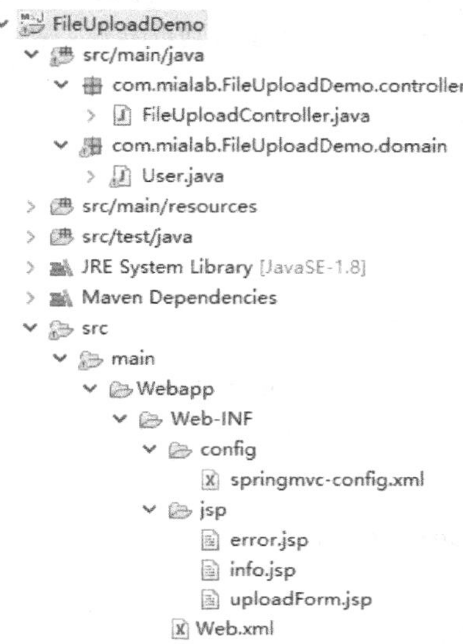

图 19-14 FileUploadDemo 工程目录层次及文件

FileUploadController.java 的主要代码如下：

```
@Controller
public class FileUploadController {
 @RequestMapping(value = "/{formName}")
 public String loginForm(@PathVariable String formName) {
     return formName;
 }
 @RequestMapping(value = "/uploadFile")
 public String uploadFile(HttpServletRequest request, @ModelAttribute User
user, Model model) throws Exception {
     List<MultipartFile> imageList = user.getImageList();
     // 判断所上传文件是否存在
     if (!imageList.isEmpty() && imageList.size() > 0) {
         // 循环输出上传的文件
         for (MultipartFile file : imageList) {
             // 获取上传文件的原始名称
             String originalFilename = file.getOriginalFilename();
             // 设置上传文件的保存地址目录
             String dirPath =
request.getServletContext().getRealPath("/upload/");
             File filePath = new File(dirPath);
             // 如果保存文件的地址不存在，则先创建目录
             if (!filePath.exists()) {
                 filePath.mkdirs();
             }
             try {
                 // 使用 MultipartFile 接口的方法将文件上传到指定位置
                 file.transferTo(new File(dirPath + originalFilename));
```

```
                } catch (Exception e) {
                    e.printStackTrace();
                    return "error";
                }
            }
            model.addAttribute("user", user); //将用户添加到 model 中
            return "info"; // 跳转到成功页面
        } else {
            return "error";
        }
    }
}
```

User.java 的主要代码如下：

```
public class User implements Serializable {
 private String userName;
 private List<MultipartFile> imageList;
 public User() {
     super();
 }
 public String getUserName() {
     return userName;
 }
 public void setUserName(String userName) {
     this.userName = userName;
 }
 public List<MultipartFile> getImageList() {
     return imageList;
 }
 public void setImageList(List<MultipartFile> imageList) {
     this.imageList = imageList;
 }
}
```

uploadForm.jsp 的<body>部分的主要代码如下：

```
<form action="uploadFile" enctype="multipart/form-data" method="post">
 <table>
     <tr>
         <td>用户名:</td>
         <td><input type="text" name="userName"></td>
     </tr>
     <tr>
         <td>请上传图片:</td>
         <td><input type="file" name="imageList" multiple="mulpiple"></td>
     </tr>
     <tr>
         <td><input type="submit" value="上传"></td>
     </tr>
 </table>
</form>
```

info.jsp 的<body>部分的主要代码如下。这里使用了 JSTL 的核心标签库。

```
<%@ page language="java" contentType="text/html; charset=UTF-8"
 pageEncoding="UTF-8"%>
<%@ taglib prefix="c" uri="http://java.sun.com/jsp/jstl/core"%>
```

```
......
<body>
 <h3>文件列表，单击文件名可下载</h3>
 <c:forEach items="${requestScope.user.imageList}" var="mFile"
     varStatus="filevst">
     文件序号：${filevst.index+1}，文件名称：
     <a href="download?filename=${mFile.originalFilename}">
         ${mFile.originalFilename}</a>
     <br>
 </c:forEach>
 <p>您的用户名是：${requestScope.user.userName}</p>
</body>
</html>
```

pom.xml 中必须添加 Commons FileUpload 依赖和 JSTL 依赖。

```
<!-- 文件上传下载 Apache Commons FileUpload 依赖 -->
<dependency>
 <groupId>commons-fileupload</groupId>
 <artifactId>commons-fileupload</artifactId>
 <version>1.3.3</version>
</dependency>
<dependency>
 <groupId>commons-io</groupId>
 <artifactId>commons-io</artifactId>
 <version>2.6</version>
</dependency>
<!-- JSTL 依赖 -->
<dependency>
 <groupId>javax.servlet</groupId>
 <artifactId>jstl</artifactId>
 <version>1.2</version>
</dependency>
```

测试结果如图 19-15 所示。（代码详见本书源码包 ch19 目录中的 FileUploadDemo 工程。）

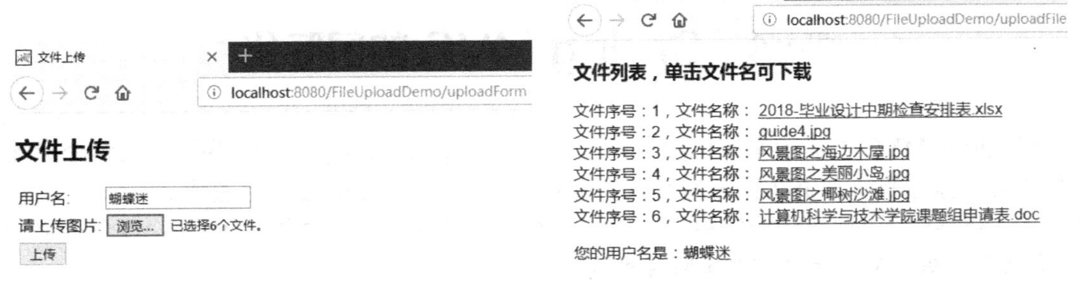

图 19-15　文件上传与上传文件列表

19.3.2　文件下载

在 Spring MVC 环境中，文件下载可分为以下两个步骤。

（1）在客户端页面中使用一个文件下载的超链接，该链接的 href 属性要指定后台文件下载的方法及文件名。可参考 FileUploadDemo 工程的 info.jsp 的代码。

（2）在后台 Controller 类中，使用 Spring MVC 提供的文件下载方法进行下载。Spring MVC

提供了一个 ResponseEntity 类型，使用它可以很方便地定义返回的 HttpHeaders 和 HttpStatus，通过对这两个对象的设置，即可完成下载文件时所需信息的配置。

在 FileUploadDemo 工程的 FileUploadController.java 中添加 download 请求方法，即可实现文件下载的功能。（代码详见本书源码包 ch19 目录中的 FileUploadDemo 工程。）

```
@Controller
public class FileUploadController {
 @RequestMapping(value = "/download")
 public ResponseEntity<byte[]> download(HttpServletRequest request,
 @RequestParam("filename") String filename,Model model)throws Exception {
     // 下载文件路径
     String path = request.getServletContext().getRealPath("/upload/");
     File file = new File(path + File.separator + filename);
     HttpHeaders headers = new HttpHeaders();
     // 下载显示的文件名，解决中文名称乱码问题
     String downloadFielName = new String(filename.getBytes("UTF-8"),
         "iso-8859-1");
     // 通知浏览器以 attachment（下载方式）打开图片
     headers.setContentDispositionFormData("attachment",
downloadFielName);
     // application/octet-stream : 二进制流数据（最常见的文件下载）
     headers.setContentType(MediaType.APPLICATION_OCTET_STREAM);
     // 201 HttpStatus.CREATED
     return                                                    new
ResponseEntity<byte[]>(FileUtils.readFileToByteArray(file), headers,
         HttpStatus.CREATED);
 }
}
```

download 处理方法接收到页面传递的文件名 filename 后，使用 Apache Commons FileUpload 组件的 FileUtils（工具类）读取项目的 upload 文件夹中的文件，将其构建成 ResponseEntity 对象并返回客户端下载。

单击图 19-15 中的上传文件列表超链接，显示文件正在下载，再保存文件即可。

19.4 Spring MVC 的国际化

在 Spring MVC 中显示本地化消息通常使用 Spring 的 message 标签。为了使用这个标签，要在使用该标签的所有 JSP 页面最前面使用 taglib 指令导入 Spring 的标签库。

```
<%@ taglib prefix="spring" uri="http://www.springframework.org/tags"%>
```

【示例】基于浏览器请求的国际化实现。

在 Eclipse 中创建 Maven Project“InternationalDemo”。InternationalDemo 工程目录层次及文件如图 19-16 所示。创建的新文件有：BookController.java、labels_en_US.properties、labels_zh_CN.properties、Book.java、bookForm.jsp 和 success.jsp。（代码详见本书源码包 ch19 目录中的 InternationalDemo 工程。）

BookController.java 的主要代码如下：

```
@Controller
public class BookController {
 @RequestMapping(value = "/book_input")
```

```
    public String inputBook(Model model) {
        model.addAttribute("book", new Book("9801","Data Structure",66.5f));
        return "bookForm";
    }
    @RequestMapping(value = "/book_save")
    public String saveBook(@ModelAttribute Book book, Model model) {
        //保存书籍
        model.addAttribute("book", book);
        return "success";
    }
}
```

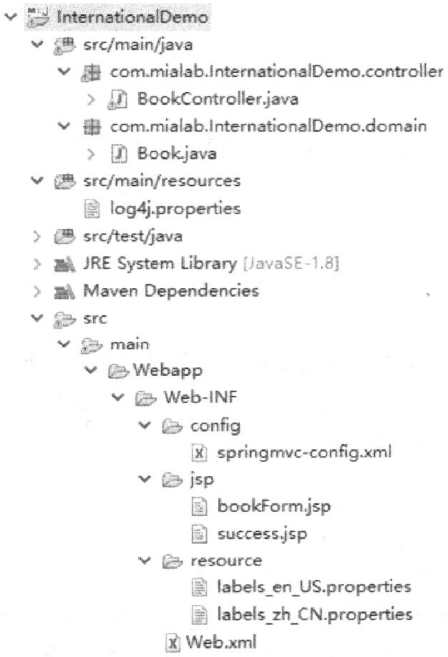

```
InternationalDemo
  src/main/java
    com.mialab.InternationalDemo.controller
      BookController.java
    com.mialab.InternationalDemo.domain
      Book.java
  src/main/resources
    log4j.properties
  src/test/java
  JRE System Library [JavaSE-1.8]
  Maven Dependencies
  src
    main
      Webapp
        Web-INF
          config
            springmvc-config.xml
          jsp
            bookForm.jsp
            success.jsp
          resource
            labels_en_US.properties
            labels_zh_CN.properties
          Web.xml
```

图 19-16 InternationalDemo 工程目录层次及文件

bookForm.jsp 的主要代码如下：

```
<%@ page language="java" contentType="text/html; charset=UTF-8"
 pageEncoding="UTF-8"%>
<%@ taglib prefix="form" uri="http://www.springframework.org/tags/form"%>
<%@ taglib prefix="spring" uri="http://www.springframework.org/tags"%>
<html>
<head>
<meta http-equiv="Content-Type" content="text/html; charset=UTF-8">
<title>Spring MVC 国际化</title>
</head>
<body>
  <!-- 使用 message 标签来输出国际化信息 -->
  <h3><spring:message code="title" /></h3>
  <form:form commandName="book" action="book_save" method="post">
      <table>
          <tr>
              <td><spring:message code="label.bookId" />: </td>
              <td><form:input path="bookId" /></td>
          </tr>
          <tr>
```

```
                <td><spring:message code="label.bookName" />: </td>
                <td><form:input path="bookName" /></td>
            </tr>
            <tr>
                <td><spring:message code="label.bookPrice" />: </td>
                <td><form:input path="bookPrice" /></td>
            </tr>
            <tr>
                <td><input id="submit" type="submit"
                    value="<spring:message code="button.submit"/>"></td>
            </tr>
        </table>
    </form:form>
</body>
</html>
```

Web-INF/resource/labels_en_US.properties 的文件内容如下：

```
label.bookId=Book Code
label.bookName=Book Name
label.bookPrice=Book Price
button.submit=Submit
title=Book Register
```

Web-INF/resource/labels_zh_CN.properties 的文件内容如下：

```
label.bookId=\u56fe\u4e66\u7f16\u53f7\u000d\u000a
label.bookName=\u56fe\u4e66\u540d\u79f0
label.bookPrice=\u56fe\u4e66\u4ef7\u683c
button.submit=\u63d0\u4ea4
title=\u56fe\u4e66\u767b\u8bb0
```

springmvc-config.xml 中需加入以下配置代码，以实现国际化。

```
<bean id="messageSource" class="org.springframework.context.support.
     ReloadableResourceBundleMessageSource">
  <property name="basenames" >
     <list>
         <value>/Web-INF/resource/labels</value>
     </list>
  </property>
</bean>
<bean id="localeResolver"

class="org.springframework.Web.servlet.i18n.AcceptHeaderLocaleResolver">
</bean>
```

<bean id="messageSource" class="…ReloadableResourceBundleMessageSource">用来告知 Spring MVC 国际化的属性文件保存在哪里。

<bean id="localeResolver" class=" …AcceptHeaderLocaleResolver">用来说明选择何种类型的 localeResolver，这是一个 ReloadableResourceBundleMessageSource 实例，它将通过读取用户浏览器的 accept-language 标题值来确定使用哪个语言区域。

这里使用的测试浏览器是 Firefox，浏览器语言设置如图 19-17 所示。如果“英语/美国”在最上面，则测试结果如图 19-18 所示；如果“汉语/中国”在最上面，则测试结果如图 19-19 所示。

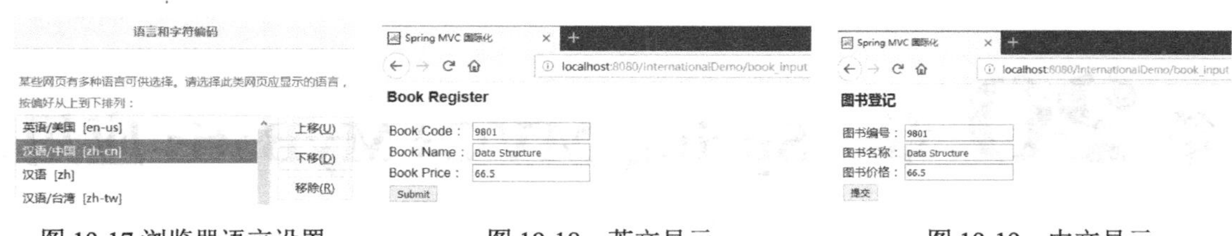

图 19-17 浏览器语言设置　　图 19-18　英文显示　　图 19-19　中文显示

19.5　本章小结

Spring 的标签库集成在 Spring Web MVC 中，因此这里的标签可以访问控制器处理命令对象和绑定数据，这使得 JSP 的开发及维护更为容易。

Spring 拥有自己独立的数据校验框架，位于 org.springframework.validation 包中。Spring 在进行数据绑定时，可同时调用校验框架来完成数据校验工作。

JSR 303 用于对 Java Bean 中的字段的值进行验证，使得验证逻辑从业务代码中脱离出来。JSR 303 的官方参考实现是 Hibernate Validator，此实现与 Hibernate ORM 没有任何关系。

Spring MVC 使用 Apache Commons FileUpload 技术实现了一个 MultipartResolver 实现类：CommonMultipartResolver。所以，Spring MVC 的文件上传还需要依赖 Apache Commons FileUpload 的组件。

若要在 Spring MVC 中选择语言区域，可以使用语言区域解析器。Spring MVC 提供了一个语言区域解析器接口 LocaleResolver，该接口的常用实现类都在 org.springframework.Web.servlet.i18n 包中。

通过配置 LocaleChangeInterceptor，可以动态改变本地语言。它会检测请求中的参数并改变地区信息。其调用 LocaleResolver.setLocal()进行配置。

在 Spring MVC 中显示本地化消息通常使用 Spring 的 message 标签。为了使用这个标签，要在使用该标签的所有 JSP 页面最前面使用 taglib 指令导入 Spring 的标签库。

可以利用 messageSource bean 告诉 Spirng MVC 国际化的属性文件保存在哪里。

习题 19

1．如何使用 Spring MVC 的表单标签库？请编程加以说明。
2．如何实现 Spring 验证和 JSR 303 验证？请编程加以说明。
3．Spring MVC 框架中如何实现文件上传和下载？请编程加以说明。
4．实现 Spring MVC 的国际化有几种方式？请编程加以说明。

第20章 Spring MVC+MyBatis应用

本章导读

健康管家 App 是一款健康类安卓客户端应用，包含运动计步、睡眠管理、饮食贴士、寻医生、找医院、查疾病、社区资讯、健康测评、自我激励、个人中心等多个模块，而健康管家管理平台是健康管家 App 数据的 Web 管理端，其实际上是一款 Web 应用。本章将讲述服务器端接口编程相关技术等知识。本章主要内容有：（1）项目简介及任务说明；（2）准备数据和总体框架；（3）登录模块及 Kaptcha 验证码组件；（4）系统管理界面；（5）系统用户管理；（6）功能模块管理。

20.1 项目总体介绍

20.1.1 项目简介及任务说明

健康管家管理平台（healthbutler-manager）的实际代码量逾万行，出于教学示范的目的，这里的示例只是代码的一部分，而且做了相当的简化。数据表也是精简过的。

任务说明：实现登录功能，登录界面如图 20-1 所示。根据不同的角色登录到不同的界面，并进行相应的数据管理。如果是一般管理员，则登录到功能模块管理界面，如图 20-2 所示；如果是超级管理员，则登录到系统管理界面，如图 20-3 所示。

图 20-1 登录界面

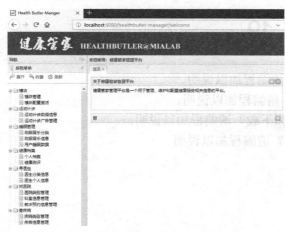

图 20-2 功能模块管理界面

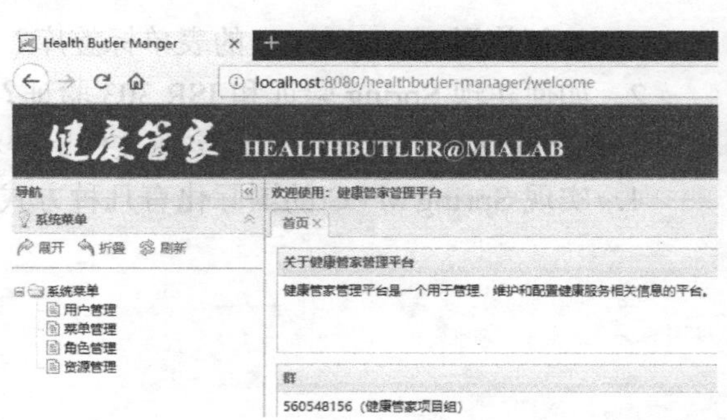

图 20-3 系统管理界面

20.1.2 准备数据

通过 pgAdmin 客户端与 PostgreSQL 数据库服务器相连，在 pgAdmin 中建立 PostgreSQL 数据库“health”，再在 health 中使用 health.backup 备份文件还原数据即可。具体操作：右击 health 数据库，在弹出的快捷菜单中选择“恢复...”选项，在弹出的“Restore database health”对话框，找到 health.backup 所在的路径，单击“恢复”按钮即可得到 account、account_group、sys_menu、module_info 等表数据。

20.1.3 总体框架

在 Eclipse 中创建 Maven Project，选择 maven-archetype-quickstart，最终完成的 healthbutler-manager 工程目录及文件如图 20-4 所示。

（代码详见 ch20 目录中的 healthbutler-manager 工程。）

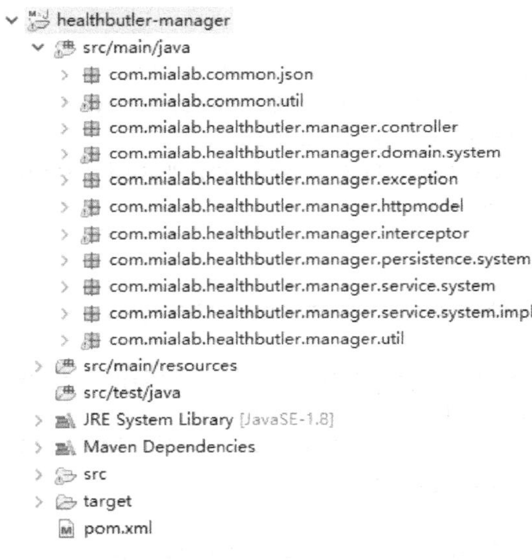

图 20-4 healthbutler-manager 工程目录及文件

20.2 典型代码及技术要点

20.2.1 登录模块及 Kaptcha 验证码组件

1．WelcomeController.java

登录表单数据是提交给控制器 WelcomeController 处理的。

WelcomeController.java 的主要代码如下：

```
@Controller
@RequestMapping("/welcome")
/*将 Model 中属性名为 Constants.USER_INFO_SESSION 的属性放到
  Session 属性列表中，以便此属性跨请求访问*/
@SessionAttributes(Constants.USER_INFO_SESSION)
public class WelcomeController {
```

```
        static Logger logger = Logger.getLogger(WelcomeController.class.getName());
        @Autowired
        private SystemUserService userService;

        @RequestMapping(method = RequestMethod.POST)
        public String login(SystemUser user, Model model, HttpSession session)
throws
        Exception {
            String sessionKey = (String) session.getAttribute
                (com.google.code.kaptcha.Constants.KAPTCHA_SESSION_KEY);
            String validCode = user.getValidCode();
            logger.info("user logining..." + user.toString());
            if (sessionKey == null || validCode == null) {
                logger.error("sessionKey=" + sessionKey + ", validCode=" +
validCode);
                return "relogin";
            }
            if (!sessionKey.equals(validCode)) {
                logger.error("验证码不正确. sessionKey=" + sessionKey + ",
                    validCode=" + validCode);
                model.addAttribute("message", "验证码不正确");
                return "relogin";
            }
            SystemUser                          user1                          =
userService.getAccountByName(user.getAccountName());
            if (user1 == null) {
                logger.error("用户不存在. sessionKey=" + sessionKey + ",
                    validCode=" + validCode);
                model.addAttribute("message", "用户不存在");
                return "relogin";
            }
            String loginPassword = user.getAccountPwd();
            String pwd = FunctionUtil.md5hashString(loginPassword, 3, false);
            String accountPwd = pwd.substring(5, 13);
            String pwdb = pwd.substring(16, 24);

            if (user == null || !user1.getAccountPwd().equals(accountPwd)
            || !user1.getPwdb().equals(pwdb)) {
                logger.error("密码错误. sessionKey=" + sessionKey + ",
                    validCode=" + validCode);
                model.addAttribute("message", "密码错误");
                return "relogin";
            }

            logger.info("user logined:" + user1.toString());
            //将名为 Constants.USER_INFO_SESSION 的属性放到 Session 属性列表中
            model.addAttribute(Constants.USER_INFO_SESSION, user1);
            return "welcome";
        }
```

```
    @RequestMapping(params = "logout")
    public String logout(HttpSession session) {
        RequestUtil.saveAccessLog(session, "logout");
        if (session != null) {
            session.invalidate();
        }
        return "login";
    }

    @RequestMapping(params = "changepassword")
    public String changepassword(SystemUser user, Model model, HttpSession
session) {
        RequestUtil.saveAccessLog(session, "changepassword");
        SystemUser  user1  =  userService.getAccountByName(user.getAccount
Name());
        if (user1 == null) {
            model.addAttribute("message", "用户不存在");
            return "passwordchange";
        }
        logger.info("user1 login:" + user1.toString());
        String loginPassword = user.getAccountPwd();
        String pwd = FunctionUtil.md5hashString(loginPassword, 3, false);
        String accountPwd = pwd.substring(5, 13);
        String pwdb = pwd.substring(16, 24);
        if (user == null || !user1.getAccountPwd().equals(accountPwd)
        || !user1.getPwdb().equals(pwdb)) {
            model.addAttribute("message", "密码错误");
            return "passwordchange";
        }
        logger.info("user login:" + user.toString());
        String newPassword = user.getNewAccountPwd();
        String newHashPassword = FunctionUtil.md5hashString(newPassword, 3,
false);
        user1.setAccountPwd(newHashPassword.substring(5, 13));
        user1.setPwdb(newHashPassword.substring(16, 24));
        int restu = userService.updatePassword(user1);
        if (restu == 1) {
            model.addAttribute("message", "密码修改成功");
            return "login";
        }
        model.addAttribute("message", "密码修改失败");
        return "passwordchange";
    }
}
```

必须注意的是，在修改密码时，写入数据库中的密码是加密的。登录验证则需要将输入的密码加密后再和数据库中的密码（已加密）进行验证。

2. login.jsp

login.jsp 中有两点需要注意：①表单提交按钮是图片形式的；②Kaptcha 验证码组件的使用。

要注意 login.jsp 中的以下关键代码：

```
<script type="text/javascript">
    $(function() {
        $('#flushcode').click(
                function() {
                    $('#kaptchaImage').attr('src',
                            'Kaptcha.jpg?' + Math.floor(Math.random() * 100));
                });
    });
</script>
…
<form action="welcome" method="post">
…
    <input type="image" name="btnLogin" id="btnLogin"
        src="<%=request.getContextPath()%>/resources/images/login.gif" … />
…
    <td width="30%" align="left">
    <a href="" id="flushcode">
    <img src="Kaptcha.jpg" id="kaptchaImage" … />
    </td>
…
    <td valign="bottom" height="12"><a href="changepwd" …>修改密码</a> </td>
…
</form>
```

<input type="image"/>定义了图像形式的提交按钮。<input type="image">往往与 src 属性、alt 属性结合使用。例如：

```
<input type="image" src="submit.gif" alt="Submit" />
```

HTML 中 image 是“创建一个图像控件，该控件被单击后将导致表单立即被提交”。<input type="image" src="xxx.gif">**本身就是一个提交按钮，和 submit 功能一样**，如果代码中再加上 onclick 就要提交两次。例如，执行以下 HTML 代码会发生表单提交两次的现象：

```
<input type="image" src="xxx.gif" onclick="return dosubmit();">
```

这样经常会造成表单元素被重复提交，数据库被写入异常！所以要谨慎使用<input type ="image">!

Kaptcha（官网地址 http://code.google.com/p/kaptcha/）是一个基于 SimpleCaptcha 的验证码开源项目。SimpleCaptcha 是一个用于随机生成验证码的 Java 框架，它为验证码提供了简单的实现。SimpleCaptcha 提供了很多图形的自定义，它对中文也提供了很好的支持。

Kaptcha 的使用比较方便，只需添加 JAR 包依赖之后简单地配置即可使用。Kaptcha 的所有配置都可以通过 Web.xml 来完成。

Kaptcha 的工作原理是调用 com.google.code.kaptcha.servlet.KaptchaServlet 生成一张图片，并在同一时刻将生成的验证码字符串放到 HttpSession 中。

如果使用 Maven 来统一管理 JAR 包，则必须在工程的 pom.xml 中添加依赖。

```
<dependency>
    <groupId>com.google.code.kaptcha</groupId>
```

```
        <artifactId>kaptcha</artifactId>
        <version>2.3.2</version>
    </dependency>
```

如果是非 Maven 管理的项目，则可直接在官网下载 Kaptcha 的 JAR 包，然后添加到项目 lib 库中，下载地址为 http://code.google.com/p/kaptcha/downloads/list。

必须在 Web.xml 中配置 Kaptcha 的 Servlet，具体如下：

```
<servlet>
    <servlet-name>Kaptcha</servlet-name>
    <servlet-class>com.google.code.kaptcha.servlet.KaptchaServlet</servlet-class>
</servlet>
<servlet-mapping>
    <servlet-name>Kaptcha</servlet-name>
    <url-pattern>/Kaptcha.jpg</url-pattern>
</servlet-mapping>
```

其中，Servlet 的 url-pattern 可以自定义。

Kaptcha 所有的参数都有默认的配置，如果不显示配置，则会采取默认的配置。

如果要显示配置 Kaptcha，则在配置 Kaptcha 对应的 Servlet 时，在初始参数 init-param 中增加响应的参数配置即可。示例如下：

```
<servlet>
    <servlet-name>Kaptcha</servlet-name>
    <servlet-class>com.google.code.kaptcha.servlet.KaptchaServlet</servlet-class>
    <init-param>
        <param-name>kaptcha.image.width</param-name>
        <param-value>200</param-value>
        <description>Width in pixels of the kaptcha image.</description>
    </init-param>
    <init-param>
        <param-name>kaptcha.image.height</param-name>
        <param-value>50</param-value>
        <description>Height in pixels of the kaptcha image.</description>
    </init-param>
    <init-param>
        <param-name>kaptcha.textproducer.char.length</param-name>
        <param-value>4</param-value>
        <description>The number of characters to display.</description>
    </init-param>
    <init-param>
        <param-name>kaptcha.noise.impl</param-name>
        <param-value>com.google.code.kaptcha.impl.NoNoise</param-value>
        <description>The noise producer.</description>
    </init-param>
</servlet>
```

（具体的配置参数参见 http://code.google.com/p/kaptcha/wiki/ConfigParameters。）

可以看到，使用 Kaptcha 能够方便地配置验证码的字体、验证码字体的大小、验证码字体的颜色、验证码内容的范围（数字、字母、中文）、验证码图片的大小、边框、边框粗细、边框颜色、验证码的干扰线（能够自己继承 com.google.code.kaptcha. NoiseProducer 来编写一个自定义的干扰线）、验证码的样式（鱼眼样式、3D、普通模糊等，也能够继承 com.google.code. kaptcha.GimpyEngine 来自定义样式）。

页面调用的示例代码如下：

```
<form action="submit.action">
    <input type="text" name="kaptcha" value="" /><img src="Kaptcha.jpg" />
</form>
```

在 submit 的 action 方法中进行验证码校验：

```
//从 Session 中取出 Servlet 生成的验证码
String kaptchaExpected = (String)request.getSession().getAttribute(com.google.code.
kaptcha.Constants.KAPTCHA_SESSION_KEY);
//获取用户页面输入的验证码
String kaptchaReceived = request.getParameter("kaptcha");
//校验验证码是否正确
if (kaptchaReceived == null || !kaptchaReceived.equalsIgnoreCase(kaptchaExpected)){
    setError("kaptcha", "Invalid validation code.");
}
```

注意：确保 JDK 设置 -Djava.awt.headless=true。

实现页面验证码刷新的示例代码如下：

```
<img src="kaptcha.jpg" width="200" id="kaptchaImage" title="看不清，单击换一张" />
<script type="text/javascript">
    $(function() {
        $('#kaptchaImage').click(function() {$(this).attr('src','kaptcha.jpg?' + Math.floor(Math.random() *
100));});
    });
</script>
<br /><small>看不清，单击换一张</small>
```

注意：为了避免浏览器的缓存，可以在验证码请求的 url 后添加随机数。

在配置文件 controllers.xml 中还要添加以下代码：

```
<!-- Maps '/' requests to the 'login' view -->
<mvc:view-controller path="/" view-name="/login" />
<mvc:view-controller path="/changepwd" view-name="passwordchange" />
<mvc:view-controller path="/report" view-name="passwordchange" />
<context:component-scan base-package="com.mialab.healthbutler.manager.controller,
    com.mialab.healthbutler.manager.persistence,
    com.mialab.healthbutler.manager.service,
    com.mialab.healthbutler.manager.exception" />
```

20.2.2 系统管理界面

以用户名 abc（超级管理员）登录到系统管理界面，如图 20-3 所示，登录密码是 123456。

1．welcome.jsp

welcome.jsp 使用了 jQuery EasyUI 布局，welcome.jsp 的<body>部分的代码如下：

```
<body id="indexLayout" class="easyui-layout" fit="true">
  <div region="north" href="layout/north.jsp" …></div>
  <div region="west" href="layout/west.jsp" title="导航" …></div>
  <div region="center" href="layout/center.jsp" …></div>
  <div region="south" href="layout/south.jsp" …></div>
</body>
```

2．west.jsp

west.jsp 的主要作用是导航，sy.bp()的作用是获得项目根路径。west.jsp 的主要代码如下：

```
<script type="text/javascript" charset="UTF-8">
 var tree;
 $(function() {
     tree = $('#tree').tree(
         {
             url : sy.bp() + '/system/menu?tree',
             animate : true,
             lines : !sy.isLessThanIe8(),
             onClick : function(node) {…
             },
             onLoadSuccess : function(node, data) {…
             }
         });
 });
 function collapseAll() {…
 }
 function expandAll() {…
 }
</script>
<div class="easyui-panel" fit="true" border="false">
 <div class="easyui-accordion" fit="true" border="false">
     <div title="系统菜单" iconCls="icon-tip">
         <div class="easyui-layout" fit="true">
             <div region="north" border="false" style="overflow: hidden;">
                 <a href="javascript:void(0);" class="easyui-linkbutton"
                     … onclick="expandAll();">展开</a><a
                     href="javascript:void(0);"
class="easyui-linkbutton"
                     … onclick="collapseAll();">折叠</a><a
                     href="javascript:void(0);"
class="easyui-linkbutton"
                     … onclick="tree.tree('reload');">刷新</a>
                 <hr style="border-color: #fff;" />
             </div>
             <div region="center" border="false">
                 <ul id="tree" …></ul>
             </div>
         </div>
     </div>
 </div>
</div>
```

由以上代码可以看出，能通过 url 得到数据，再对 JSP 页面进行渲染，把数据绑定到#tree 上。实际上，这里是通过调用控制器 SystemMenuController 的 tree()方法来得到数据的。url 映射的是 SystemMenuController 的 tree()方法。

3．north.jsp

north.jsp 的主要代码如下：

```
<%@ page language="java" pageEncoding="UTF-8"%>
<script type="text/javascript" charset="UTF-8">
 function logout(b) {
     window.location.href="welcome?logout";
```

```
    }
    </script>
    <div style="position: absolute; right: 0px; bottom: 0px; ">
     <a href="javascript:void(0);" class="easyui-menubutton"
         menu="#layout_north_zxMenu" iconCls="icon-back">注销</a>
    </div>
    <div id="layout_north_zxMenu" style="width: 100px; display: none;">
         <div onclick="logout();">退出系统</div>
    </div>
```

north.jsp 的主要作用是通过选择页面右上角的“注销→退出系统”选项，来调用控制器 WelcomeController 的 logout()方法注销当前用户，并重新回到登录页面。

4. center.jsp

center.jsp 的主要代码如下：

```
    <%@ page language="java" pageEncoding="UTF-8"%>
    <script type="text/javascript" charset="UTF-8">
     var centerTabs;
     function addTabFun(opts) {
         var options = $.extend({
             title : '',
             content : '<iframe src="' + opts.src + '" frameborder="0" ...></iframe>',
             closable : true,
             iconCls : ''
         }, opts);
         if (centerTabs.tabs('exists', options.title)) {
             centerTabs.tabs('close', options.title);
         }
         centerTabs.tabs('add', options);
     };
     $(function() {
         centerTabs = $('#centerTabs').tabs({
             border : false,
             fit : true
         });
         setTimeout(function() {
             centerTabs.tabs('add', {
                 title : '首页',
                 content:'<iframesrc = "layout/home.jsp "frameborder = "0" ...>
</iframe>',
                 closable : true,
                 iconCls : ''
             });
         }, 0);
     });
    </script>
    <div id="centerTabs"></div>
```

由以上代码可以看出，在 center.jsp 页面中能动态增加标签页。

5. south.jsp

south.jsp 的主要代码如下：

```
    <%@ page language="java" pageEncoding="UTF-8"%>
    <div align="center">版权所有，推荐分辨率 1024*768+，使用谷歌内核浏览器...</div>
```

20.2.3　系统用户管理

在如图 20-3 所示的系统管理界面中，在左侧的导航栏中选择“用户管理”，在 center.jsp 页面中便会相应地增加“用户管理”标签页，如图 20-5 所示。在“用户管理”标签页中，可对用户信息进行“增加”、“删除”、“编辑”及“重置密码”等操作，默认情况下，重置密码为“111111”。

图 20-5　用户管理

1. SystemUserController.java

SystemUserController.java 的主要代码如下：

```
@Controller
@RequestMapping("/system/user")
public class SystemUserController {
 @Autowired
    private SystemUserService userService;

    @RequestMapping(params = "user")
    public String user() {
        return "/admin/user";
    }

    @RequestMapping(params = "datagrid")
    @ResponseBody
    public EasyuiDataGridJson datagrid(DataGridModel dg, SystemUser bean){
        return userService.datagrid(dg, bean);
    }

    @RequestMapping(params = "add")
    @ResponseBody
    public int add(SystemUser bean, HttpSession session) {
        RequestUtil.saveAccessLog(session, "user.add " + bean.toString());
        bean.setCreateTime(Calendar.getInstance().getTime());
        bean.setEntryNumber(0);
        String pwd = FunctionUtil.md5hashString(bean.getAccountPwd(), 3,
```

```
false);
            String accountPwd = pwd.substring(5, 13);
            String pwdb = pwd.substring(16, 24);
            bean.setAccountPwd(accountPwd);
            bean.setPwdb(pwdb);
            return userService.add(bean);
        }

        @RequestMapping(params = "edit")
        @ResponseBody
        public int edit(SystemUser bean, HttpSession session) {
            …
        }

        @RequestMapping(params = "resetpwd")
        @ResponseBody
        public Json resetpwd(String ids, HttpSession session) {
            …
        }

        @RequestMapping(params = "del")
        @ResponseBody
        public Json del(String ids, HttpSession session) {
            …
        }
    }
```

sys_menu 表数据如图 20-6 所示。因为 sys_menu 表数据要与 EasyUI 树形菜单控件绑定，sys_menu 表字段 pid 的值决定了这个菜单的目录层次，如 pid=“999999”表示“系统菜单”是 EasyUI 树形菜单控件的根；pid=“1”表示“用户管理”、“角色管理”等是 EasyUI 树形菜单控件的一级目录，以此类推。

id	pid	text	iconcls	src	seq
1	999999	系统菜单	(Null)	(Null)	100
2	1	用户管理	(Null)	/system/user?user	8
4	1	角色管理	(Null)	/system/role?role	10
5	1	资源管理	(Null)	/system/resource?resource	11
3	1	菜单管理	(Null)	/system/menu?menu	9

图 20-6 sys_menu 表数据

字段 src 的值存储在 EasyUI 树形菜单控件的节点中。在 west.jsp 中单击树形菜单控件的节点会有以下代码：

```
    onClick : function(node) {
     …
     href = sy.bp() + node.attributes.src;
     …
     addTabFun({
         src : href,
         title : node.text
     });
    }
```

addTabFun()方法被调用，把 sy.bp() + node.attributes.src 传给了 addTabFun()方法参数 **src**。

在 center.jsp 中定义的 addTabFun()方法代码如下：

```
function addTabFun(opts) {
 var options = $.extend({
     title : '',
     content : '<iframe src="' + opts.src + '" frameborder="0" …></iframe>',
     closable : true,
     iconCls : ''
 }, opts);
 …
 centerTabs.tabs('add', options);
};
```

例如，单击左侧导航树形菜单的“用户管理”，就会调用 addTabFun()方法，把 sy.bp() + node.attributes.src 传给了 addTabFun()方法参数 **src**。

sy.bp() + node.attributes.src 的值就是/healthbutler-manager/system/user?user，由前面的 SystemUserController.java 代码可以知道/healthbutler-manager/system/user?user 请求得到的就是视图“/admin/user”，实际上就是/healthbutler-manager/admin/user.jsp。

也就是说，在 center.jsp 中动态生成的“用户管理”标签页实际上就是 user.jsp。

2. user.jsp

user.jsp 中有以下代码：

```
<head>
<script type="text/javascript" charset="UTF-8">
 var datagrid;
 …
 $(function() {
     datagrid = $('#datagrid').datagrid({
         url : sy.bp() + '/system/user?datagrid',
         toolbar : […],
         onLoadSuccess:function(){
             $('#datagrid').datagrid('clearSelections');
         },
         …
     });
 });
 …
</script>
</head>
<body class="easyui-layout" fit="true">
 <div region="center" border="false" …>
     <table id="datagrid"></table>
 </div>
 <div id="userDialog" …>
     <form id="userForm" method="post">
         <table class="tableForm">…</table>
     </form>
 </div>
</body>
```

在 user.jsp 中，通过 url 得到数据，再对 JSP 页面进行渲染，把数据绑定到# datagrid 上。实际上，这里是通过调用控制器 SystemUserController 的 datagrid ()方法来得到数据的。url 映射的就是 SystemUserController 的 datagrid ()方法。

3. 业务层 SystemUserServiceImpl.java

SystemUserServiceImpl.java 的主要代码如下：

```
@Service("accountService")
public class SystemUserServiceImpl implements SystemUserService {
 @Autowired
   private SystemUserMapper userMapper;

   @Override
   publicEasyuiDataGridJsondatagrid(DataGridModeldg,SystemUser record){
      EasyuiDataGridJson grid = new EasyuiDataGridJson();
      Map<String, Object> para = new HashMap<String, Object>();
      para.put("dg", dg);
      para.put("record", record);
      List<SystemUser> list = userMapper.find(para);
      String count = userMapper.findCount(para);
      grid.setTotal(TextUtils.parseLong(count));
      grid.setRows(list);
      return grid;
   }

   @Override
   @Transactional
   public int add(SystemUser bean) {
      userMapper.save(bean);
      if (bean.getGroupId() == null) {
         bean.setGroupId(0);
      }
      System.out.println(bean);
      return userMapper.saveAccountGroup(bean);
   }

   @Override
   @Transactional
   public int edit(SystemUser bean) {
      if (bean.getGroupId() == null) {
         bean.setGroupId(0);
      }
      userMapper.updateAccountGroup(bean);
      return userMapper.update(bean);
   }

   @Override
   @Transactional
   public int resetpwd(String accountId) {
      SystemUser user = new SystemUser();
      user.setAccountId(Integer.valueOf(accountId));
      String pwd = FunctionUtil.md5hashString("111111", 3, false);
      String accountPwd = pwd.substring(5, 13);
```

```
        String pwdb = pwd.substring(16, 24);
        user.setAccountPwd(accountPwd);
        user.setPwdb(pwdb);
        return userMapper.update(user);
    }

    @Override
    @Transactional
    public int remove(String ids) {
        for (String id : ids.split(",")) {
            userMapper.remove(Integer.valueOf(id));
            userMapper.removeAccountGroup(Integer.valueOf(id));
        }
        return 1;
    }

    @Override
    public SystemUser getAccountByName(String accountName) {
        return userMapper.findByName(accountName);
    }

    @Override
    public int updatePassword(SystemUser bean) {
        return userMapper.update(bean);
    }
}
```

业务逻辑是由 SystemUserServiceImpl.java 来实现的，这个业务组件负责使用 MyBatis 框架同数据库交互，完成用户信息的“增加”、“删除”、“编辑”、“重置密码”等功能。

20.2.4　功能模块管理

以用户名 ppp（一般管理员）登录到功能模块管理界面，如图 20-2 所示，登录密码是 111111。在如图 20-2 所示的功能模块管理界面中，在左侧的导航栏中选择“模块管理”，在 center.jsp 页面中会相应地增加“模块管理”标签页，如图 20-7 所示。在“模块管理”标签页中，可对模块信息进行“增加”、“删除”、“编辑”及“查询”等操作。

可参考源码包 ch20 目录中的 healthbutler-manager 工程中的 ModuleInfoController.java、ModuleServiceImpl.java、ModuleInfoMapper.java、ModuleInfoMapper.xml、moduleInfo.jsp、Web.xml、root-context.xml、servlet-context.xml 和 controllers.xml 等文件。

限于篇幅，在此不再赘述。

图 20-7 “模块管理”标签页

20.3 本章小结

本章主要介绍了基于 Spring MVC + MyBatis + Maven 这一组合的应用实现——健康管家 Web 管理端。其中，重点介绍了健康管家 Web 管理端（healthbutler-manager）的登录模块及 Kaptcha 验证码组件、系统管理界面、系统用户管理模块以及功能模块管理中的模块信息管理的编程实现。

习题 20

1．使用 Spring MVC + MyBatis + Maven 组合改写“网络商城”，“网络商城”是基于 JSP + Servlet + JavaBean 框架的 Web 应用。

2．编写 mytree.jsp 页面，显示 EasyUI 树形菜单，要求：树形菜单的目录层次有 3 层；EasyUI 树形菜单控件绑定的数据来源于服务器端接口，表数据可仿照本章应用的 sys_menu 表。

3．如何使用 Kaptcha 验证码组件？给出应用实例。

4．如何在表单中使用图像形式的提交按钮？请编程加以说明。

5．使用 Spring MVC + MyBatis + Maven 组合改写“CMS 内容管理系统”，“CMS 内容管理系统”是使用 Struts 2 + Hibernate + Spring 框架实现的 Web 应用。

参 考 文 献

[1] 杨开振，周吉文，梁华辉，等．Java EE 互联网轻量级框架整合开发[M]．北京：电子工业出版社，2017．
[2] 李刚．轻量级 Java EE 企业应用实战：Strus 2+Spring 4+Hibernata[M]．4 版．北京：电子工业出版社，2014．
[3] 疯狂软件．Spring+MyBatis 企业应用实战[M]．北京：电子工业出版社，2017．
[4] 杨开振．深入浅出 MyBatis 技术原理与实战[M]．北京：电子工业出版社，2016．
[5] （加）Budi Kurniawan，（美）Paul Deck．Servlet、JSP 和 Spring MVC 初学指南[M]．林仪明，俞黎敏译．北京：人民邮电出版社，2016．
[6] 明日科技（中国）有限公司．Java Web 从入门到精通[M]．北京：清华大学出版社，2012．
[7] （美）Craig Walls． Spring 实战[M]．3 版．耿渊，张卫滨译．北京：人民邮电出版社，2013．
[8] 计文柯．Spring 技术内幕：深入解析 Spring 架构与设计原理 [M]．2 版．北京：机械工业出版社，2010．
[9] 郝佳．Spring 源码深度解析[M]．北京：人民邮电出版社，2013．
[10] QST 青软实训．Java Web 技术及应用[M]．北京：清华大学出版社，2015．
[11] 何致亿．SCWCD 认证专家应考指南 [M]．北京：电子工业出版社，2004．
[12] 黑马程序员．Java Web 程序设计任务教程[M]．北京：人民邮电出版社，2017．
[13] 孙鑫．Java Web 开发详解[M]．北京：电子工业出版社，2006．
[14] 黑马程序员．Java EE 企业级应用开发教程[M]．北京：人民邮电出版社，2017．
[15] （美）Bryan Basham，Kathy Sierra，Bert Bates．Head First Servlets & JSP（中文版）：通过 SCWCD 考试之路[M]．苏钰函，林剑译．北京：中国电力出版社，2006．
[16] 贾蓓，镇明敏，杜磊．Java Web 整合开发实战[M]．北京：清华大学出版社，2013．
[17] 郭克华．Java Web 程序设计[M]．北京：清华大学出版社，2011．
[18] 施铮．SUN 国际认证 SCWCD 应试指南[M]．北京：科学出版社，2007．
[19] Servlet 教程． http://www.runoob.com/servlet/servlet-tutorial.html
[20] Spring 基础． http://www.runoob.com/w3cnote/basic-knowledge-summary-of-spring.html
[21] Spring 教程． https://www.w3cschool.cn/wkspring/
[22] Spring MVC 教程． https://www.yiibai.com/spring_mvc/
[23] MyBatis 教程． https://www.yiibai.com/mybatis/
[24] jQuery EasyUI 中文网． http://www.jeasyui.net/．